CHEMISTRY *for* ENGINEERS

CHEMISTRY
for
ENGINEERS

For

B.E./B.Tech. I and II Semesters

Strictly as per the latest syllabus prescribed by M.D.U. Rohtak

By

Dr. B.K. AMBASTA

Professor and Head
Department of Chemistry
Haryana College of Technology and Management
Kaithal, Haryana

UNIVERSITY SCIENCE PRESS

(An Imprint of Laxmi Publications Pvt. Ltd.)

An ISO 9001:2008 Certified Company

BANGALORE • CHENNAI • COCHIN • GUWAHATI • HYDERABAD
JALANDHAR • KOLKATA • LUCKNOW • MUMBAI • RANCHI • NEW DELHI
INDIA • USA • GHANA • KENYA

CHEMISTRY FOR ENGINEERS

Printed and bound in India
Typeset at : Goswami Associates, Delhi
First Edition : 2005, *Second Edition* : 2006, *Third Edition* : 2008, *Reprint* : 2010
Fourth Edition : 2012, *Reprint* : 2013, 2014, 2015
UCE-9497-250-CHEMISTRY FOR ENGINEERS-AMB
Price: ₹ 250.00

Published in India by
UNIVERSITY SCIENCE PRESS
(An Imprint of Laxmi Publications Pvt. Ltd.)
An ISO 9001:2008 Certified Company
113, GOLDEN HOUSE, DARYAGANJ, NEW DELHI-110002, INDIA
Telephone : 91-11-4353 2500, 4353 2501
Fax : 91-11-2325 2572, 4353 2528
www.laxmipublications.com info@laxmipublications.com

Branches

Bangalore	080-26 75 69 30
Chennai	044-24 34 47 26
Cochin	0484-237 70 04, 405 13 03
Guwahati	0361-254 36 69, 251 38 81
Hyderabad	040-27 55 53 83, 27 55 53 93
Jalandhar	0181-222 12 72
Kolkata	033-22 27 43 84
Lucknow	0522-220 99 16
Mumbai	022-24 91 54 15, 24 92 78 69
Ranchi	0651-220 44 64

C—9038/014/08
Printed at : Ajit Printing Press, Delhi.

Dedicated

To

My Mother

Contents

Foreword

I have gone through the manuscript, especially the chapter on Polymers in 'Chemistry for Engineers' and have found it satisfactory and catering to the prescribed syllabus. The author, who is an experienced polymer chemist himself, has taken pains in making his points look easy and comprehensible.

Polymers are important industrial materials and a technologist is required to familiarise himself well with their chemistry in order to correlate the polymers with their properties and manifold applications. This criterion is well served from the text developed in 'Polymers and Polymerization'.

(C.S. Pande)
Professor Emeritus
H.P. University
Shimla–171005

Preface to the Fourth Edition

I am very glad to express my sincere thanks to teachers as well as students for accepting the third edition of this book. Their suggestions and comments have encouraged me to bring out the present fourth edition.

Since the syllabus and pattern of examinations of B.Tech students of M.D.U. Rohtak has been changed w.e.f. 2009–10. So on keeping this view the fourth edition is strictly based on the new prescribed syllabus of M.D.U. Rohtak, as a new one. Some corrections and alterations have been made in each chapter. A new chapter "The Catalyst" has been introduced in place of Thermodynamics. I express my since thanks to Dr. V.K. Syal (Formerly Professor of Chemistry, H.P. University, Shimla now professor of Chemistry K.I.I.T. College of Engineering, Maruti Kunj, Gurgaon). Dr. A. Paul (Professor of Chemistry, Kurukshetra University, Kurukshetra) Dr. Sanjeev Arora (Associate Professor, Deptt. of Chemistry, Kurukshetra University, Kurukshetra), Dr. D.P. Tiwari (Professor of Chemical Engineering C.R.S.E. University, Murthal) Dr. S.K. Verma (Professor of Chemistry, TERI, Kurkshetra) and Dr. K.S. Sinha formerly Professor of Chemistry, Magadh University, Bodh-Gaya). I also express my heartiest gratitude and sincere thanks to Dr. D.P. Gupta, Principal, Haryana College of Technology and Management, Kaithal, for helpful discussions, moral encouragement and suggestions.

I sincerely hope that the new edition of this book will be very much helpful to students of different engineering colleges of M.D.U. Rohtak in Haryana State. However suggestions and comments for further improvement of this book will be most welcome.

—Author

Preface to the First Edition

I really feel pleasure and satisfaction in presenting the first edition of my book "Chemistry for Engineers" for engineering students of different universities and specially for M.D.U., Rohtak, Haryana. The book has been written strictly in accordance with the latest syllabus of M.D.U., Rohtak.

During teaching of engineering students, I have found that there is hardly any single book that fully covers the chemistry syllabus of B.Tech./B.E. students of any university. This has motivated me to bring out this book.

Salient feature of the book:

(*i*) It covers all the topics of new syllabus of M.D.U., Rohtak, in a very simple and systematic way.

(*ii*) The language of the book is quite simple.

(*iii*) For students practice, a number of solved and unsolved problems have been included in the book.

(*iv*) Model sets of probable question papers has also been added for the benefit of students.

(*v*) The subject matter of the book is based on the lecture notes and the long teaching experience.

—Author

Acknowledgement

I express my deepest gratitude towards Dr. C.S. Pande, Emeritus Professor of UGC, Department of Chemistry, H.P. University, Shimla, who spared some time out of his busy schedule to go through the book and then writing foreword of this book. It has, of course, enhanced its value.

In the preparation of this book in the present form a number of Senior Colleagues have assisted me a great deal whose help I wish to acknowledge here. They are Dr. V.K. Syal (formerly Professor of Chemistry, H.P. University, Shimla now Professor of Chemistry, KIIT College of Engineering, Maruti Kunj, Gurgaon), Dr. S.C. Chaudhary (Professor of Chemistry, H.P. University, Shimla), Dr. K.S. Sinha (formerly Professor of Chemistry, Magadh University Bodh-Gaya, Bihar), Dr. A.K. Singh (Professor of Chemistry, Magadh University Bodh-Gaya), Dr. R.K. Verma (Professor of Chemistry, Magadh University Bodh-Gaya), Dr. A. Paul (Professor of Chemistry, Kurukshetra University Kurukshetra), Dr. Sanjiv Arora, Associate Professor, Department of Chemistry, Kurukshetra University Kurukshetra), Dr. S.K. Verma (Professor of Chemistry, TERI Kurukshetra), Dr. S.C. Chauhan (Department of Chemistry, M.L.S.M. College, Sundernagar, H.P.), Dr. C.P. Kaushal (Deptt. of Chemistry, M.L.S.M. College, Sundernagar, H.P.), Dr. D.P. Tiwari (Professor of Chemical Engineering, C.R.S.E. College, Murthal).

I also express my sincere thanks to my elder brother Prof. (Dr.) S.K. Verma, (Department of Mechanical Engineering, NIT, Patna) for important discussions and suggestions.

I am also highly thankful to my colleagues and friends for their valuable suggestions and discussions from time to time.

I express my sincere thanks to college authorities Er. Ranjan Agrawal (Chairman H.C.T.M. Kaithal), Er. G.K. Sethi (Managing Director, H.C.T.M. Kaithal), Professor P.N. Vijayvergiya (Formerly Principal of HCTM, Kaithal), Professor M.L. Grover (Formerly vice principal of HCTM, Kaithal) and Dr. D.P. Gupta (Principal, H.C.T.M. Kaithal) for moral encouragement. Above all, I must express my deep reverence to my father late Sri Kamleshwari Prasad and mother late Smt. Rajkumari Devi, without blessing it could not have been possible for me to complete the book. I am also highly thankful to my wife Ms. Renu Sinha and son Chitrak (Minku) for cooperation in completing the project.

I hope that the present edition of the book will be liked by both students as well as teachers. Any suggestion for further improvement in the book will be highly appreciated.

—Author

Syllabus

ENGINEERING CHEMISTRY—CH-101F

B.Tech I/II Semester

L	T	P		
			Sessional	50 Marks
L	T	P	Exam	100 Marks
3	1	0	Total	150 Marks
			Duration of exam	3 Hrs.

Note: Examiner will set 9 questions in total, with two questions from each section and one question covering all sections which will be Q.1. This Q.1 is compulsory and of short answer type. Each question carries equal mark (20 marks). Student have to attempt 5 questions in total, selecting at least one question from each session.

SECTION A

Phase Rule: Terminology, One component system (H_2O system and CO_2 – system), two components system, simple eutectic system (Pb – Ag), system with congruent melting point (Zn – Mg), system with incongruent melting point (Na_2SO_4 – H_2O), Cooling curves.

Catalysis: Characteristics of catalytic reactions, Homogeneous, heterogeneous, and enzymatic. Mechanism of catalytic action (Intermediate compound formation theory and Adsorption theory). Concepts of promoters, inhibitors and poisioners, Enzymatic catalysis : its characteristics, factors affecting, Mechanism (lock and key hypothesis and induced fit hypothesis).

SECTION B

Water and its Treatment: Part-I: Sources of water, impurities in water, hardness of water and its determination (EDTA method), units of hardness, alkalinity of water and its determination, Related numerical problems, scale and sludge formation (Composition properties and methods of prevention) Boiler corrosion and caustic embrittlement. Priming and foaming.

Water and its Treatment: Part-II: Treatment of water for domestic use, coagulation, sedimentation, filtration and disinfection. Water softening: Lime-Soda treatment, Zeolite, lon – exchange process, mixed bed demineralization, Desalination (Reverse Osmosis, electro dialysis) and related numericals.

SECTION C

Corrosion and its Prevention: Mechanism of Dry and wet corrosion (rusting of iron), types of corrosion galvanic corrosion, differential aeration corrosion, stress corrosion. Factors affecting corrosion, preventive measures (Proper design, Cathodic and Anodic protection, Electroplating, tinning, galvanization.). Soil Corrosion, Microbiological Corrosion.

Lubrication and Lubricants: Introduction, mechanism of lubrication, classification of lubricants, Liquid, Grease (Semi-Solid) and Solid (MoS_2, Graphite). Additives for lubricants, Properties of lubricants (Flash and Fire point, Sapnification number, Iodine value, Acid value, Viscosity and Viscosity index Aniline point, Cloud point and pour point). Numerical problems based of viscosity index. Biodegradable lubricants.

SECTION D

Polymers and Polymerization: Introduction and Classification of polymers, mechanism of polymerization (Addition, condensation and co-ordination), effect of structure on properties of polymers, Bio polymerization, Bio degradable polymerization, preparation, properties and technical applications of thermoplastic (PVC, PVA, Teflon) and thermo sets (PF, UF), Natural elastomers and synthetic rubber (SBR, GR–N). Silicones, introduction to polymeric composites.

Instrumental Methods of Analysis: Principle and application of Thermal methods of Analysis. (TGA, DTA, DSC), Basic concepts of spectroscopy, Lambert and Beers law. Absorption and Emission spectroscopy Different spectroscopic Techniques (UV- Visible and IR spectroscopy), elementary discussion on Flame photometry.

Chapter **1**

PHASE EQUILIBRIA (PHASE RULE)

1.1 INTRODUCTION

1.1.1 Explanation of the Terms Involved in Phase Equilibria

1.1.1.1 Phase (P)

A phase is a homogeneous, physically distinct and mechanically separable portion of the heterogeneous system, which is separated from other parts of the system by well-defined boundary surface. It is denoted by 'P'.

Examples:

1. Water exists in three forms—ice, water and vapours so it is a three phase system and represented as

$$\text{Solid (ice)} \rightleftharpoons \text{Liquid (water)} \rightleftharpoons \text{Gas (water vapour)}$$

Each phase is separated by a phase boundary known as interface.

2. A gas or a gaseous mixture is a single phase because there is no interface between one gas and another. *e.g.*, air.

3. Let us consider the decomposition of $CaCO_3$ as

$$CaCO_3(s) \rightleftharpoons CaO(s) + CO_2(g)$$

Here are three phases. Two phases are of $CaCO_3$ (*s*) and CaO(*s*) and third phase is of $CO_2(g)$. All phases are separated by interface.

4. All totally miscible liquids exist in only one liquid phase. But since each liquid has its vapour above, so the total number of phases in a system of totally miscible liquids is two, one for liquid and the other for vapour.

Required Conditions for Phase

A phase must fulfill the following three conditions:

1. It should be physically homogeneous.
2. In case of more than one phase, all phases must be separated from each other by surface of contact (interface) at equilibrium.
3. A dynamic equilibrium must be established between phases of the system through exchange of chemical species.

1.1.1.2 Components (C)

The number of components 'C' of a system at equilibrium is the smallest number of independently variable chemical constituents by means of which the compositions of all the phases present in the system can be expressed either directly or in the form of a chemical equation.

Examples:

1. Water exists in three phases as

$$\underset{\text{(ice)}}{\text{Solid}} \rightleftharpoons \underset{\text{(water)}}{\text{Liquid}} \rightleftharpoons \underset{\text{(vapour)}}{\text{Gas}}$$

   Each phase can be represented by H_2O. Thus, the number of components is one.

2. Consider aqueous solution of sugar. Here the composition of this solution is described by specifying the presence of sugar and water. Thus, the number of components is two.

3. Consider the decomposition of ammonium chloride as

$$NH_4Cl(s) \rightleftharpoons NH_3(g) + HCl(g)$$

   Actually this equilibrium exists as

$$\underset{\text{Phase 1}}{NH_4Cl(s)} \rightleftharpoons \underset{\text{Phase 2}}{NH_4Cl(g)} \rightleftharpoons \underset{\text{Phase 3}}{NH_3(g) + HCl(g)}$$

   If the reaction is carried out in a closed vessel or in vacuum then the number of components is one inspite of the dissociation of the $NH_4Cl(g)$ into $NH_3(g)$ and $HCl(g)$. Because in the gaseous phase both HCl and NH_3 are always present in equal amount and represents $NH_4Cl(g)$. However, if this reaction is carried out in excess of NH_3 or HCl gas then the composition of the vapour differs from that of the solid NH_4Cl. Hence the number of components of the system will be two (2) not one (1).

4. In case of chemically reactive system where various chemical reactions take place between the species, the number of components is calculated by the relation $C = S - R$
   where S = number of chemical species present in the system.
   and R = number of independent chemical reactions taking place among the chemical species.

   For example, (*i*) Thermal decomposition of calcium carbonate in a sealed tube, represented as

$$CaCO_3(s) \rightleftharpoons CaO(s) + CO_2(g)$$

   There are three phases, three species and one relation.

   Here $S = 3, R = 1$

   $\therefore$ $C = S - R = 3 - 1 = 2$

Hence the number of components is two.

If the number of moles of two species is altered, the number of moles of third species is fixed by the equilibrium condition

$$K = \frac{[CaO][CO_2]}{[CaCO_3]}$$

Since it is a two component system, hence all the phases in equilibrium can be expressed by choosing only any two constituents.

If we choose the constituents CaO and CO_2 then

Phase expressed		Composition
$CaCO_3$	$\longrightarrow$	$CaO + CO_2$
CaO		$CaO + 0CO_2$
CO_2		$0CaO + CO_2$

If we choose the constituents $CaCO_3$ and CO_2 then

Phase expressed	Composition
$CaCO_3$	$CaCO_3 + 0CO_2$
CaO	$CaCO_3 - CO_2$
CO_2	$0CaCO_3 + CO_2$

If we choose the constituents $CaCO_3$ and CaO then

Phase expressed	Composition
$CaCO_3$	$CaCO_3 + 0CaO$
CaO	$0CaCO_3 + CaO$
CO_2	$CaCO_3 - CaO$

If ions are also present in the system then the condition of electroneutrality (number of positive and negative charged ions should be equal) is also considered. Hence the number of components may be calculated by the modified relation as $C = S - (R + 1)$.

For example, (*i*) In the system $NaBr - KCl - H_2O$, the number of component is calculated as follow:

Here numbers of species are nine NaBr, KCl, NaCl, KBr, H_2O, Na^+, Br^-, K^+, Cl^- (The dissociation of water is ignored). The number of independent reactions (relations) is four.

$$NaBr \rightleftharpoons Na^+ + Br^-$$

$$KCl \rightleftharpoons K^+ + Cl^-$$

$$Na^+ + Cl^- \rightleftharpoons NaCl$$

$$K^+ + Br^- \rightleftharpoons KBr$$

$\therefore$ Number of components $C = S - (R + 1) = 9 - (4 + 1) = 4$.

Hence it is a four component system.

(*ii*) A dilute solution of sulphuric acid in water may be represented as:

$$H_2SO_4 + H_2 \rightleftharpoons H_3O^+ + HSO_4^-$$

$$HSO_4^- \rightleftharpoons H^+ + SO_4^{2-}$$

Here the number of species is five H_2SO_4, H_2O, H_3O^+, HSO_4^- and SO_4^{2-}.

The number of relations is two.

∴ Number of components C = S – (R + 1) = 5 – (2 + 1) = 2

Hence it is a two component system.

1.1.1.3 Degree of Freedom (F)

It is the least number of independent variables (temperature, pressure and concentration), which may be altered without changing the number of phases. It is also called variance. These variables describe the state of the system.

Examples:

1. We consider a system containing a gas enclosed in a cylinder provided with a movable piston. It is a single phase system having one component. To describe the system completely, we need only temperature and pressure because the volume is automatically known from the equation of state. So we may change only temperature and pressure. Hence it is bivariant *i.e.*, degree of freedom is two(2). F = 2.

2. Consider one component having two phase system.

$$\text{water} \rightleftharpoons \text{vapour}$$

Since it is one component system, so the states of vapour phase depends upon the temperature and pressure not on the concentration. So there is a fixed value of vapour pressure at a particular temperature. The vapour pressure changes with change in temperature. We cannot alter both the variables without disturbing the equilibrium. Hence we have to mention only one variable either pressure or temperature. Thus the system is said to be univariant (monovariant) *i.e.*, F = 1.

If we add a small amount of salt in the above system, the vapour pressure changes with concentration of the salt in the solution. Thus, if the temperature of system is fixed then the vapour pressure will also be fixed. Thus, only two variables can be altered without disturbing the equilibrium. Hence the system will be bivariant *i.e.*, F = 2.

3. We consider a one component system having three phases in equilibrium.

$$\text{ice } (s) \rightleftharpoons \text{water } (l) \rightleftharpoons \text{vapour } (g)$$

All three phases exist in equilibrium at a particular temperature (0.0098°C) and a pressure (4.58 mm of Hg). In this case none of the variables can be altered without disturbing the system. For example, on raising the temperature, ice will melt and on lowering the temperature, the vapour will condense. Thus by changing any of the variables, the number of phase will be reduced to two or one. Hence, none of the variables can be changed for existing all three phases in equilibrium. Thus, the system is non-variant or invariant *i.e.*, degree of freedom F = 0 (zero)

1.1.1.4 True Equilibrium

A system is said to be in state of true equilibrium under a given set of conditions if the same state can be realised by approach from either direction by following any possible procedure.

Example: The equilibrium exists between ice and water at 1 atm pressure and 273 K.

$$\text{ice} \rightleftharpoons \text{water}$$

It is true equilibrium because it can be attained by partial melting of ice or by partial freezing of water.

1.1.1.5 Metastable Equilibrium

A system is said to be in a state of metastable equilibrium under a given set of conditions if a state can be realised only from one direction by a careful change of conditions.

Example: It is possible to cool water slowly and very carefully to 271 K (– 2°C) or even lower temperature without the appearance of ice. Hence water at 271 K is said to be in a state of metastable equilibrium. Metastable state may be preserved, if the system is not disturbed by addition of solid phase or stirring etc. However, if stirring is done or if a small piece of ice is added, solidification sets in and the temperature rises to 0°C.

Criteria for Phase Equilibria

The temperature, pressure and chemical potential must be same throughout the system at the equilibrium. If we consider the equilibrium between two phases in an isolated system *i.e.,*

$$\text{Phase I} \rightleftharpoons \text{Phase II}$$

If T_1 and T_2 are temperatures

P_1 and P_2 are pressures

and μ_1 and μ_2 are chemical potentials then

At constant temperature

$$T_1 = T_2 \text{ (Thermal equilibrium)}$$

At constant pressure

$$P_1 = P_2 \text{ (Mechanical equilibrium)}$$

At constant composition

$$\mu_1 = \mu_2 \text{ (Chemical equilibrium).}$$

1.1.1.6 Eutectic Mixture and Eutectic Point

Eutectic mixture is a mixture of two or more components without chemically reacted in solution state having lowest freezing or melting point among all possible ratio of mixing that components at a particular temperature. Such type of system where eutectic mixture is formed is called the eutectic system and the lowest melting point of that eutectic mixture is called the eutectic point.

1.1.1.7 Triple Point

Triple point is the point at which three phases co-exist in equilibrium. The degree of freedom of one component system is zero.

1.2 PHASE RULE

It states that if the equilibrium in a heterogeneous system is not influenced by electrical, magnetic or gravitational forces, the number of degrees of freedom (F) of the system, is related to the number of components (C) and the number of phases (P) of the system as

$$F = C - P + 2$$

where the digit '2' represents the temperature and pressure. It is applicable to all systems in equilibrium without exception.

It is derived thermodynamically. Let us consider a heterogeneous system in equilibrium consisting of C-components (C_1, C_2, C_3 ......, C_C) distributed in P-phases (P_1, P_2, P_3, ..., P_P) as in Fig. 1.1.

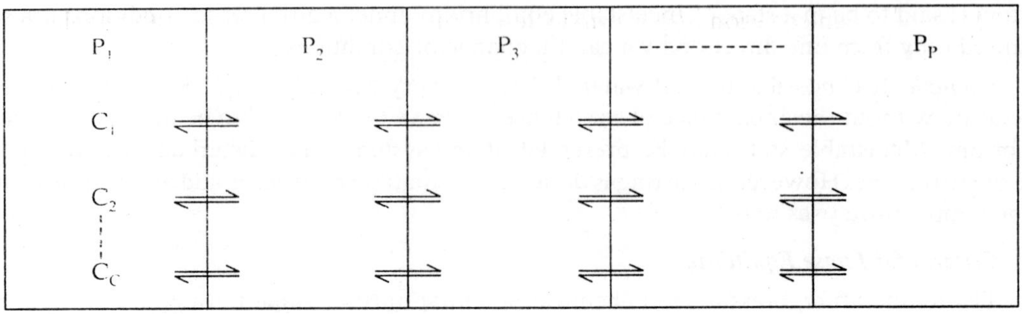

FIGURE 1.1 **Distribution of C-components in P phases**

By definition, the degree of freedom of a system in equilibrium is the least number of variables, which must be specified to define a system completely.

(*a*) To find the total number of independent variables

1. Temperature: Same for all phases (one variable)
2. Pressure: Same for all phases (one variable)
3. Concentration:

Independent concentration variables for 1 phase in respect of C-components

$$= C - 1 \quad \text{(The concentration of last component is not independent)}$$

∴ Independent concentration variables for P phase in respect of C-components

$$= P\,(C - 1)$$

The total number of independent variables is given by

$$V = P\,(C - 1) + 2 \quad \text{('2' stands for temperature and pressure variables)}$$

(*b*) To find the number of relations (equations) of equilibrium:

When a heterogeneous system is in equilibrium at constant temperature and pressure, the chemical potential of any component will have the same value in all P phases.

i.e.,

For Component 1	$\mu_1(a)$	= $\mu_1(b)$ =	......	$\mu_1(P)$
For Component 2	$\mu_2(a)$	= $\mu_2(b)$ =	......	$\mu_2(P)$
For Component 3	$\mu_3(a)$	= $\mu_3(b)$ =	......	$\mu_3(P)$
....	......	......		......
....	......	......		......
For Component C,	$\mu_C(a)$	= $\mu_C(b)$ =	......	$\mu_C(P)$

For each component, the relationships (equations) for 'P' phases = (P – 1)

∴ For 'C' component, the relationships (equations) for 'P' phases = C(P – 1).

Thus, the total number of independent equations is

$$E = C(P - 1)$$

The number of degree of freedom

= Total number of variables – The number of equations between them.

(Because each equation or relation decreases the number of independent variables by one)

$$\therefore \quad F = V - E$$

$$= [P(C - 1) + 2] - [C(P - 1)]$$

$$F = C - P + 2 \qquad ...(A)$$

Equation (A) represents the Gibbs phase rule.

If one of these two variables (temperature and pressure) does not affect on equilibria then degree of freedom for such a system will be reduced by one and in this case phase rule is called the reduced phase rule and is represented as

$$\boxed{F' = C - P + 1}$$

Uses

1. It provides a simple basis for the classification of equilibrium state of systems.

2. It is applicable to macroscopic systems and hence it is not necessary to have information about molecular structures.

3. It predicts the behaviour of systems when subjected to changes in the governing variables.

4. It is applicable to physical as well as chemical phase reactions.

5. It is helpful in deciding whether, under a given set of conditions, a number of substances put together would remain as such in equilibrium or not.

Limitations

1. It applies to a single equilibrium state and does not tell about the number of other equilibria possible in the system.

2. It takes into account the number of phases and not their quantities, even the small quantity of a phase when presents counts to the number phases present in the system in equilibrium.

3. Attainment of equilibrium state is an essential prerequisite for the application of phase rule. It is thus not applicable for such a system which is slow in reaching the equilibrium state.

4. This rule does not give any information regarding the time taken for the system to attain equilibrium.

5. Besides the concentration variables applicable to the system the other variables for the system could only be temperature and pressure. If the number of variables other than the concentration variables is different from 2, this factor of the phase rule has to be adjusted accordingly.

6. The liquid and solid phases should not be so finely subdivided as to change their vapour pressures from normal values.

1.3 PHASE DIAGRAMS

It is the graphical representation giving the conditions of temperatures and pressure under which the various phases are capable or stable existence and transform into another phase. Hence the relation between the solid, liquid and gaseous states of a given substance as a function of the temperature and pressure can be summarized on a single graph known as a phase diagram.

1.4 ONE COMPONENT (WATER) SYSTEM

Water system is a one component system as water is the only chemical compound involved in this system. Water exists in three phases in equilibrium as

$$\underset{\text{(solid)}}{\text{ice}} \rightleftharpoons \underset{\text{(liquid)}}{\text{water}} \rightleftharpoons \underset{\text{(gas)}}{\text{vapour}}$$

These three phases may occur in the three possible combinations of the two phases in equilibrium as

(I) $\underset{\text{(liquid)}}{\text{water}} \rightleftharpoons \underset{\text{(gas)}}{\text{vapour}}$

(II) $\underset{\text{(solid)}}{\text{ice}} \rightleftharpoons \underset{\text{(gas)}}{\text{vapour}}$

(III) $\underset{\text{(solid)}}{\text{ice}} \rightleftharpoons \underset{\text{(liquid)}}{\text{water}}$

The phase diagram of water system is represented as Fig. 1.2.

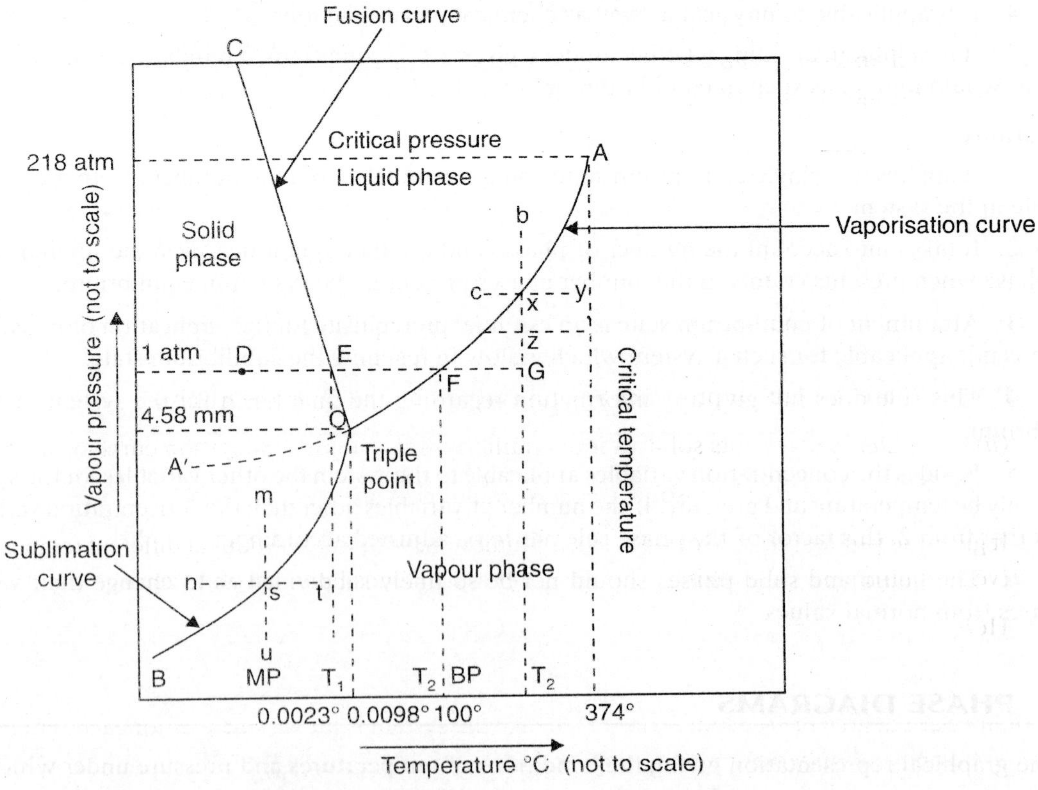

FIGURE 1.2 **Phase diagram of water system**

There are three curves OA, OB and OC, three areas AOC, BOC and below AOB, a triple point O and only one metastable curve OA′.

1.4.1 Study of Curves

(*i*) *Curve OA.* It represents liquid-vapour equilibrium.

$$\text{liquid} \rightleftharpoons \text{vapour}$$

The curve start from 'O' and extends up to critical temperature (374°C) at critical pressure (278 atm)

Here $\quad P = 2, \text{ and } \quad C = 1$

Applying Phase rule $\quad F = C - P + 2$

$$= 1 - 2 + 2 = 1$$

Since the degree of freedom is one (1) hence the system is univariant. At 100°C, the vapour pressure of water is equal to the 1 atmosphere (760 mm of Hg). This temperature is called the boiling point of water. Beyond the point 'A' (Critical temperature) liquid phase of water does not exist.

The slope of the curve OA is positive *i.e.*, the vapour pressure of water increases with temperature. It is also predicted by the Clausius-Clapeyron equation as

$$\frac{dP}{dT} = \frac{\Delta H_V}{T(V_g - V_l)} = +ve \qquad (\because \; V_g > V_l)$$

where ΔH_v = change in molar heat of vaporization

V_g = molar volume of water vapours

V_l = molar volume of water (liquid)

T = boiling point of water.

On integrating, we get

$$\log \frac{P_2}{P_1} = \frac{\Delta H_V}{2.303\,R}\left[\frac{T_2 - T_1}{T_1 T_2}\right]$$

where P_1 and P_2 are the vapour pressures at temperatures T_1 and T_2, respectively.

(*ii*) *Curve OB.* It represents solid-vapour equilibrium *i.e.*, it is a sublimation curve of ice

$$\text{solid} \rightleftharpoons \text{vapour}$$

It gives the vapour pressure of solid ice in equilibrium with water vapour at different temperatures. The curve starts from O, the freezing point of water and extends up to absolute zero.

Here $\quad P = 2, \quad C = 1$

Therefore, $\quad F = C - P + 2 = 1$

Since the degree of freedom is one (1), hence the system is univariant, *i.e.*, for each temperature there can be one and only one pressure and vice versa.

Here the slope of the curve OB is positive and also predicted by the Clausius-Clapeyron equation as

$$\frac{dP}{dT} = \frac{\Delta H_S}{T(V_g - V_S)} = +ve \qquad (\because \quad V_g > V_S)$$

where, ΔH_S = change in molar heat of sublimation.

On integrating, we get

$$\log \frac{P_2}{P_1} = \frac{\Delta H_S}{2.303\ R} \left[\frac{T_2 - T_1}{T_1 T_2} \right]$$

(*iii*) *Curve OC.* It represents solid-liquid equilibrium *i.e.*, it is a freezing point (fusion) curve of ice

$$\text{solid} \rightleftharpoons \text{liquid}$$

It shows how melting temperature of ice (freezing point of water) varies with pressure.

Here $\quad P = 2, \quad C = 1$

Therefore, $\quad F = C - P + 2 = 1$

Since the degree of freedom is one (1), hence the system is univariant. The slope of the curve is negative *i.e.*, the melting point of ice is lowered by increase of pressure. It is also predicted by Clausius-Clapeyron equation as

$$\frac{dP}{dT} = \frac{\Delta H_f}{T(V_l - V_S)} = -ve \qquad (\because \quad V_S > V_l)$$

where ΔH_f = change in molar heat of fusion.

As the $\frac{dP}{dT}$ is negative, to counter effect this the curve OC is sloping towards pressure axis.

The freezing point of water is lowered by 0.0075° by 1 atmosphere increase of pressure.

Metastable curve OA′. It represents the liquid water-water vapour in metastable equilibrium. Sometimes it is possible to cool liquid water below its freezing point without the separation of ice as shown by dotted line curve OA′. Below the freezing point the liquid is in the supercooled state. This state is known as metastable state. It is an unstable state and can be kept as such indefinitely if the presence of any other solid phase (ice) is avoided. The curve OA′ lies above the curve OB, it means the vapour pressure of the metastable supercooled water is higher than the vapour pressure of the stable solid phase at the same temperature. Therefore, as soon as the equilibrium is disturbed by adding small pieces of ice, the entire liquid will solidify.

Here $\quad P = 2, \quad C = 1$

Therefore, $\quad F = C - P + 2 = 1$

The degree of freedom is one (1) hence the system is univariant.

1.4.2 Points

Triple point 'O'. It is the point at which all the three phases ice, water and vapour coexist in equilibrium. Here the three curves OA, OB and OC intersect each other at point 'O'. This point 'O' is called triple point in the phase diagram.

Here $P = 3, \quad C = 1$

Therefore, $F = C - P + 2$

$= 1 - 3 + 2$

$= 0$ (zero)

Since the degree of freedom is zero, hence the system is invariant. The temperature and pressure corresponding to this equilibrium are 0.0098°C and 4.58 mm respectively. It is important to note that three phases can coexist in equilibrium only under one set of conditions. Any change in the temperature and pressure variable even slightly, the equilibrium will shift and the three phases cannot coexist.

1.4.3 Areas or Regions between the Lines

(*i*) *Area AOC.* The area above the curve OA *i.e.*, area AOC consists of liquid phase only.

i.e., $P = 1 \quad C = 1$

$\therefore$ $F = C - P + 2$

$= 1 - 1 + 2 = 2$

The degree of freedom is two, hence the system is bivariant. Let the state of equilibrium be represented by a point '*x*' on the curve OA. Here two phases liquid and vapour are in contact with each other. By keeping the temperature constant and increasing the pressure, the vapours are compressed wholly and changed into liquid state. It is shown by dotted line *xb*. When pressure is kept constant and temperature is decreased, the vapours will again be converted into liquid. It is represented by dotted line *xc*.

(*ii*) *Area below AOB (below the curve OA).* This area represents only vapour phase. Hence the degree of freedom is two *i.e.*, it is a bivariant system. If the temperature is increased along '*xy*' keeping the pressure constant, the liquid will change completely into vapours. Similarly, on decreasing pressure along '*xz*' at constant temperature, the liquid will again change into vapours. Thus, the area below the curve OA represents the vapour phase only.

(*iii*) *Area BOC.* It represents only solid phase. Degree of freedom is two, so it is also a bivariant system. Consider a point '*s*' on the curve OB. On increasing pressure along '*sm*' at constant temperature or on decreasing temperature along '*sn*' at constant pressure, vapour will completely change into solid. Similarly on decreasing pressure along '*su*' at constant temperature or on increasing temperature along '*st*' at constant pressure, the solid will completely change into vapours.

Since all areas are bivariant system, hence to locate any point in the region, temperature and pressure must be specified. Salient features of water system may be represented as in Table 1.1.

Table 1.1. Salient Features of Water System

Curve/Area/ Point	*Name of the System*	*Phase Equilibrium*	*No. of Phase (P)*	*Degree of Freedom (F)*
Curve OA	Vaporization curve	liquid $\rightleftharpoons$ vapour	02	01 (univariant)
Curve OB	Sublimation curve	solid $\rightleftharpoons$ vapour	02	01 (univariant)
Curve OC	Fusion curve	solid $\rightleftharpoons$ liquid	02	01 (univariant)
Curve OA′	Metastable vaporization curve	liquid $\rightleftharpoons$ vapour	02	01 (univariant)
Area AOC	—	liquid	01	02 (bivariant)
Area BOC	—	solid	01	02 (bivariant)
Area AOB (below the curve OA)	—	vapour	01	02 (bivariant)
Point 'O'	—	solid $\rightleftharpoons$ liquid $\rightleftharpoons$ vapour	03	zero (invariant)

Phase Diagram of Water at High Pressure or Polymorphism

At very high pressure if the ice is heated at a particular temperature then many new solid phases are obtained. These new solid phases are stable at only very high pressure and corresponding to ice with different crystalline forms having different physical properties. This phenomenon is known as polymorphism.

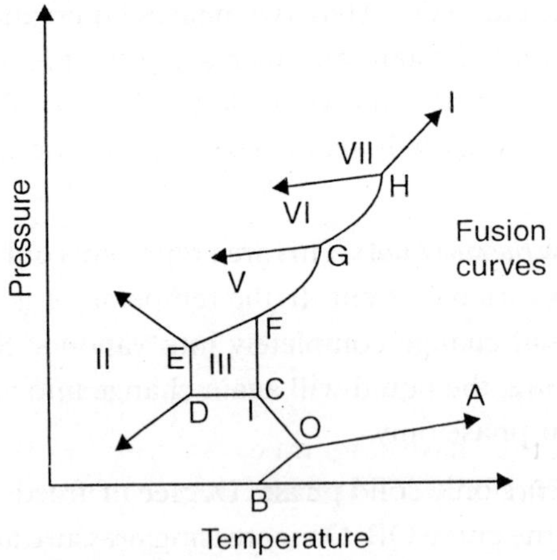

FIGURE 1.3 **Polymorphism of water**

The effect of temperature and pressure may be studied by considering again phase diagram of water. We consider the system of heating ice at constant pressure (at 1 atmosphere). Let us consider a point 'D' at 1 atm pressure below the corresponding temperature T_1. This point 'D' is in solid phase. On increasing the temperature slowly at constant pressure, the system moves along DE and ice begins to melt. The temperature remains constant till complete fusion. During fusion the system will remain bivariant (solid and liquid phases). After complete melting the ice if again temperature is raised, it follows the line EF in liquid region. In between E and F only change is an increase in temperature of the liquid. At point 'F' vapourization takes place and the temperature will remain constant till complete vapourization. After this, the system will move along FG in vapour phase with higher temperature.

At very high pressure (more than 2000 atm pressure) the polymorphism of water may be represented as in Fig. 1.3. Due to the condensed nature of the scale on the pressure axis, the line OA and OB in phase diagram merge into the temperature axis. Ice I is the ordinary ice while ice II, III, V, VI and VII are the different forms which are stable at very high pressures. Ice IV is unstable form, it could not be confirmed by later work. So it is absent amongst the types. It is very interesting fact that ice VI exists between 0.16°C and 81.6°C. At very high pressure (25000 atm.) the melting point of ice is about 100°C and the melting point can be raised to 190°C when the pressure is raised to 40,000 atmospheres. These ices may be regarded as hot ices. Ices III, V, VI and VII are denser than water and do not melt by increasing pressure on them. Polymorphism of water is represented as in Table 1.2.

Table 1.2. Polymorphism of Water

Triple Point	*Phases*	*Temperature (°C)*	*Pressure in kg/cm²*
O	I, L, V	+ 0.0098	4.58 mm of Hg
C	I, III, L	– 22.0	2115
D	I, II, III	– 34.7	2170
E	II, III, V	– 24.3	3510
F	III, V, L	– 17.0	3530
G	V, VI, L	+ 0.16	6380
H	VI, VII, L	+ 81.6	23150

Importance

The phase diagram of water is helpful in explaining the examples like ice skating, flow of glacier etc.

(*i*) *Ice skating.* From the phase diagram of water it is clear that on increasing the pressure by the skater, freezing point of ice is lowered. Hence, ice melts at 0°C and skating becomes easier because water now act as lubricant.

(*ii*) *Flow of glacier.* From the phase diagram of water it is evident that density of ice is less than that of liquid water. Hence ice floats over water and glacier flows.

1.5 CARBON DIOXIDE SYSTEM

Carbon dioxide system is an example of one component system because only one chemical compound CO_2 is involved in this system. Carbon dioxide exists in three phases in equilibrium as

$$\underset{\text{(solid, dry ice)}}{CO_2(s)} \rightleftharpoons \underset{\text{(liquid)}}{CO_2(l)} \rightleftharpoons \underset{\text{(gas)}}{CO_2(g)}$$

CO_2 exists in liquid form at very high pressure about 5.2 atm.

All these three phases may occur in the three possible combinations of the two phases in equilibrium as

$$CO_2(s) \rightleftharpoons CO_2(l)$$
$$CO_2(s) \rightleftharpoons CO_2(g)$$
$$CO_2(l) \rightleftharpoons CO_2(g)$$

The phase diagram of carbon dioxide system may be represented as in Fig. 1.4.

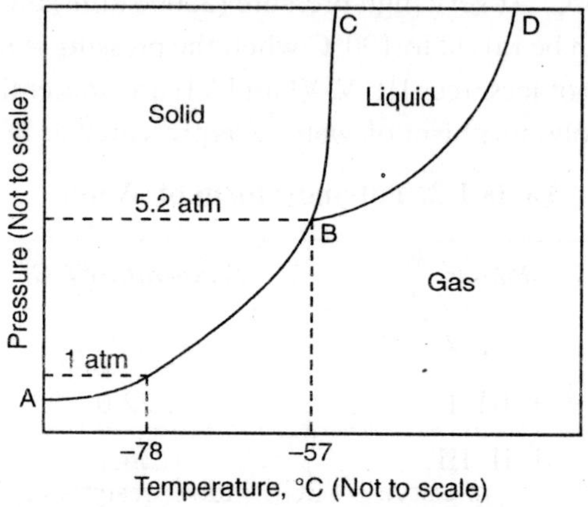

FIGURE 1.4 **The phase diagram for the carbon dioxide system**

The whole phase diagram of CO_2 system may be studied in their components *i.e.*, curves, areas and triple point as follow:

1. Curves

(*i*) *Curve AB:* It represents the sublimation curve and represents the solid-liquid equilibria as

$$CO_2(s) \rightleftharpoons CO_2(g)$$

Here, $P = 2$ and $C = 1$

$\therefore$ Degree of freedom $F = C - P + 2$
$= 1 - 2 + 2 = 1$

Hence, the system is monovariant (univariant) in nature along this curve.

The slope of the curve AB is positive *i.e.*, the pressure of CO_2 increases with rise in temperature. It is also predicted by the Clausius-Clapeyron equation as

$$\frac{dp}{dt} = \frac{\Delta H_s}{T(V_g - V_s)} = +\text{ve} \qquad (\because \ V_g > V_s)$$

where ΔH_s = change in molar heat of sublimation
V_g = molar volume of CO_2 gas
V_s = molar volume of CO_2 solid
T = sublimation point of CO_2 gas.

(*ii*) *Curve BC:* It is the fusion curve and represents the solid-liquid equilibria as

$$CO_2(s) \rightleftharpoons CO_2(l)$$

The solid CO_2 can exist in equilibrium with its liquid only at a very high pressure equal to 5.2 atm.

Here, P = 2, and C = 1

$\therefore$ Degree of freedom F = C – P + 2

= 1 – 2 + 2 = 1

Hence, the system is monovariant in nature along this curve. The nature of slope is positive *i.e.*, the melting point of solid CO_2 is increased with increase of pressure. It is also predicted by Clausius-Clapeyron equation as

$$\frac{dp}{dt} = \frac{\Delta H_f}{T(V_l - V_s)}$$

Since specific volume of liquid CO_2 is greater than that of solid CO_2.

$$\therefore \quad \frac{dp}{dt} = +\text{ve} \qquad (\because \ V_l > V_s)$$

where, ΔH_f = change in molar heat of fusion

V_l = molar volume of CO_2 liquid

V_s = molar volume of CO_2 solid

T = melting point of CO_2 solid

(*iii*) *Curve BD:* It is the vaporisation curve of CO_2 and represents liquid-vapour equilibria as

$$CO_2(l) \rightleftharpoons CO_2(g)$$

Here, P = 2 and C = 1

$\therefore$ Degree of freedom F = C – P + 2

= 1 – 2 + 2 = 1

Hence, the system is monovariant in nature along this curve. The nature of slope is positive *i.e.*, the pressure of CO_2 increases with the rise in temperature. It is also predicted by Clausius-Clapeyron equation

$$\frac{dp}{dt} = \frac{\Delta H_v}{T(V_g - V_l)} = +\text{ve} \qquad (\because \ V_g > V_l)$$

2. Areas

(*i*) *Area above ABC:* The area above ABC consists of solid phase only.

i.e., P = 1 and C = 1

$\therefore$ Degree of freedom F = C – P + 2

= 1 – 1 + 2 = 2

Hence, the system is bivariant in nature, in this region. Consider a point '*s*' on the curve AB. On increasing pressure along '*sm*' at constant temperature or on decreasing temperature along '*sn*' at constant pressure, vapour will completely change into solid. Similarly, on decreasing pressure along '*su*' at constant temperature or on increasing temperature along '*st*' at constant pressure, the solid will completely change into vapours.

(*ii*) *Area CBD:* It consists of liquid phase only

i.e., P = 1 and C = 1

$\therefore$ Degree of freedom $F = C - P + 2$

$$= 1 - 1 + 2 = 2$$

Hence, the system is bivariant in nature, in this region.

(*iii*) *Area below ABD.* It consists of gas phase only. *i.e.*, P = 1 and C = 1

$\therefore$ Degree of freedom $F = C - P + 2$

$$= 1 - 1 + 2 = 2$$

Hence the system is bivariant in nature, in this region.

3. Triple Point 'B'

B is the triple point at which all the three phases (solid, liquid and gas) of CO_2 coexist in equilibrium with one another *i.e.*,

$$CO_2(s) \rightleftharpoons CO_2(l) \rightleftharpoons CO_2\ (g)$$

Here P = 3, and C = 1

$\therefore$ Degree of freedom $F = C - P + 2$

$$= 1 - 3 + 2 = \text{zero}$$

Hence, the system is invariant in nature. The temperature of triple point of CO_2 is – 57°C at 5.2 atm. A slight variation in temperature or pressure at triple point may change the phase. For example, on increasing in temperature, the solid phase will disappear and the equilibrium will shift along the curve BD. Similarly the liquid phase may disappear by decreasing the temperature and the equilibrium will shift along BA. If the pressure is increased at constant temperature, the gaseous phase will disappear and the equilibrium will shift along the curve BC.

A significant feature of phase diagram of CO_2 system is that CO_2 gas can be directly solidified without the appearance of liquid phase on cooling at – 78°C at 1 atm. pressure. The solid form of carbon dioxide is known as dry ice and has a great importance.

Salient features of CO_2 system may be represented as in Table 1.3.

Table 1.3. Salient Features of CO_2 System

Curve/Area/ Point	*Name of the System*	*Phase Equilibrium*	*No.of Phase (P)*	*Degree of Freedom (F)*
Curve AB	Sublimation curve	Solid $\rightleftharpoons$ vapour	02	01 (univariant)
Curve BC	Fusion curve	Solid $\rightleftharpoons$ liquid	02	01 (univariant)
Curve BD	Vaporisation curve	Liquid $\rightleftharpoons$ gas	02	01 (univariant)
Area ABC	—	Solid	01	02 (bivariant)
Area CBD	—	Liquid	01	02 (bivariant)
Area below ABD	—	Gas	01	02 (bivariant)
Triple point 'O'	—	Solid $\rightleftharpoons$ liquid $\rightleftharpoons$ gas	03	zero (invariant)

Comparison between the Phase Diagram of CO_2 and H_2O

Both phase diagrams (CO_2 system and H_2O system) are of one component system. The phase diagram of CO_2 system resembles that of water in many respects.

The main differences are following:

(*i*) The slopes of fusion curve of CO_2 system is away from the pressure axis. This indicates that increase of pressure raises the melting point of solid CO_2. The factor $\frac{dp}{dT}$ of the Clausius-Clapeyron equation is positive whereas the slope of fusion curve of water system is towards the pressure axis. It indicates that the melting point of ice is lowered by increase of pressure. The factor $\frac{dp}{dT}$ of the Clausius-Clapeyron equation is negative.

(*ii*) The solid CO_2 exist in equilibrium with its liquid phase only at a very high pressure equal to 5.2 atm. pressure whereas water can exist in equilibrium with its liquid phase at very lower pressure equal to 4.58 mm of Hg.

(*iii*) The vapour pressure of solid CO_2 is very high even at extremely low temperature, whereas the vapour pressure of ice is very low at low temperature.

1.6 TWO COMPONENT SYSTEM

When we express the composition of all the phases in terms of two substances then the system is said to be 'Two Component System'. Such type of system may have as many as three degrees of freedom.

i.e., $P = 1, \quad C = 2.$

$$\therefore \text{ Degree of freedom (F)} = C - P + 2$$

$$= 2 - 1 + 2 = 3$$

Hence the system is trivariant *i.e.*, three variables—temperature, pressure and concentration will express the phase of a system. For graphical representation, a three dimensional model is required which is more complex and difficult to understand. For simplicity we consider the third variable is constant and a simple plane diagram with two variables only is considered. Hence in actual practice three types of phase diagrams are possible.

1. Pressure-temperature diagram
2. Temperature-composition diagram
3. Pressure-composition diagram.

1.6.1 Lead-Silver System

Both lead and silver are completely miscible and form homogeneous mixture. The melting point of pure lead is 327°C. It is lowered by addition of silver. Similarly the melting point of pure silver (961°C) is lowered by addition of lead. The phase diagram of lead-silver system is shown in Fig. 1.5.

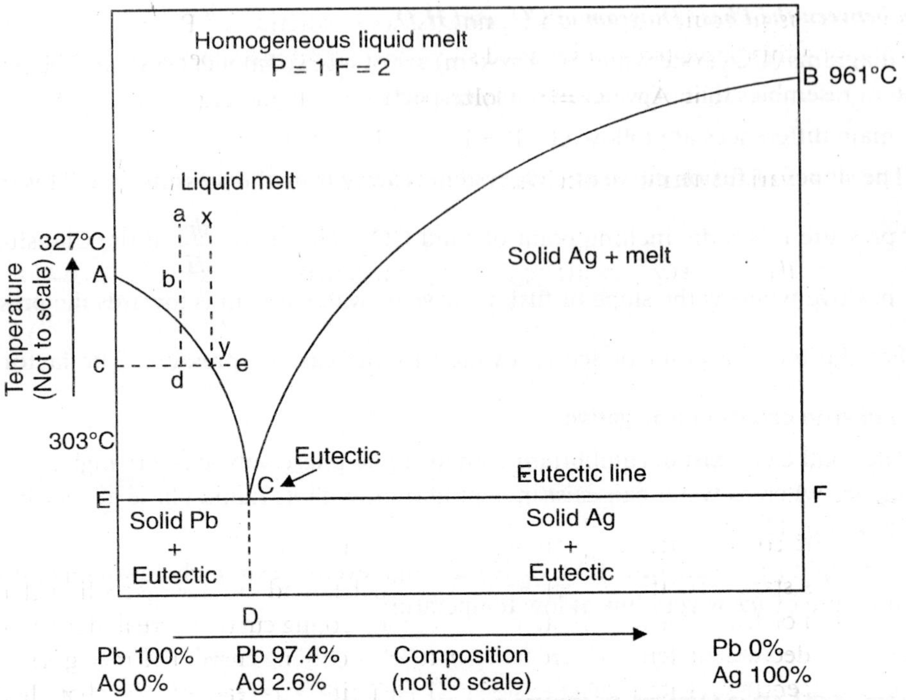

***FIGURE 1.5* Phase diagram for lead-silver system**

Since it is solid-liquid equilibrium system which does not have gaseous phase is called condensed system. Since the pressure does not have any effect on this type of equilibria, hence the degree of freedom for such a system will be reduced by one. Here the reduced phase rule

$$F' = C - P + 1 \text{ is applicable.}$$

This system has four possible phases (1) solid Ag, (II) solid Pb, (III) solution of Ag + Pb and (IV) vapour.

The phase diagram contains following, curves, areas and points.

1. Curves

(*i*) *Curve AC.* It represents the melting point or freezing point curve of Pb by addition of small amount of Ag. All along this curve, the silver which is added goes into solution while the separation of solid Pb takes place. Along this curve solid Pb is in equilibrium with liquid melt (*i.e.*, solution of Ag in liquid Pb) of varying composition. Along this curve two phases solid lead and liquid melt coexist.

$$\text{Solid Pb} \rightleftharpoons \text{Solution of Ag in liquid Pb}$$

$$F' = C - P + 1$$

$$= 2 - 2 + 1 = 1$$

The system is univariant along this curve.

(*ii*) *Curve BC.* It represents the melting point or freezing point curve of Ag by addition of small amount of Pb. The melting point of Ag gets lowered gradually by the addition of lead into it. All along this curve, the Pb which is added goes into solution while the separation of solid Ag takes place. Along this curve solid Ag is in equilibrium with liquid melt (*i.e.*, solution of Pb in liquid Ag), of varying composition. Along this curve two phases solid silver and liquid melt coexist.

$$\text{Solid Ag} \rightleftharpoons \text{Solution of Pb in liquid Ag}$$

$$F' = C - P + 1 = 2 - 2 + 1 = 1$$

i.e., the system is univariant along this curve.

2. Areas

(*i*) *Area above the curve ACB.* In this region only one phase *i.e.*, liquid alloy (melt of Ag and Pb) coexist.

$$\therefore \qquad F' = C - P + 1 = 2 - 1 + 1 = 2$$

i.e., the system is bivariant in this region (temperature and composition).

Cooling of Melt in the Area Above Curve ACB

When a melt of certain composition at a point (say '*a*') in the area above the curve ACB is allowed to cool, it follows Newton's law of cooling and represented as in Fig. 1.6. '*pq*' represents the cooling of liquid and corresponds to '*ab*' in the phase diagram.

Lead crystals start separating when point '*b*' is reached and the rate of cooling changes due to latent heat of fusion of lead. It is represented by '*q*' in the cooling curve. More and more lead goes on separating with the decrease in temperature from '*q*' to '*r*' in cooling curve. The melt goes on becoming more and more concentrated with respect to silver till the point '*r*' is reached when both lead and silver start freezing out. At point 'C' in phase diagram lead and silver freeze out simultaneously. At this point, the composition of the melt is the same as that of the solid mixture of lead and silver. The liquid melt freezes out as if it was a pure compound. The whole of the melt freezes out at constant temperature, represented by horizontal line '*rs*' in the cooling curve. This constant temperature corresponds to point 'C' in phase diagram. This point 'C' is known as eutectic point (low melting) and the solid mixture is called eutectic mixture which has a characteristic composition for each system.

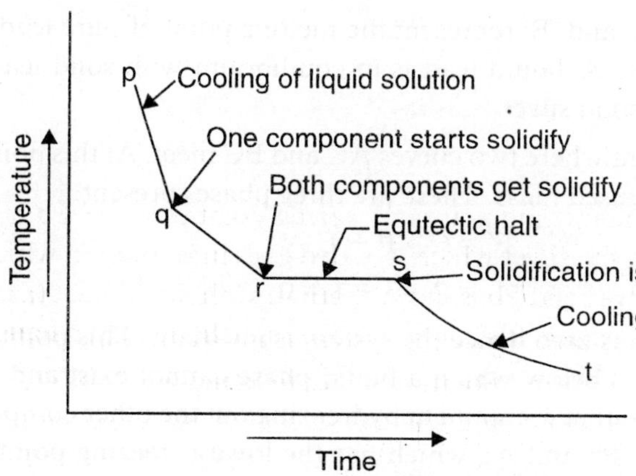

***FIGURE 1.6* Cooling curve of liquid melt**

The eutectic though melts at a constant temperature, is not a chemical compound of lead and silver as (*i*) its composition does not correspond to any simple chemical formula (*ii*) its composition depends on temperature. Below point C only solid phase exists and the cooling of this solid eutectic is represented by a dotted line CD.

(*ii*) *Area ACE.* In this area solid lead and liquid melt coexists. The composition of the liquid melt at any temperature, can be obtained from the curve AC. At point '*d*' in this area solid lead is in equilibrium with a liquid melt of composition which corresponds to the point '*e*' on the curve AC. A horizontal line passing through '*d*' is known as tie line. The amount of solid lead and the amount of saturated solution (melts) is given by the equation.

$$\frac{\text{Amount of lead}}{\text{Amount of melt}} = \frac{cd}{ed}$$

$$\text{Degree of freedom (F}') = C - P + 1$$
$$= 2 - 2 + 1 = 1$$

Therefore, the system is univariant. A point near '*d*' indicates all the solid lead sample and a point near '*e*' implies that nearly all the sample is a melt.

(*iii*) *Area BCF.* In this area solid silver and melt exist. The composition of the liquid melt at any temperature can be obtained from the curve BC as similar in case of area ACE. It is also univariant system as the degree of freedom is one.

(*iv*) *Area below the line ECF.* In this area two phases, solid lead and solid silver exists. In the region (area) represented by solid Pb + eutectic, we find crystals of lead and solid eutectic crystals whereas in the region (area) represented by solid Ag + eutectic, we find crystals of silver and solid eutectic crystals.

$$F' = C - P + 1$$
$$= 2 - 2 + 1 = 1$$

i.e., the system is univariant in this area.

3. Points

(*i*) *Points 'A' and 'B'.* Point 'A' and 'B' represent the melting point of pure lead (327°C) and pure silver (961°C) respectively. At point 'A' liquid lead is in equilibrium with solid lead and at point 'B' liquid sliver is in equilibrium with solid silver.

(*ii*) *Eutectic point C.* It is a point where two curves AC and BC meet. At this point solid mass (lead + silver) is in equilibrium with the fused mass. There are three phases present.

$$\therefore \quad F' = C - P + 1$$
$$= 2 - 3 + 1 = 0$$

Since the degree of freedom is zero hence the system is invariant. This point 'C' represents the lowest possible temperature (303°C) below which a liquid phase cannot exist and beyond which the liquid phase cannot be enriched in either component by freezing out the other component. Such type a liquid mixture of two components Pb and Ag which has the lowest freezing point correspond to all other liquid mixtures is called eutectic mixture and the temperature corresponding to this mixture is called eutectic temperature. This temperature is always less than the melting point of the constituents.

Desilverisation of Lead

Desilverisation of lead is generally done by Pattinson's process. The argentiferrous lead contains a very small percentage of silver. The argentiferrous lead is heated to a temperature above its melting point so that the system consists of only liquid phase. Let us consider the point '*x*' in the phase diagram. It is then allowed to cool gradually along the line '*xy*' without any change of concentration till the point '*y*' is reached. As soon as the point '*y*' is reached, lead starts separating and the solution will contain increasing amount of silver. Further cooling will shift the system along the curve '*y*C'. Lead continues to separate out and the melt continues to be richer and richer in Ag till the eutectic point 'C' is reached. At this point Ag remains 2.6% only in alloy. The alloy is then further refined by the process of cupellation techniques. This process of increasing the relative proportion of silver in the alloy is known as Pattinson's process and is represented as in Fig. 1.7.

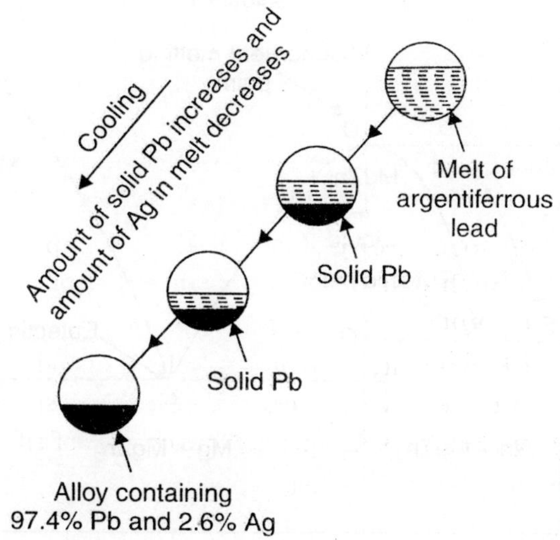

FIGURE 1.7 **Illustrating the gradual increase in the proportion of Ag on cooling of argentiferrous lead**

1.7 SYSTEM HAVING CONGRUENT MELTING POINTS

A system (compound) is said to possess congruent melting point, if it melts sharply at a constant temperature, having the same composition as in solid state. It has been observed that in case of binary alloys, the components undergo chemical combination in some stoichiometric ratio and form one or more compounds. These compounds melt at a particular temperature without changing the ratio of composition. Hence such types of compounds are termed as compounds (systems) having congruent melting point. For examples, Zinc-Magnesium system, Gold-Tin system, Ferric Chloride-water system, Phenol-Aniline system, Aluminium-Magnesium system, Mercury-Thallium system, etc.

1.7.1 Zinc-Magnesium System

Zinc-magnesium binary alloys forms a two component system having a congruent melting point. The melting point of zinc and magnesium is 419°C and 650°C respectively. Both metals chemically combined together in the ratio of 1:2 to form a compound $MgZn_2$. This compound melts at 590°C without changing the composition. It prevails that the congruent melting point of the system is 590°C. Zn-Mg system has four phases.

1. Solid zinc
2. Solid magnesium
3. Solid $MgZn_2$
4. Liquid solution of Zn and Mg.

The phase diagram of the Zn-Mg system may be shown as in Fig. 1.8.

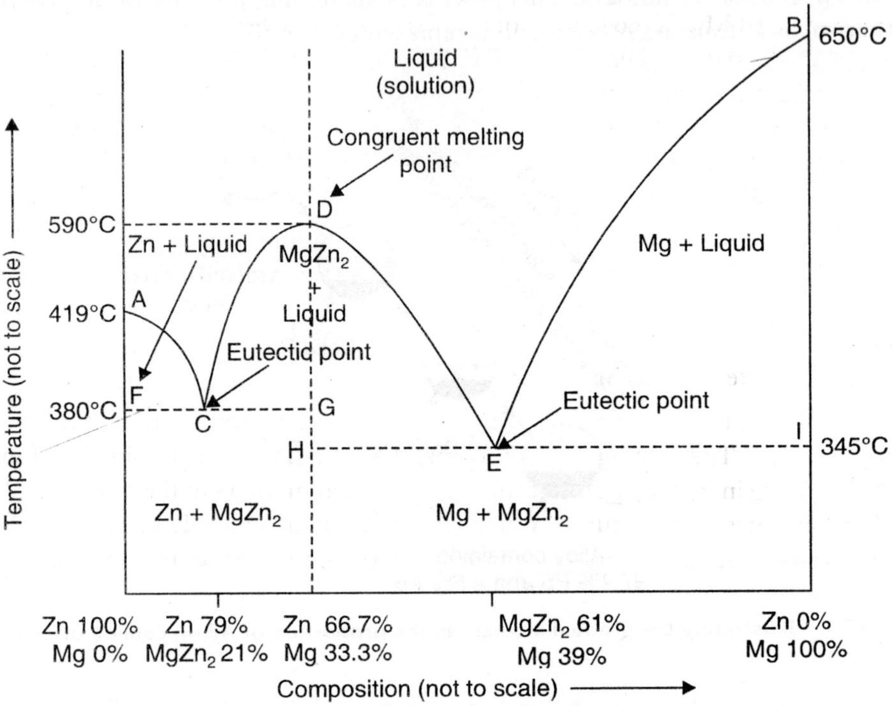

FIGURE 1.8 **The phase diagram of Zn-Mg system**

The phase diagram of this system may be considered as equivalent to two phase diagrams of Pb-Ag type placed side-by-side. The left hand side consists of Zn and $MgZn_2$ system while the right hand side consists of $MgZn_2$ and Mg. Since the pressure does not have any effect on this type of equilibria hence the degree of freedom for such a system will be reduced by one. That's why the reduced phase rule is applicable *i.e.*, $F' = C - P + 1$.

The phase diagram contains the following curves, points and areas.

1. Curves

(*i*) *Curve AC.* It represents the melting point or freezing point curve of zinc (Zn) by addition of small amount of magnesium (Mg). The melting point of Zn gets lowered slowly by the addition of increasing amount of Mg. The lowering in melting point continues till the minimum point C is reached. All along this curve the magnesium is added goes to the solution while the separation of solid Zn takes place. At point 'C' a new solid phase $MgZn_2$ separates out. The equilibrium gets set up along this curve between solid Zn and Zn-Mg solution.

$$\text{Solid Zn} \rightleftharpoons \text{Solution of Mg in liquid}$$

Here $P = 2, C = 2$

$\therefore$ $F' = C - P + 1 = 2 - 2 + 1 = 1$

The system is univariant along this curve.

(*ii*) *Curve CDE.* The curve CDE has a maxima at a point D. This point 'D' represents the melting point (590°C) of compound $MgZn_2$. It is known as congruent melting point of the system because at this point compound $MgZn_2$ coexit in both state *i.e.*, solid as well as liquid state without changing the composition. The curve has been divided into two parts CD and DE. CD portion of the curve represents the increase in freezing or melting point by adding of magnesium in the liquid. DE portion of the curve represents the lowering of melting point of the system till the eutectic point E is reached by adding the amount of magnesium. The equilibrium gets set up along this curve (except at the point D) between solid and liquid $MgZn_2$.

$$\text{Solid } MgZn_2 \rightleftharpoons \text{Liquid } MgZn_2$$

Here $P = 2, C = 2$

$\therefore$ $F' = C - P + 1 = 2 - 2 + 1 = 1$

Thus, the system is univariant along this curve.

(*iii*) *Curve BE.* It represents the melting point or freezing point curve of Mg by addition of small amount of zinc (Zn). The melting point of Mg gets lowered slowly by the addition of increasing amount of Zn. The lowering in melting point continues till the minimum point E is reached. All along this curve the zinc is added goes to the solution while the separation of solid Mg takes place. At point 'E' a new solid phase $MgZn_2$ separates out. The equilibrium gets set up along this curve between solid Mg and Zn-Mg solution.

$$\text{Solid Mg} \rightleftharpoons \text{Solution of Zn in liquid}$$

Here $P = 2, C = 2$

$\therefore$ $F' = C - P + 1 = 2 - 2 + 1 = 1$

Thus, the system is univariant along this curve.

2. Points

(*i*) *Point A and B.* Points A and B represent the melting point of pure Zn (419°C) and pure Mg (650°C) respectively. At point A liquid Zn is in equilibrium with solid Zn and at point B liquid Mg is in equilibrium with solid Mg.

(*ii*) *Point C.* It is the eutectic point (380°C) of the system (Zn-$MgZn_2$ system). At this point three phases are coexists *i.e.*, solid Zn, solid $MgZn_2$ and liquid $MgZn_2$.

$\therefore$ $C = 2, P = 3$

$\therefore$ $F' = C - P + 1 = 2 - 3 + 1 = 0$

Thus, the system is invariant at point C.

(*iii*) *Point D.* It is the congruent melting point (590°C) of the system. The point D is that point at which the compound $MgZn_2$ has a sharp melting point without changing its composition *i.e.*, it has only one component ($MgZn_2$).

$\therefore$ $P = 2, C = 1$

$\therefore$ $F' = C - P + 1 = 1 - 2 + 1 = 0$

Thus, the system is invariant (non-variant) at point D.

(*iv*) *Point E.* It is the eutectic point (345°C) of the system (Mg-$MgZn_2$). At this point three phases coexists *i.e.*, solid Mg, solid $MgZn_2$ and liquid $MgZn_2$.

$\therefore$ $P = 3, C = 2$

$\therefore$ $F' = C - P + 1 = 2 - 3 + 1 = 0$

Thus, the system is non-variant at point E and the composition is 61% $MgZn_2$ and 39% of Mg.

3. Areas

(*i*) *Area ACF.* It consists of two phases zinc and liquid.

Here $P = 2, C = 2$

$\therefore$ $F' = C - P + 1 = 2 - 2 + 1 = 1$

Thus, the system is univariant.

(*ii*) *Area DCG.* It consists of two phases $MgZn_2$ and liquid.

Here $P = 2, C = 2$

$\therefore$ $F' = 1$

Thus, the system is univariant.

(*iii*) *Area DHE.* It consists of two phases *i.e.*, $MgZn_2$ and liquid.

Here $P = 2, C = 2$

$\therefore$ $F' = 1$

Thus, the system is univariant.

(*iv*) *Area BEI.* It consists of two phases Mg and liquid.

Here $P = 2, C = 2$

$\therefore$ $F' = 1$

Thus, the system is univariant.

(*v*) *Area below FCG.* It consists of two phases Zn and $MgZn_2$.

Here $P = 2, C = 2$

$\therefore$ $F' = 1$

Thus, the system is univariant.

(*vi*) *Area below HEI.* It consists of two phases Mg and $MgZn_2$.

Here $P = 2, C = 2$

$\therefore$ $F' = 1$

Thus, the system is univariant.

(*vii*) *Area above ACD.* It consists of only one phase *i.e.*, liquid phase. In this area liquid Zn and liquid $MgZn_2$ coexist.

Here $P = 1, C = 2$

$\therefore$ $F' = C - P + 1 = 2 - 1 + 1 = 2$

Thus, the system is bivariant.

(*viii*) *Area above DEB.* It consists only one phase *i.e.*, liquid Mg and liquid $MgZn_2$.

Here $P = 1, C = 2$

$\therefore$ $F' = 2$

Thus, the system is bivariant.

1.7.2 System having Incongruent Melting Points

A system (compound) is said to possess incongruent melting point, if it decomposes much below its melting point and forms a new solid phase and a solution having different composition from solid state. It has no any sharp melting point.

$$\text{Original Solid} \rightleftharpoons \text{New Solid + Solution (melt)}$$

The decomposition at this temperature is known as transition reaction or meritectic or peritectic reaction and the temperature (the incongruent melting point) is known as transition temperature or meritectic or peritectic temperature. For examples, sodium-potassium system, gold-antimony system, potassium chloride-copper chloride system, picric acid-benzene system, sodium sulphate-water system etc. In such type of systems the components undergo chemical combination and forms a new compound which is unstable and on heating it decomposes much below its melting point to form a new solid and a solution or melt.

1.8. SODIUM SULPHATE-WATER SYSTEM

Sodium Sulphate-water system is a two component system having incongruent melting point. Decahydrated sodium sulphate decomposes at 32.4°C into the anhydrous rhombic sodium sulphate and water. It prevails that the peritectic temperature of this system is 32.4°C. There are seven phases exist in the system.

(*i*) $Na_2SO_4.10H_2O$ (*ii*) $Na_2SO_4.7H_2O$

(*iii*) Anhydrous rhombic Na_2SO_4 (*iv*) Anhydrous monoclinic Na_2SO_4

(*v*) Solid ice (*vi*) Water

(*vii*) Water vapour.

The phase diagram representing temperature versus composition at constant vapour pressure may be represented as in Fig. 1.9.

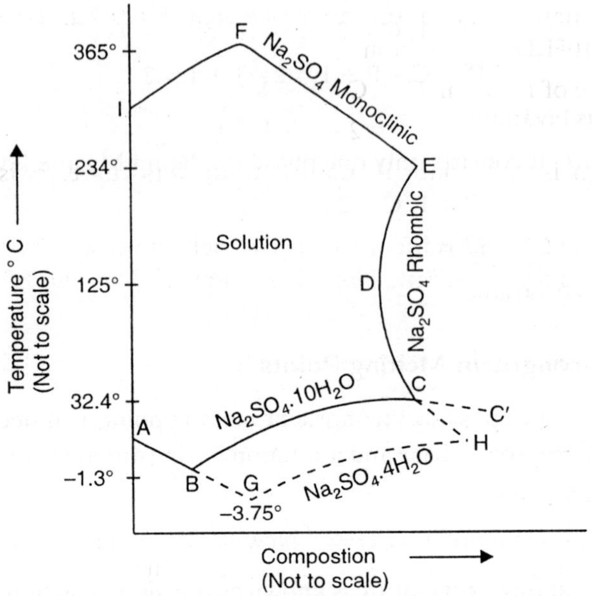

FIGURE. 1.9 **The phase diagram for the $Na_2SO_4.H_2O$ system**

The phase diagram of $Na_2SO_4.H_2O$ system may be explained as follow

(*i*) *Point A and Curve AB:* A is the melting point of ice. Ice and water coexist in equilibrium at 0°C and 1 atm. pressure.

At this point degree of freedom $F' = C - P + 1$

$$= 1 - 2 + 1 = \text{zero}$$

i.e., the system is invariant at this point.

When we add sodium sulphate, a solution of salt in water of varying composition is obtained and the equilibrium shifts along the curve AB. The curve AB represents the equilibrium between liquid solution and solid ice. It is the fusion curve of ice.

$$\text{Solid ice} \rightleftharpoons \text{liquid solution}$$

Here P = 2, C = 2

∴ Degree of freedom $F' = C - P + 1$

$$= 2 - 2 + 1 = 1$$

Hence, the system is univariant in nature along this curve.

(*ii*) *Point B and curve BC:* If we add Na_2SO_4 continuously, a point B is reached at – 1.3°C where $Na_2SO_4.10\,H_2O$ separates out. Here three phases coexist.

$$Na_2SO_4.10H_2O \rightleftharpoons \text{ice} \rightleftharpoons \text{Solution}$$

Here P = 3, C = 2

∴ Degree of freedom $F' = C - P + 1$

$$= 2 - 3 + 1 = \text{zero}$$

i.e., the system is invariant at this point. The point B is the eutectic point of the system.

If we increase the temperature at this point, ice melts rapidly and the system will contain only two phases *i.e.*, $Na_2SO_4.10H_2O$ and solution

In this case, degree of freedom $F' = C - P + 1$

$$= 2 - 2 + 1 = 1$$

Hence the system is univariant in nature along this curve. It is the solubility curve of $Na_2SO_4.10H_2O$.

(*iii*) *Point C and curve CDE:* C is the incongruent melting point (32.4°C) where $Na_2SO_4.10H_2O$ decomposes into the anhydrous rhombic sodium sulphate and solution.

$$Na_2SO_4.10H_2O \rightleftharpoons Na_2SO_4 + 10H_2O.$$

Here the composition of solution phase is different from that of solid phase.

Here P = 3, C = 2

∴ Degree of freedom $F' = C - P + 1$

$$= 2 - 3 + 1 = \text{zero}$$

Hence the system is invariant in nature at this point.

On increasing the temperature, all the decahydrates disappear. In this case only two phases will coexist.

∴ Degree of freedom $F' = C - P + 1$

$$= 2 - 2 + 1 = 1$$

Hence the system is univariant in nature along this curve CDE. Along this curve CDE, the solution is in contact with anhydrous rhombic Na_2SO_4. It is the solubility curve of the anhydrous rhombic Na_2SO_4. The solubility of anhydrous rhombic Na_2SO_4 first decreases upto the point D (125°C) and then increases with rise of temperature upto 234°C (point E).

(*iv*) *Point E and curve EF:* E is the point where rhombic anhydrous Na_2SO_4 changes into monoclinic form. Here three phases rhombic anhydrous Na_2SO_4, monoclinic anhydrous Na_2SO_4 and solution coexist.

Here P = 3, C = 2

∴ Degree of freedom $F' = C - P + 1$

$$= 2 - 3 + 1$$

$$= \text{zero}$$

Hence the system is invariant in nature at this point.

On continue heating at this point E, rhombic form is completely changed into monoclinic form and only two phases coexist. In this case curve EF will obtain.

Here, degree of freedom $F' = C - P + 1$

$$= 2 - 2 + 1 = 1$$

Hence the system is univariant in nature along this curve EF. It is the solubility curve of anhydrous monoclinic Na_2SO_4. The solubility decreases with rise of temperature.

(*v*) *Point F and curve FI:* Point F at 365°C represents the slight solubility of monoclinic sodium sulphate in the vapour phase. It should be noted that it does not corresponds to zero solubility. Curve FI represents the solubility of monoclinic Na_2SO_4 in vapour phase.

(*vi*) *Metastable curves CC′ and CH:* Curve CC′ represents the metastable phase of $Na_2SO_4.10H_2O$ in equilibrium with the solution. It indicates that $Na_2SO_4.10H_2O$ can exist beyond the transition temperature (32.4°C) without changing into the anhydrous rhombic sodium sulphate.

Here P = 2, and C = 2

$\therefore$ Degree of freedom $F' = C - P + 1$

$= 2 - 2 + 1 = 1$

Hence, the system is univariant in nature along this curve.

Curve CH represents the equilibrium of anhydrous rhombic Na_2SO_4 with the solution. It indicates that if a solution of anhydrous rhombic sodium sulphate in water is cooled quickly below the transition temperature 32.4°C, the solid may not change into the decahydrate salt *i.e.*, $Na_2SO_4.10H_2O$.

Here P = 2, and C = 2

$\therefore$ Degree of freedom $F' = C - P + 1$

$= 2 - 2 + 1 = 1$

Hence, the system is univariant in nature along this curve.

1.9 APPLICATIONS

It has wide applications in industries, pharmaceutical science, medical science, etc. Some important applications are the following:

1.9.1 Freeze Drying (Lyophilization)

'Freeze drying' is a special technique used for the dehydration of food and several applications in pharmaceutical industries. Actually this technique 'Freeze Drying' is accomplished by sublimation of ice in vacuum. It is also known as Lyophilization. It is actually a preservation process that is a combination of science and creative design, which was originally created for the pharmaceutical and food industry. Almost all botanicals, fruits and vegetables can be freeze-dried. Sublimation means the conversion of solid ice directly to the vapour state. It is represented by sublimation curve in the phase diagram of water as in Fig. 1.2. This technique become famous after the World War II. Now a days it is used in pharmaceutical industry, camping/hiking food processors, museums, taxidermy, and the floral industry.

The basic principle of floral freeze drying is the removal of water as water vapour from frozen botanicals and collected as ice in a condenser. This process is done under a deep vacuum and below a pressure of about 1 mm of Hg. During the process shrinkage of flowers are eliminated or minimized. The material or tissue is left almost as a skeleton and original matter can be obtained by adding water. Once the flowers are deeply frozen, a vacuum is applied to pull all of the water out in the vapour form. Sublimation occurs when this frozen liquid is directly converted into a gaseous state. This water vapour is transferred to a separate chamber (ice bank) as ice form at – 40°F or lower and is removed. After removing all of the water the freeze drying chamber is slowly returned to room temperature. The whole process takes approximately 12–14 days, and cannot be rushed.

When the freeze dried flower industry began in 1988, the flowers were fragile, shattered easily, and had a tendency to change colour. These problems have all been solved through Chemistry. The pre and post-treating chemicals used today meet the needs for colour stabilization, durability and longevity in freeze dried flowers.

1.9.2 Solders

A solder is an alloy having lower melting point than that of the metal pieces which have to be joined together. Eutectic mixture can be used for this purpose but there is one great disadvantage that eutectic mixture freezes sharply at its freezing point. Solders in actual use, have compositions some what different from the eutectics so that the freezing occurs over a range of temperatures. The capacity of solder depends upon the formation of a surface alloy between the solder and parts of metals being soldered. The selection of solder alloy is based upon the melting point desired and the pieces of metals to be joined. Generally soft (tinner's) solder which is an alloy of Pb and Sn is used. Another alloy is 'Plumber alloy' containing Pb = 67% and Sn = 33%. It is commonly used for soldering the pipes. The alloy in which Pb and Sn are equally shared (*i.e.*, Pb = 50% and Sn = 50%) is very common and famous solder in market. It is called 'half-half' solder. It provides a bright surface finish after soldering. It has very good soldering properties but is very expensive. Also due to high contents of tin, it is not widely applicable for several appliances. The alloy containing about 60% Pb is used as solders in electrical wiring. Hence, we can conclude that different types of solders are useful in different types of soldering.

A good solder has following characteristics:

(*i*) Its melting point should be lower than the soldering metals.

(*ii*) It should spread in liquid form.

(*iii*) It should possess good quality of wetting the soldering metals.

(*iv*) It should form homogeneous mixture with the soldering metals.

1.9.3 Safety Plug

Safety plugs or safety fuse is an alloy having low melting point, used to ensure the safe working and avoid accidents. Safety fuses are used in buildings to protect them against fires. One such alloy is woods metal (Bi = 50%, Pb = 25%, Sn = 12.5% and Cd = 12.5%). This alloy melts at 65°C. It is used for plugging water sprayers in a building. In case of a fire in a room the plug melts off and the automatic spraying of water puts off the fire. Similarly safety fuses can be used in the working of electric alarms also. Safety plugs are fitted in steam boilers, pressure cookers etc. to prevent the accidents. In case of blockage of excess steam in the boiler or cooker the safety plugs automatically melts down and the pressure of the system decreases.

1.9.4 Freezing Mixture

A mixture of ice and salt is known as freezing mixture. It is used to obtain low temperatures. It has been observed that the addition of salt to ice results in considerable lowering of temperature. A good freezing mixture should satisfy the following conditions:

1. It should have a low cryohydric temperature *i.e.*, salt must be highly soluble.
2. The heat of solution of the salt should be high *i.e.*, solubility of the salt should increase rapidly on increasing the temperature.
3. The salt should be cheap.
4. Components used should be such that they form an intimate mixture on cooling.

In actual practice, a mixture of ice and common salt is used as a freezing mixture. Although common salt is very cheap and easily available but it is not a good component for freezing mixture because the heat of solution of this salt is very low and the heat absorbed is almost due to the heat of fusion of ice. Calcium chloride hexahydrate and ice form an excellent freezing mixture because of very low cryohydric point and high heat of solution. Some important freezing mixtures along with their eutectic temperature and percentage eutectic composition have been listed in Table 1.4.

Table 1.4. Freezing Mixtures

System	*Composition (% of Salt in the Mixture)*	*Eutectic Temperature (°C)*
NH_4Cl and Ice	20.1	– 16.0
NH_4NO_3 and Ice	43.0	– 18.0
$NaNO_3$ and Ice	33.3	– 18.1
$NaCl\ 2H_2O$ and Ice	23.0	– 22.0
KI and Ice	52.0	– 23.0
$CaCl_2\ 6H_2O$ and Ice	15.2	– 55.9

SOLVED NUMERICALS

1. *Calculate the number of components and degrees of freedom for the following equilibria:*

(*i*) *An aqueous solution of glucose*

(*ii*) *An aqueous solution of NaCl.*

Solution.

(*i*) The number of components (C) = 2 (glucose and water)

$$F = C - P + 2$$
$$= 2 - 1 + 2$$
$$= 3$$ (temperature, pressure, and concentration of solution)

(*ii*) Number of components = 2 (NaCl and water)

$$F = C - P + 2$$
$$= 2 - 1 + 2$$
$$= 3$$ (temperature, pressure, concentration of solution)

2. *Is it possible to have a quadruple point on a phase diagram for one component system?*

Solution.

For a quadruple point the number of phases in equilibrium with one another must be four.

According to the phase rule equation for one component system

$$F = C - P + 2 \text{ (Here } C = 1, P = 4)$$
$$= 1 - 4 + 2 = -1$$

which is ridiculous. Hence, we cannot have a quadruple point on a phase diagram for one component system.

3. *Write down the number of component, number of phases and calculate the degrees of freedom for the following equilibria:*

(*i*) $N_2(g) + 3H_2(g) \rightleftharpoons 2NH_3(g)$

(*ii*) $Fe(s) + H_2O(g) \rightleftharpoons FeO(s) + H_2(g)$

(*iii*) $H_2O(g) \rightleftharpoons H_2(g) + \frac{1}{2}O_2(g)$

Solution.

(*i*) Chemical species present are

$$N_2(g),\ H_2(g) \text{ and } NH_3(g)$$

Number of component $C = 2$ ($\because$ $C = S - R = 3 - 1 = 2$)

Number of phases $P = 1$

Degree of freedom $F = C - P + 2$

$= 2 - 1 + 2 = 3.$

(*ii*) Chemical species present are

$$Fe(s),\ FeO(s),\ H_2O(g),\ H_2(g)$$

$\therefore$ Number of components $C = 3$ ($\because$ $C = S - R = 4 - 1 = 3$)

Number of phases $P = 3$ (Two solids and one gaseous phase)

Degree of freedom $F = C - P + 2$

$= 3 - 3 + 2 = 2$

(*iii*) Number of species present are

$$H_2O(g),\ H_2(g) \text{ and } O_2(g)$$

$\therefore$ Number of components $C = S - (R + 1)$ ($\because$ $P_{H_2} = 2P_{O_2}$)

$= 3 - (1 + 1) = 1$

Number of phases $P = 1$

$\therefore$ Degree of freedom $F = C - P + 2$

$= 1 - 1 + 2 = 2.$

4. *Write down the number of components, number of phases and evaluate the degrees of freedom for the following equilibria:*

(*i*) $N_2O_4(g) \rightleftharpoons 2NO_2(g)$

(*ii*) *Solid carbon in equilibrium with gaseous CO, CO_2 and O_2 at 373 K.*

Solution.

(*i*) Number of component $= 1$

Number of phases $= 1$

Degree of freedom $F = C - P + 2$

$= 1 - 1 + 2 = 2$

(*ii*) Number of components $= 2$ $(\because \quad C = S - R = 4 - 2 = 2)$

Number of phases $= 2$

$\therefore$ Degree of freedom $F = C - P + 2$

$= 2 - 2 + 2 = 2$

5. *1000 kg of a sample of argentiferrous lead containing 0.1% silver is melted and then allowed to cool. If eutectic contains 2.6% Ag, what mass of:*

(*i*) *eutectic will be formed, and*

(*ii*) *mass of lead will separate out?*

Solution. (*i*) Mass of Ag in 1000 kg argentiferrous lead

$$= \frac{0.1}{100} \times 1000 \text{ kg} = 1 \text{ kg}$$

Mass of eutectic $= \dfrac{1 \text{ kg}}{2.6} \times 100 = 38.46$ kg

(*ii*) Mass of Pb separated $= [1000 - 38.46]$ kg $= 961.54$ kg.

6. *KCl—NaCl—H_2O is a three-component system while NaBr—KCl—H_2O is a four-component system. Explain it.* **(K.U.K. 2003)**

Solution.

The first system consists of six species: KCl, NaCl, Na^+, K^+, Cl^-, H_2O (dissociation of H_2O is ignored). The number of independent equilibrium reactions (R) are two.

$$NaCl \rightleftharpoons Na^+ + Cl; \; KCl \rightleftharpoons K^+ + Cl^-.$$

Condition of electroneutrality is one. Thus $C = S - (R + 1)$ gives:

$$C = 6 - (2 + 1) = 3$$

$$(S = 6, R = 2)$$

In the second system there are nine species:

KCl, NaCl, NaBr, KBr, Na^+, K^+, Cl^-, Br^-, H_2O. The number of independent reactions is four:

$$KCl \rightleftharpoons K^+ + Cl^- \; ; \; NaBr \rightleftharpoons Na^+ + Br;$$

$$Na^+ + Cl^- \rightleftharpoons NaCl; \; K^+ + Br^- \rightleftharpoons KBr;$$

$C = S - (R + 1)$ (Here S = 9, R = 4)

$\therefore$ $C = 9 - (4 + 1) = 4.$

EXERCISE

1. State phase rule and explain the significance of the terms involved. Illustrate your answer with suitable examples.
2. Define the terms with suitable example:
   (*i*) Phase
   (*ii*) Component
   (*iii*) Degree of Freedom. ***(K.U.K. June 2006, K.U.K. 2003, Jan. 2007, Jan. 2008)***
3. (*a*) Explain the following terms:
   (*i*) Eutectic mixture
   (*ii*) Eutectic point
   (*iii*) Triple point
   (*iv*) Condensed phase rule. ***(K.U.K. Jan. 2008)***
   (*b*) What do you understand by congruent melting point and incongruent melting point? Discuss Zn-Mg system in details.
4. Derive phase rule equation *viz.* F = C – P + 2. Why reduced phase rule is applicable in Pb-Ag system.
5. Draw and discuss the phase diagram of lead-silver system. Explain the eutectic point and Pattinson's process of desilverization of lead. ***(K.U.K. Jan. 2007)***
6. Explain the terms:
   (*i*) Stable equilibrium (*ii*) Metastable equilibrium.
7. (*a*) Draw and discuss the phase diagram of water system from the stand point of the phase rule and give special emphasis on the following fact:
   (*i*) Why is solid-liquid line almost vertical and slightly tilted towards the left?
   (*ii*) Why is solid-vapour line more steep than liquid-vapour line at the triple point?
   (*iii*) What is the upper limit of the liquid-vapour equilibrium line?
   (*b*) What is meant by peritectic point?
   (*c*) Explain why the fusion curve of ice has a negative slope whereas sublimation curve has a positive slope in the phase diagram of water? ***(K.U.K. June 2009)***
8. (*a*) How is the phase diagram of water helpful in explaining:
   (*i*) ice skating, and (*ii*) flow of glaciers?
   (*b*) Draw the complete phase diagram of water system and prove that the conclusion in regard to the degree of freedom as derived from the diagram are the same as the deduction from the phase rule.
9. (*a*) What is triple point? Explain triple point with reference to water system.
   (*b*) Write a note on 'Reduced phase rule'.
   (*c*) State Gibb's phase rule and explain the terms involved.
10. (*a*) Write a short note on:
   (*i*) Freezing mixture, (*ii*) Solders, and (*iii*) Freeze drying.
   (*b*) Distinguish between congruent and incongruent melting point. Discus sodium sulphate-water system in details.
11. Justify the statement, "The eutectic is a mixture and not a compound." ***(K.U.K. June 2004)***
12. What is the principle of sublimation? Discuss its application in Freeze Drying. ***(K.U.K. June 2004)***

13. (*a*) Differentiate between 'true equilibrium' and 'metastable equilibrium'. ***(K.U.K. Jan. 2004)***
(*b*) What do you understand by Reduced Phase Rule? ***(K.U.K. Jan. 2004)***

14. What are the criteria for phase equilibria? How is supercooled water an example of metastable equilibrium?

15. Define triple point. How is it differ from eutectic point? Explain clearly why there is no choice of triple point in water system? What is the upper limit of liquid-vapour equilibrium line in this system?

16. State and explain any two terms with suitable examples frequently involved in phase rule. Describe the metastable curve in H_2O system. Write the mathematical form of reduced phase rule equation. ***(K.U.K. Jan. 2005)***

17. What is meant by eutectic point? Explain how can the eutectic point be calculated? Discuss the Pb-Ag system. ***(K.U.K. Jan. 2005)***

18. Discuss the general phase diagram of two components system forming compounds in the solid state with congruent melting point. ***(K.U.K. Jan. 2005)***

19. Define the various curves involved in water system with a neat and sketched diagram. Why is the fusion curve in the phase diagram of water system inclined towards the pressure axis? Explain. ***(K.U.K. June 2006, K.U.K. June 2005)***

20. Describe with net, sketched diagram a two component system with incongruent melting point. ***(K.U.K. June 2005)***

21. State and explain the reduced phase rule equation. Write its physical significance. ***(K.U.K. Jan. 2006, M.D.U. Rohtak June 2008)***

22. Draw a net, cleaned and labelled sketch of the phase diagram of water system. Discuss all the equilibrium involved in it. ***(K.U.K. Jan. 2006)***

23. Calculate P, C and F in the following cases:
(*i*) $NH_3(g)$ at 42°C
(*ii*) An emulsion of oil in water at 2 atm and 70°C.
(*iii*) $S_P \rightleftharpoons S_M$ at the transition temperature.
(*iv*) Pure crystals of $CuSO_4 . 5H_2O$
(*v*) Water system at 4.578 mm of Hg and at 0.0098°C. ***(K.U.K. Jan. 2006)***

24. Draw a neat, cleaned, labelled sketch of the phase diagram of carbon dioxide system. Explain the curves, points and areas of the phase diagram.

25. What is phase rule? Draw a neat, cleaned phase diagram of Zn-Mg system. Explain all points, curves and areas in detail.

26. Explain the phase diagram of sodium sulphate-water system in details.

27. Draw a neat, cleaned and labelled sketch of the phase diagram of sodium sulphate-water system. Discuss all the equilibrium involved in it.

28. What do you understand by one component and two component phase diagrams. How the phase diagram of carbon dioxide is differ from the phase diagram of water?

29. What is dry ice? Explain the phase diagram of CO_2 system in details.

30. Explain a two component system involving the formation of a eutectic solid. What is an important application of this system. ***(K.U.K. June 2009)***

31. Derive an expression for Gibb's phase rule equation. ***(K.U.K. June 2009)***

32. Differentiate with suitable examples the congruent and incongruent melting solids. ***(K.U.K. June 2009)***

33. Define phase, component, degree of freedom, eutectic point, freezing point.

(K.U.K. June 2009, Jan. 2009, June 2008, M.D.U. Rohtak June 2008)

34. What are phase rule and reduced phase rule? Discuss Pb-Ag system helpful in desilverisation of lead.

(K.U.K. Jan. 2009)

35. What is the composition of eutectic in Pb-Ag system? Write its significance in industries.

(K.U.K. June 2009)

36. An alloy of tin and lead contains 73% tin. Calculate the mass of eutectic in 1 kg of solid alloy, if the eutectic contains 64% of tin.

(K.U.K. Jan. 2008)

[Hint: 73% tin present in alloy means 730 g of tin and 270 g of lead are present in 1 kg alloy. Eutectic contains 64% tin means 640 g of tin and 360 g of lead is present in 1 kg of eutectic.

Therefore, mass of tin present in eutectic = $\frac{640 \times 270}{360} = 480$ g.

$\therefore$ Total mass of eutectic in an alloy = 480 g + 270 g = 750 g.]

37. Differentiate between triple point and eutectic point. Mention two limitations of phase rule.

(K.U.K. Jan. 2009)

38. Draw a labelled diagram of Pb-Ag system explaining the significance of all the curves and the intersections. Also discuss the utility of this diagram in desilverisation of lead. *(K.U.K. Jan. 2002)*

Chapter 2 CATALYSIS

2.1 INTRODUCTION

A catalyst is a substance which alters the rate of a chemical reaction without undergoing any chemical change and can be recovered at the end of the reaction. The phenomenon of acceleration or retardation of the rate of a chemical reaction by using catalyst is called catalysis. If a catalyst accelerate the rate of chemical reaction is called a positive catalyst and the phenomenon is called positive catalysis. On the other hand, if a catalyst retards the rate of chemical reaction is called a negative catalyst and the phenomenon is called negative catalysis.

Examples of positive catalysis:

(*i*) Decomposition of $KClO_3$ by using MnO_2 catalyst

$$2KClO_3 \xrightarrow{MnO_2, \Delta} 2\,KCl + 3O_2$$

(*ii*) Oxidation of SO_2 by using Pt catalyst

$$2SO_2 + O_2 \xrightarrow{Pt, \Delta} 2\,SO_3$$

(*iii*) Hydrogenation of vegetable oil by using finely divided nickel catalyst

$$\text{Vegetable oil} + H_2 \xrightarrow{Ni} \text{Vegetable ghee}$$

Examples of negative catalysis:

(*i*) Decomposition of H_2O_2 by using H_3PO_4 as a catalyst

$$2H_2O_2 \xrightarrow{H_3PO_4, \Delta} 2H_2O + O_2$$

(*ii*) In the contact process, the rate of combination of SO_2 and O_2 is slowed down by using arsenic compound or V_2O_5 as a catalyst.

$$2SO_2 + O_2 \xrightarrow{V_2O_5} 2SO_3$$

(*iii*) Decomposition of H_2O_2 is retarded by using small amount of glycerol or acetanilide as a catalyst.

$$2H_2O_2 \xrightarrow{\text{acetanilide}} 2H_2O + O_2$$

2.2 GENERAL CHARACTERISTICS OF CATALYTIC REACTIONS

Following are the general characteristics of catalytic reactions:

1. **The catalyst does not initiate the reaction:** The catalyst only accelerates the rate of reaction. In the absence of the catalyst, the reaction is already occurring very slowly. When we use catalyst the reaction takes place some alternative path which requires much lower energy of activation. Hence it accelerates the rate of reaction.
2. **Only a small quantity of the catalyst is generally required:** It has been observed that in many reactions only a small quantity of the catalyst is generally required to convert very large amount of reactants to the products. For example, one gram of platinum is quite sufficient for the decomposition of 10^8 litres of hydrogen peroxide. However, there are some reactions in which the rate of reactions increases with the increase in concentration of the catalyst.
   For example, the inversion of cane sugar in the presence of dil HCl. Here HCl acts as a catalyst. The rate of reaction increases with the increase in the concentration of catalyst.
3. **A catalyst remains unchanged chemically at the end of the reaction:** The amount of the catalyst recovered at the end of the reaction although its physical state like size, colour etc may change.
   For example, MnO_2 which is used as a catalyst in the decomposition of $KClO_3$ changes its physical state from coarse to fine powder at the end of the reaction.
4. **A catalyst is specific in its action:** A catalyst can catalyse only a specific reaction. For example, MnO_2 may catalyze the decomposition of $KClO_3$ but not of $KClO_4$. Similarly enzymes are also specific in their action. On the other hand it has been observed that same reactant yield different products with different catalysts.
   For example, Ethyl alcohol undergoes dehydration in the presence of alumina (catalyst) in hot condition and gives ethylene whereas the same ethyl alcohol undergoes dehydration in the presence of copper (catalyst) in hot condition and yields acetaldehyde.

$$C_2H_5OH \xrightarrow{\text{hot } Al_2CO_3} C_2H_4 + H_2O$$

$$C_2H_5OH \xrightarrow{\text{hot Cu}} CH_3CHO + H_2$$

5. **A catalyst does not affect the state of equilibrium in a reversible reaction:** According to Thermodynamics point, for a reversible reaction whether it occurs in the presence or absence of a catalyst, the free energy of the process is the same. Since a catalyst accelerates both the forward and backward reaction and help to attain the equilibrium quickly. Hence it does not affect the equilibrium but it is true only in the cases where the catalyst used is in small

quantity. Those reactions where large amount of the catalyst is required the value of equilibrium constant may be affected. For example, the value of equilibrium constant changes for the hydrolysis of ethyl acetate using large amount of HCl as a catalyst.

6. **A catalyst has an optimum temperature:** A catalyst work effectively at a particular temperature for a particular reaction. For example, the optimum temperature for the oxidation of SO_2 to SO_3 in the presence of Pt catalyst is 450°C to 500°C.

$$2SO_2 + O_2 \xrightarrow[450-500°C]{Pt} 2SO_3$$

7. **Activity of a catalyst is increased by the presence of promotors:** The activity of a catalyst is generally increased in the presence of small quantity of promotor or activator. For example, in the manufacture of ammonia by Haber's process, finally divided iron supported over molybdenum. Here molybdenum acts as a promotor and iron acts as a catalyst.
8. **A catalyst is poisoned by the presence of certain substances:** The substances which make the catalyst inactive are called catalytic poisons. Their trace amount is enough to deactivate the catalyst. Some powerful catalytic poisons are HCN, H_2S, CO, AS_2O_3 etc. For example, the iron catalyst is poisoned by H_2S gas during the synthesis of ammonia by Haber's process.

2.3 TYPES OF CATALYSIS

Generally catalysis is classified into two categories:

1. Homogeneous catalysis
2. Heterogeneous catalysis

1. **Homogeneous catalysis:** Homogeneous catalysis are those types in which the reactants and the catalyst exist in the same phase.

Examples:

(*i*) Oxidation of SO_2 to SO_3 in the presence of nitric oxide as a catalyst in lead chamber process during the manufacturing of sulphuric acid.

$$2SO_2(g) + O_2(g) \xrightarrow{NO(g)} 2SO_3(g)$$

(*ii*) Decompositions of acetaldehyde in the presence of iodine.

$$CH_3\ CHO(g) \xrightarrow{I_2\ vapours} CH_4(g) + CO(g)$$

(*iii*) Inversion of cane sugar by acid catalyst.

$$\underset{\text{(cane sugar)}}{C_{12}H_{22}O_{11}(l)} + H_2O(l) \xrightarrow{H^+(aq)} \underset{\text{(glucose)}}{C_6H_{12}O_6(l)} + \underset{\text{(fructose)}}{C_6H_{12}O_6(l)}$$

There are several examples of homogeneous catalysis amongst which the acid-base catalysis and enzyme catalysis are very important.

2.3.1 Acid Base Catalysis

There are large number of acid-base catalysis reactions which are catalysed by H^+ or OH^-. For example, inversion of cane sugar, hydrolysis of esters, mutarotation of glucose etc.

$$\underset{\substack{\text{Cane sugar}\\ \text{(Dexo-rotatory)}}}{C_{12}H_{22}O_{11}} + H_2O \xrightleftharpoons{H^+ \text{ or } H_3O^+} \underset{\substack{\text{Glucose}\\ \text{(Dexo rotatory)}}}{C_6H_{12}O_6} + \underset{\substack{\text{Fructose}\\ \text{(Laevo rotatory)}}}{C_6H_{12}O_6}$$

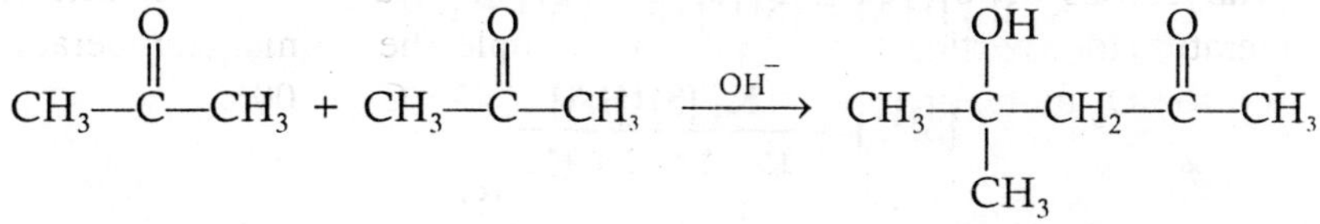

Acetone　　　　Diacetonyl alcohol

The reactions which are catalysed by H^+ or H_3O^+ are known as specific acid catalysis and the reactions which are catalysed by OH^- are known as specific base catalysis.

Recently it has been observed that many reactions which are catalysed by H^+ or H_3O^+ or any other substance which is an acid according to Bronsted concept (Proton donor) can act as a catalyst and are called general acid catalysis. Similarly reactions which are catalysed by all bases (proton acceptor) are called general base catalysis.

2.3.1.1 Mechanism (Kinetics) of Acid-Base Catalysis

In this reaction transfer of proton takes place from an acid to a substrate molecule (S) or from substrate molecule to the base. Let us consider a general reaction.

$$\underset{\text{(Substrate)}}{S} + \underset{\text{(Acid)}}{HA} \underset{K_{-1}}{\overset{K_1}{\rightleftharpoons}} SH^+ + A^-$$

$$SH^+ + H_2O \xrightleftharpoons{K_2} \text{Product}$$

where S = Substrate

HA = Acid

P = Product

K_1, K_{-1} and K_2 are rate constants for various reactions.

Rate of formation of product is given by

$$r = \frac{d\,[p]}{dt} = K_2\,[SH^+] \quad \text{...(1)}$$

Rate of formation of intermediate is given by

$$\frac{d\,[SH^+]}{dt} = K_1\,[S]\,[HA] - K_{-1}\,[A^-]\,[SH^+] - K_2\,[SH^+] \quad \text{...(2)}$$

When the reaction is started the concentration of the species SH^+ may be considered as steady state concentration, it means the concentration of intermediate which does not change with time *i.e.* the rate of formation of intermediate will be equal to the rate of its disappearance. Hence,

$$\frac{d\,[SH^+]}{dt} = 0 \quad \text{...(3)}$$

From equation (2) and (3) we get

$$K_1 [S] [HA] - K_{-1} [A^-] [SH^+] - K_2 [SH^+] = 0 \quad ...(4)$$

or, $$K_1 [S] [HA] = K_{-1} [A^-] [SH^+] + K_2 [SH^+]$$

or, $$K_1 [S] [HA] = [SH^+] (K_{-1} [A^-] + K_2)$$

or, $$[SH^+] = \frac{K_1 [S] [HA]}{K_{-1} [A^-] + K_2} \quad ...(5)$$

Here are two cases:

Case I: When $K_2 >> K_{-1} [A^-]$

In this case $K_{-1}[A^-]$ may be considered negligible as compared to K_2.

i.e., $$K_{-1}[A^-] + K_2 \cong K_2$$

Hence equation (5) becomes

$$[SH^+] = \frac{K_1 [S] [HA]}{K_2} \quad ...(6)$$

Under these conditions, the reaction is subjected to general acid catalysis. The overall rate of reaction is the rate of formation of the product.

On putting the value of $[SH^+]$ in equation (1) we get

Rate of formation of product $$r = \frac{K_2 . K_1 [S] [HA]}{K_2}$$

or, $$r = K_1 [S] [HA]$$

Hence the rate of reaction is directly proportional to the product of concentration of the substrate and the acid.

Case II: When $K_2 << K_{-1} [A^-]$

In this case K_2 may be considered as negligible as compared to $K_{-1} [A^-]$

i.e., $$K_{-1} [A^-] + K_2 \cong K_{-1} [A^-]$$

Hence equation (5) becomes

$$[SH^+] = \frac{K_1 [S] [HA]}{K_{-1} [A^-]} \quad ...(7)$$

Now, consider the dissociation of the acid HA as

$$HA \rightleftharpoons H^+ + A^-$$

The dissociation constant is given by

$$K_a = \frac{[H^+][A^-]}{[HA]}$$

or, $$\frac{[HA]}{[A^-]} = \frac{[H^+]}{Ka} \quad ...(8)$$

On putting the value of $\frac{[HA]}{[A^-]}$ in equation (7) we get

$$[SH^+] = \frac{K_1 [S][H^+]}{K_{-1} . Ka} \quad ...(9)$$

Hence equation (1) becomes

$$\text{rate of formation of product } r = \frac{K_1K_2 [S][H^+]}{K_{-1} . Ka}$$

or,
$$r = \left(K_2 . \frac{K_1}{K_{-1}} . \frac{1}{Ka} \right) [S] [H^+]$$

or,
$$r = K' [S] [H^+] \quad ...(10)$$

Hence, the rate of the reaction is directly proportional to the concentration of substrate and H^+ even in the presence of HA and A^-. It is also a case of specific acid catalysis.

2.3.2 Enzyme Catalysis

It is a type of homogeneous catalysis. Chemical reactions which are catalysed by enzymes are called Enzyme catalysis. Enzyme catalysis generally occurs in living organisms. Enzymes are complexes of nitrogeneous organic substances called proteins having very high molecular weight. It has been observed that enzyme catalysis is intermediate between homogeneous and heterogeneous catalysis. Hence, sometimes it is also called micro-heterogeneous. Since enzymes are secreted by living cells and capable of catalysing the biochemical reactions in and outside the cells in living organisms without themselves undergoing any change, so they are also called biocatalysts. The reactants in an enzyme catalysed reaction are called the enzyme substrates. All enzymes are synthesised by amino acids. Name of some enzymes are following:

Lipase, ATP-ase, Urease, Ribonuclease, Yeast, Sucrase etc. 'Ribonuclease' is the first enzyme which have been synthesised in laboratory and used as a catalyst. It is very important question arises here that how enzymes speed up reactions ?

We know that reacting molecules must possess a certain amount of energy for initiating the chemical reactions. The required amount of energy is called Activation Energy. When substrate (reactant) molecules acquire the required amount of energy of activation, they across the energy barrier and form the products. It is a difficult task to acquire a large amount of activation energy for the substrate molecules. The enzyme used as a catalyst lowers the level of energy barrier so that a large number of substrate molecules may cross the energy barrier. When a catalyst is added, a new reaction path having a lower energy barrier is provided as in Fig. 2.1.

Since the energy barrier is reduced in magnitude, a large number of reacting molecules cross the energy barrier and form the product. It is believed that the enzymes combine with the substrate molecules and bring them close together which favours their collisions in the most suitable directions and locations for the reactions. As a result, number of effective collisions are increased and hence it increases the rate of reactions.

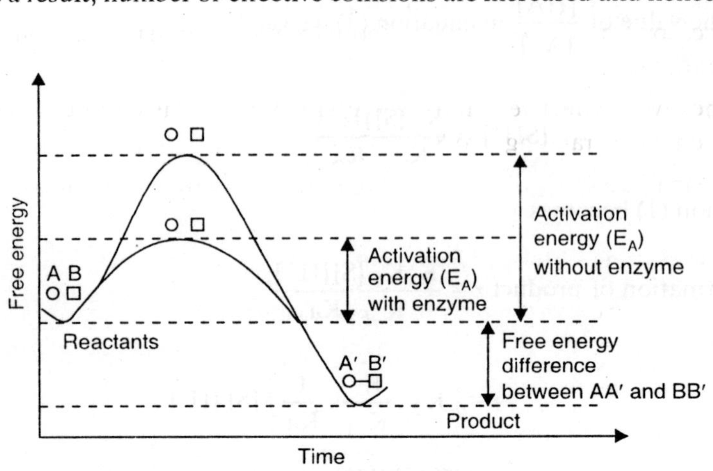

FIGURE 2.1 **Activation energy requirement of noncatalysed and enzyme-catalysed reactions. Reactants A and B absorb energy from surroundings to climb the hill of activation energy (E_A) and reach the unstable transition state. Enzyme speeds the reaction by reducing the uphill climb to the transition state. In this state, the reactants are in an unstable condition and reaction can occur**

2.3.2.1 Characteristics of Enzyme Catalysis

Following are important characteristics of enzyme catalysis:

(*i*) Enzyme catalyse most of the biochemical processes like digestion, biosynthesis, respiration etc.

(*ii*) The catalytic action of each enzyme is specific.

(*iii*) Each enzyme catalyses a particular reaction.

For examples,

(*a*) The enzyme invertase can break up sucrose into glucose and fructose but fails to break a similar disaccharide maltose. Maltose can break up by the enzyme maltase.

$$\underset{\text{(Sucrose)}}{C_{12}H_{22}O_{11}} + H_2O \xrightarrow{\text{Invertase}} \underset{\text{(Glucose)}}{C_6H_{12}O_6} + \underset{\text{(Fructose)}}{C_6H_{12}O_6}$$

$$\underset{\text{(Maltose)}}{C_{12}H_{22}O_{11}} + H_2O \xrightarrow{\text{Maltase}} \underset{\text{(Glucose)}}{2\ C_6H_{12}O_6}$$

(*b*) Urea undergoes hydrolysis by the enzyme urease.

$$\underset{\text{Urea}}{NH_2CONH_2} + H_2 \xrightarrow{\text{Urease}} 2NH_3 + CO_2$$

(*iv*) The catalytic efficiency of enzyme is maximum at a particular temperature known as optimum temperature. Above this temperature the enzyme becomes denatured and it loses its activity. Below the optimum temperature the rate of reaction is slow.

(*v*) Enzyme reactions are very sensitive towards the catalytic poisons like H_2S, HCN, CS_2 etc.

(*vi*) Enzymes lose their activity when exposed to the UV-radiations, γ-radiatons or in the presence of electrolytes.

(*vii*) Enzymes work effectively at a particular pH known as optimum pH. The optimum pH of mostly enyzmes ranging 6 to 8.

(*viii*) A very small quantity of enzyme is required to catalyse the reaction.

2.3.2.2 Kinetics of Enzyme Catalysed Reactions

The mechnism of enzyme catalysis was proposed by Michaelis and Menton. Hence it is also known as Michaelis and Menton mechanism. The mechanism may be understood as follow.

Let us consider a general reaction

$$E + S \underset{K_{-1}}{\overset{K_1}{\rightleftharpoons}} ES \qquad \text{(fast)}$$

$$ES \xrightarrow{K_2} P + E \qquad \text{(slow)} \qquad ...(11)$$

Here, E = enzyme

S = substrate (A reactant in an enzyme catalysed reaction is called substrate)

ES = enzyme substrate complex

and P = product

K_1, K_{-1} and K_2 are rate constants for various reactions.

Rate of formation of the product is given by

$$r = \frac{d\,[P]}{dt} = K_2\,[ES] \qquad ...(12)$$

Rate of formation of enzyme substrate complex is given by

$$\frac{d\,[ES]}{dt} = K_1\,[E]\,[S] - K_{-1}\,[ES] - K_2\,[ES]$$

or,

$$\frac{d\,[ES]}{dt} = K_1\,[E]\,[S] - (K_{-1} + K_2)\,[ES] \qquad ...(13)$$

The enzyme is consumed in the step 1 *i.e.,* fast step and regenerated in the step 2 *i.e.,* slow step. It may be explained either the equilibrium approximation or the steady state approximation. According to experimental evidences the true equilibrium is not achieved in the fast step because the subsequent slow reaction is constantly removing the intermediate complex *i.e.,* enzyme-substrate (ES) complex. Generally the concentration of enzyme is very less than the concentration of substrate *i.e.,* [E] << [S] so that [ES] << [S]. Hence we can use the steady state approximation for the intermediate ES. The rate of formation of [ES] complex will be equal to the rate of its disappearance. Hence,

$$\frac{d\,[ES]}{dt} = 0 \qquad ...(14)$$

From equation (13) and (14) we get

$$K_1 [E] [S] - (K_{-1} + K_2) [ES] = 0 \qquad ...(15)$$

or,
$$[ES] = \frac{K_1 [E] [S]}{(K_{-1} + K_2)}$$

or,
$$[ES] = \frac{[E] [S]}{\frac{K_{-1} + K_2}{K_1}}$$

or,
$$[ES] = \frac{[E] [S]}{K_m} \qquad ...(16)$$

where $K_m = \frac{K_{-1} + K_2}{K_1}$ and K_m is known as Michaelis constant. (*It should be noted that K_m is not an equilibrium constant.*)

Now [E] cannot be measured experimentally. The equilibrium between the free and bounded enzyme is given by the enzyme conservation equation as

$$[E]_0 = [E] + [ES]$$

where, $[E]_0$ = the total enzyme concentration and is measured

[E] = free enzyme concentration

[ES] = bounded (reacted) enzyme substrate concentration

∴
$$[E] = [E]_0 - [ES] \qquad ...(17)$$

Substituting the value of E in equation (16) we get

$$[ES] = \frac{([E]_0 - [ES]) . [S]}{K_m}$$

or,
$$K_m [ES] = [E]_0 [S] - [ES] [S]$$

or,
$$[ES] (K_m + [S]) = [E]_0 [S]$$

or,
$$[ES] = \frac{[E]_0 [S]}{K_m + [S]} \qquad ...(18)$$

Substituting the value of [ES] from equation (18) in equation (12) we get

Rate of formation of the product

$$r = \frac{d [P]}{dt} = \frac{K_2 [E]_0 [S]}{K_m + [S]} \qquad ...(19)$$

Equation (19) is known as **Michaelis-Menton equation.**

Now we consider when all the enzyme has reacted with the substrate at high concentration, the reaction will be going at maximum rate. In this case no free enzyme will remain so that $[E]_0 = [ES]$.

Hence equation (12) may be written as

$$r_{max} = V_{max} = K_2 [E]_0 \quad ...(20)$$

where V_{max} = maximum rate of reaction and called enzymology and K_2 is called turnover number of the enzyme. The turnover number of enzyme may be defined as the number of reacting molecules converted into the product in unit time (1 sec) by one molecule of enzyme. For example, one molecule of enzyme urease break down 30,000 molecules of urea into the products (CO_2 and NH_3) in one second. Hence its turnover is 30,000.

On putting the value of $K_2 [E]_0$ from equation (20) into equation (19) we get the Michaelis-Menten equation $r = \dfrac{V_{max} [S]}{K_m + [S]}$ ...(21)

Now there are three cases arise:

Case I: When the concentration of substrate is very low

i.e., $$K_m >> [S]$$

In this case [S] can be neglected in the denominator of equation (21). Hence equation (21) becomes

$$r = \frac{V_{max} [S]}{K_m}$$

$$= K' [S]$$

where $K' = \dfrac{V_{max}}{K_m}$

$$\therefore \quad \text{Rate } (r) \propto [S]$$

Hence every enzyme catalysed reaction is of the first order.

Case II: When the concentration of the substrate is very high.

i.e., $$[S] >> K_m$$

$$\therefore \quad K_m + [S] \cong [S]$$

In this case K_m can be neglected in the denominator of equation (21). Hence equation (21) becomes

$$r = \frac{V_{max} [S]}{[S]} = V_{max} = \text{constant}$$

Case III: In an intermediate case when $[S] = K_m$

Hence equation (21) becomes

$$r = \frac{V_{max} [S]}{2 [S]} = \frac{1}{2} V_{max}$$

Hence the rate of reaction is independent of [S].

When the rate of enzyme catalysed reaction is equal to half of its maximum value, Michaelis constant (K_m) becomes equal to the concentration of the substrate [S].

Kinetics of an enzyme-catalysed reaction my be represented in Fig. 2.2.

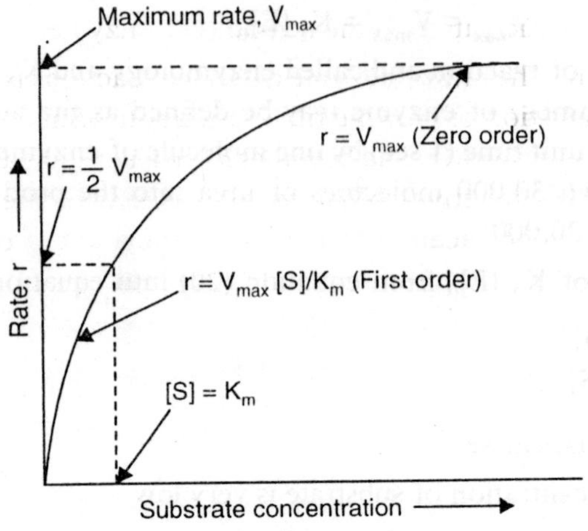

FIGURE 2.2 **Kinetics of an enzyme-catalysed reaction**

It is very interesting question arises here why the reaction rate of an enzyme–catalysed reaction changes from first order to zero order as the substrate concentration is increased?

It can be explained as the enzyme molecule has one or more active sites at which the substrate must be bound in order that the catalytic action may occur. In case of low concentration of substrate, most of the active sites remain vacant at any time. On the other hand when the concentration of substrate is high, the number of active sites which are occupied increases and hence the rate of reaction also increases. However, at very high concentration of substrate, virtually all the active sites are occupied at any time so that further increase in substrate concentration cannot further increase the formation of enzyme-substrate complex.

It is difficult to calculate V_{max} (and hence K_m) directly from the plot of r against [S]. It can be determine by rearranging equation (21) we get

$$\frac{1}{r} = \frac{K_m}{[S]\,V_{max}} + \frac{1}{V_{max}}$$

Lineweaver-Burk plotted a graph of $1/r$ vs $1/[S]$ gives a straight line whose intercept on the x-axis and y-axis are $-1/K_m$ and $1/V_{max}$ respectively and slope gives the value of K_m/V_{max} as in Fig. 2.3.

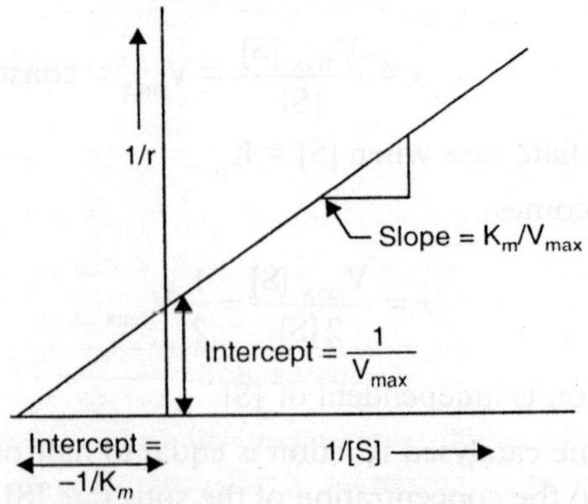

FIGURE 2.3 **The Lineweaver-Burk method**

2.3.2.3 Factors Affecting the Activity of Enzyme Catalyst During Enzyme Catalysis

Following are the various factors which affected the activities of enzyme:

1. **Enzyme concentration:** The rate of enzyme catalysed reactions is initially increases with an increase in the concentration of enzyme till a saturation point is attained. After saturation point the rate increases very slowly and becomes almost constant as in Fig. 2.4. This is due to the fact that number of effective collisions increases with the increase in the enzyme concentration initially. The vacant active sites of enzyme are occupied by the substrate molecule and achieved the saturation point. After saturation point if we increase the concentration of enzyme the number of collisions between the enzyme molecules take place and the rate of reactions become almost constant.

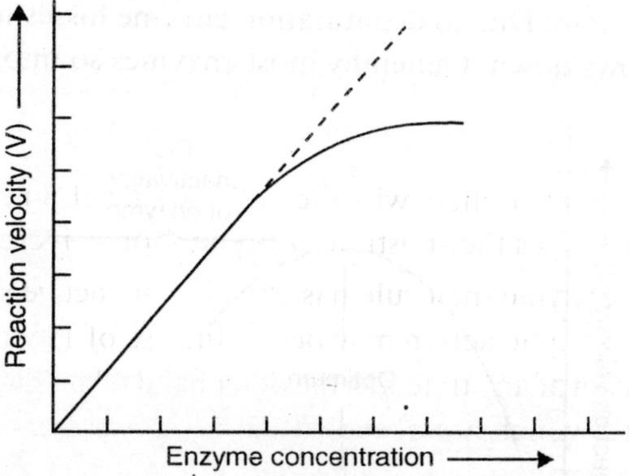

FIGURE 2.4 **Effect of enzyme concentration on the velocity of enzyme action**

2. **Substrate concentration:** The rate of enzyme catalysed reaction increases with an increase in the substrate concentration till a saturation effect is attained as in Fig. 2.5. Initially the rate of reaction rapidly increases with the increase in substrate concentration and is represented by $\frac{1}{2} V_{max}$ in graph. The line AB represents it. After that the rate of reaction increases

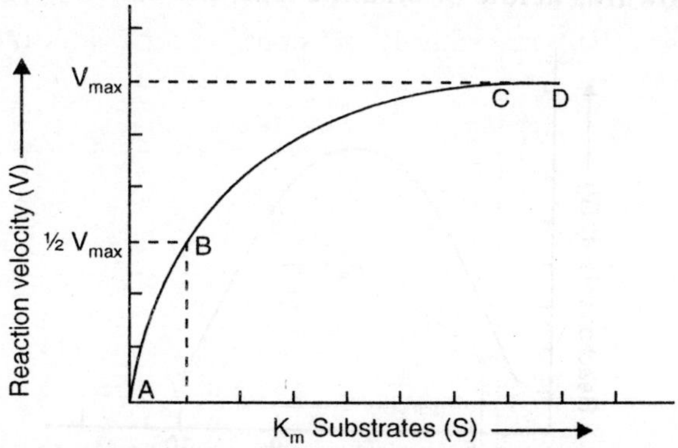

FIGURE 2.5 **Effect of substrate concentration on the velocity of enzyme action S = Substrate, V = Reaction velocity, V_{max} = Maximum velocity, K_m = Michaclis mention constant**

gradually represented by BC in curve in graph and the rate of reaction reaches to a maximum point C and is represented by V_{max}. After this point, increasing in the substrate concentration does not affect the rate of reaction. At this stage, the active sites of all the available enzyme molecules are occupied by the substrate molecules.

3. **Temperature:** Temperature play very important rôle in the enzyme catalysis. As the temperature rises the effectiveness of enzyme increases upto a certain optimum and then decreases thereafter as in Fig. 2.6. The activity of enzymes stop altogether at about 70–80°C and also at 0°C. Initially rise in temperature increases the kinetic energy of the molecule. Therefore, a large number of molecules have the required activation energy and take part in chemical reaction. As a result the rate of reaction is increased. At very high temperatures, the kinetic activity of molecules in an enzyme becomes very strong and is enough to break the weak hydrogen bonds that maintain the tertiary structure of the enzyme. It is called denaturation of protein. Due to denaturation enzyme losses its catalytic activity. Hence the rate of reaction slows down. Generally most enzymes so maximum work at 25 to 40°C.

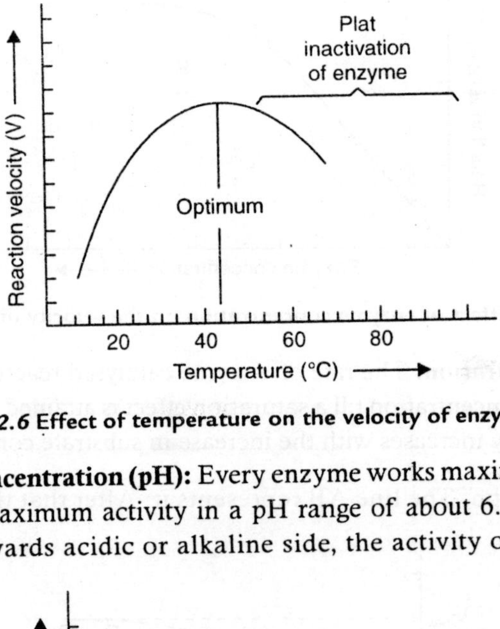

FIGURE 2.6 **Effect of temperature on the velocity of enzyme action**

4. **Hydrogen ion concentration (pH):** Every enzyme works maximum at a particular pH. Most enzymes show maximum activity in a pH range of about 6.0 – 7.5 as in Fig. 2.7. When pH is shifted towards acidic or alkaline side, the activity of enzyme decreases rapidly

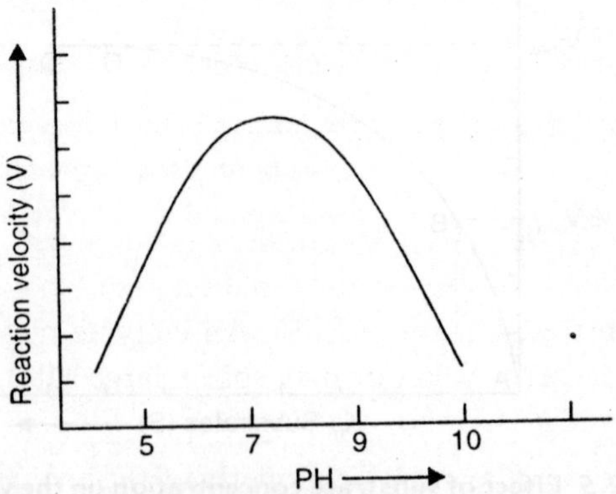

FIGURE 2.7 **Effect of pH on the velocity of enzyme action**

due to the denaturation of the enzyme molecule. It has been noticed that some enzymes act best in an acidic medium or basic medium. For example, pepsin enzyme of gastric juice has its optimum activity at pH 2 (highly acidic) and trypsin enzyme of pancreatic juice shows maximum activity at pH 8.8 (highly alkaline).

5. **Product concentration:** Since enzyme catalysed reactions are reversible in nature hence accumulation of the product lowers the enzyme activity. So, the product should be quickly removed during reactions.
6. **Poisonous substance and radiation:** Poisons like cyanide etc or radiations destroy the tertiary structure of the enzymes, making them passive.

2.3.2.4 Mechanism of Enzyme Catalysis

There are two mechanisms have been proposed for enzyme action:

(*i*) Lock and key hypothesis and (*ii*) Induced-fit hypothesis.

It is better to know the proximity and orientation for understanding these mechanisms.

Proximity and Orientation: It is the basic concept of enzyme catalysis. An enzyme is believed to bind the substrate molecule in especially favourable position in such a way that the susceptible bond is not only in close proximity to the catalytic group on the enzyme but also is precisely oriented to it. Thus, the maximum probability is that the enzyme-substrate (E–S) complex will enter the transition state as in Fig. 2.8. This phenomenon is also called 'Orbital Steering' and has qualitative merit as a contributing factor to enzyme catalysis.

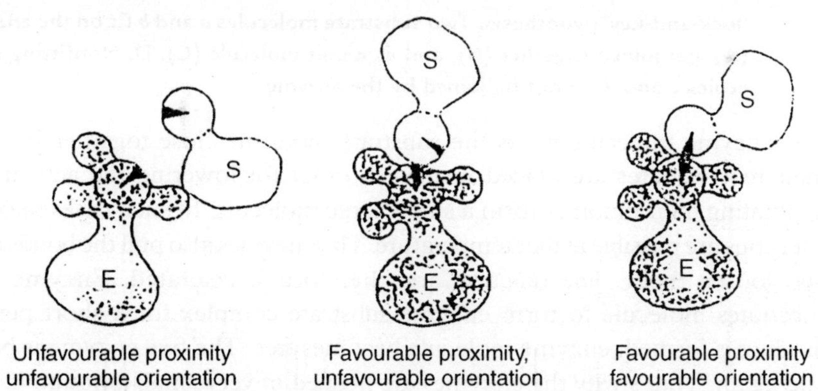

FIGURE 2.8 **Effect of proximity and orientation in the interaction of the substrate molecule with a catalytic group on the enzyme (E) active site**

1. **Lock and key hypothesis:** This hypothesis was suggested by Emil Fischer. The enzyme molecule operates chemically for the combination itself (enzyme) with the substrate molecule to form enzyme-substrate complex (E–S). An enzyme may combine two small substrate molecules into one larger product or may split a large substrate molecule into two smaller product molecules.

   (*a*) *Combination reactions:* The catalytic property of the enzymes is located at specific regions on their surface. These reasons are called active sites or active centres. Enzymes acquire 3-dimensional forms having some small packets or grooves like structure. The size, shape and electrical charge of amino acid at the active site determines the substrate that can fit there. Similarly the substrates have also reactive sites at their surface in specific areas. The active sites of enzyme molecule and reactive sites of substrate molecule should fit like a lock and key as in Fig. 2.9.

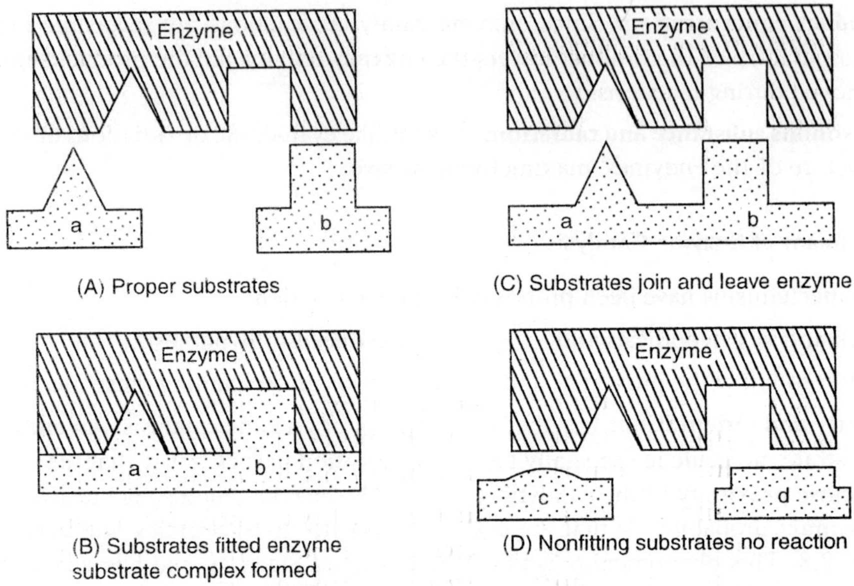

FIGURE 2.9 **Schematic representation of enzyme and substrate interaction according to 'lock-and-key' hypothesis. Two substrate molecules *a* and *b* fit on the enzyme (A), get joined together (B), and as a unit molecule (C). D. Nonfitting molecules *c* and *d* cannot be joined by the enzyme**

The enzyme molecule holds the substrate molecule close together in such a way that their reactive sites are placed near each other for lowering the activation energy and facilitating their union to form a single large molecule. In such cases maximum effective collisions are possible at that temperature. The enzymes also pull the bonds of the substrates and loosen them. The reaction are, therefore, accelerated. Enzyme combines with substrates molecule to form enzyme substrate complex for a short period. When the product is formed, enzyme molecule becomes free. The free enzyme is being used again and again. That's why the enzymes are needed in very small amount.

(*b*) *Splitting reactions:* Enzymes also split large molecules into smaller one as in Fig. 2.10. The enzyme molecule combines with the substrate molecule at different places at their active site and holds the substrate molecules in a way that strains in molecular bonds and facilitates their breaking. As a result, the product is formed and enzyme is released which is reused.

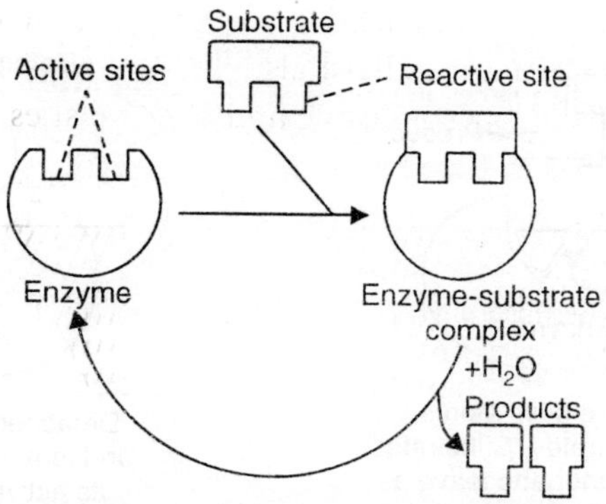

FIGURE 2.10 **Breakdown reaction catalysed by an enzyme according to lock-and-key hypothesis**

2. **Induced fit hypothesis:** It is the new hypothesis for enzymatic catalysis. This hypothesis was suggested by Daniel E. Koshland *et. al.* The main points are following:
   (*i*) There is an intermediate condition (transition state) between the substrate and products. In this stage old chemical bonds in substrate break and new chemical bonds are formed in the product. The transition state is highly unstable and persists for a very short period.
   (*ii*) The active sites do not have a rigid lock and key conformation for the substrates. As the substrates enter the active sites, they induce the enzyme to change its shape slightly so that the active sites fit more closely around the substrates.
   (*iii*) The induced fit is possible because of the flexibility of the protein molecules. The enzyme pulls the bond in substrate and lowers the activation energy. This stage is called transition state.
   (*iv*) During the transition state, the enzyme pulls atoms from the substrate and forms product.
   (*v*) The product that fits the last conformation of the active site leaves the enzyme molecule.
   (*vi*) The detached atoms combine and leaves the enzyme. The free enzyme is again involved with the new set of substrate molecule and the reaction is continued.

It may be represented as in Fig. 2.11.

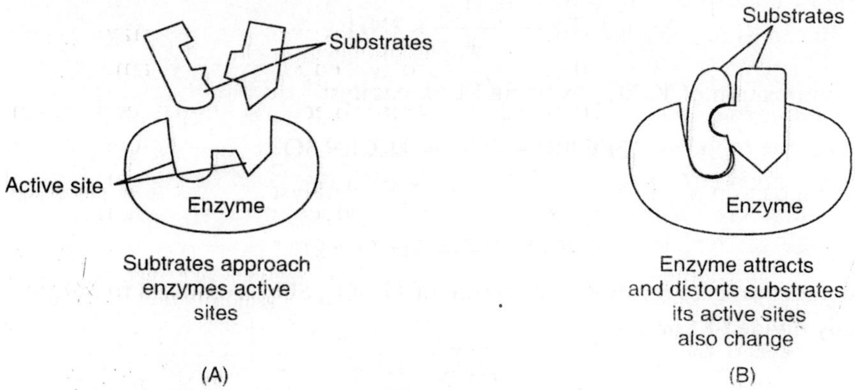

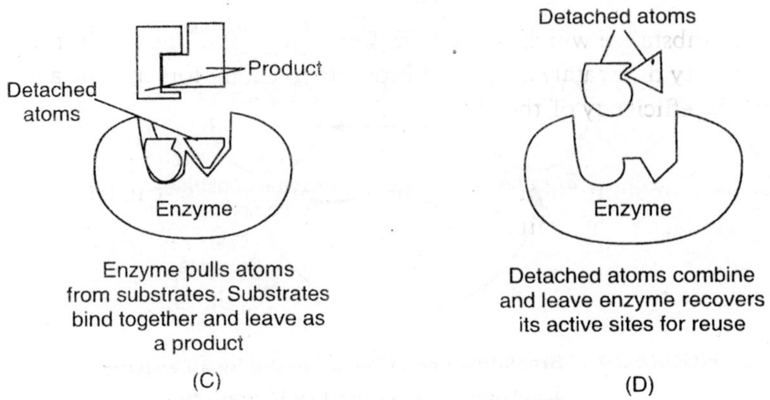

FIGURE 2.11 **Working of an enzyme according to the induced fit hypothesis**

2.3.3 Heterogeneous Catalysis

Heterogeneous catalysis are those type in which the catalyst and the reactants are present in different phases. Such type of reactions generally occur through adsorptions of reactants on the active sites at the surface of the catalyst. Active sites are the types of lattice defects which change their positions continuously due to the motion of the excited electrons in the lattice. Generally solid catalysts like Pt, Cu, Ni, Fe, Fe_2O_3, Al_2O_3, ZnO etc. are used and reactants are mostly gases or sometimes liquid. Catalysts which are used in this reaction are generally in fine divided powder form so as to have a large surface area for maximum reaction.

Examples:

1. Combination of H_2 and O_2 in the presence of Pt catalyst

$$2H_2 + O_2 \xrightarrow{Pt} 2H_2O$$

2. During the manufacturing of ammonia in the Haber's process, combination of N_2 and H_2 in the ratio 1 : 3 is catalysed by the catalyst finally divided iron supported by a promoter molybdenum.

$$N_2 + 3\,H_2 \xrightarrow[450^\circ\text{–}500^\circ C]{Fe/Mo} 2NH_3$$

3. Decomposition of $KClO_3$ by using M_nO_2 catalyst

$$2KClO_3 \xrightarrow{M_nO_2} 2KCl + 3O_2$$

4. Decomposition of H_2O_2 by using Pt or M_nO_2 catalyst

$$2H_2O_2 \xrightarrow{Pt} 2H_2O + O_2$$

5. In contact process for the manufacture of H_2SO_4, SO_2 is oxidised to SO_3 in the presence of finally divided Pt or V_2O_5.

$$2SO_2 + O_2 \xrightarrow[450^\circ C]{Pt\ or\ V_2O_5} 2SO_3.$$

2.4 CONCEPTS OF PROMOTORS, INHIBITORS AND POISONERS

Promotors: A substance which is added in small amount to the catalyst during its preparation for improving the activity of a catalyst is called Promotors. Promotors may be a non catalyst or feeble catalyst. It increase the efficiency of the catalyst.

Examples:

(*i*) Molybdenum or aluminium oxide promotes the activity of catalyst Fe in the Haber process for the manufacture of ammonia.

$$N_2 + 3\,H_2 \xrightleftharpoons{Fe/Mo} 2\,NH_3$$

Here Mo acts as a promotor.

(*ii*) Chromic oxide promotes the activity of catalyst ZnO in the reduction of CO.

$$CO + H_2 \xrightarrow{ZnO/Cr_2O_3} CH_3OH$$

Here Cr_2O_3 acts as a promotor.

It is beleived that the promotor increases the activity of catalyst by following way:

(*i*) It increases the lattice spacing of the catalyst. Due to this the bond between the adsorbed atoms of a molecule weakenes and hence cleared more easily thereby making the reaction go faster.

(*ii*) It stabilizes the surface of catalyst by inhibiting its loss during use.

(*iii*) It improves the diffusion properties of the catalyst.

(*iv*) It increases the number of active sites at the surface of the catalyst. As a result the reactant molecules absorbs more and more on the interface between the promotor and the catalyst, where the free valencies are crowded. Increased adsorption results in higher rate of reaction.

Inhibitors: A substance which is added in small amount during the preparation of the catalyst decreases the activity of the catalyst is called inhibitors.

Examples:

(*i*) Fe_2O_3 reduces the catalytic oxidation of naphthalene into phthalic anhydride by adding V_2O_5.

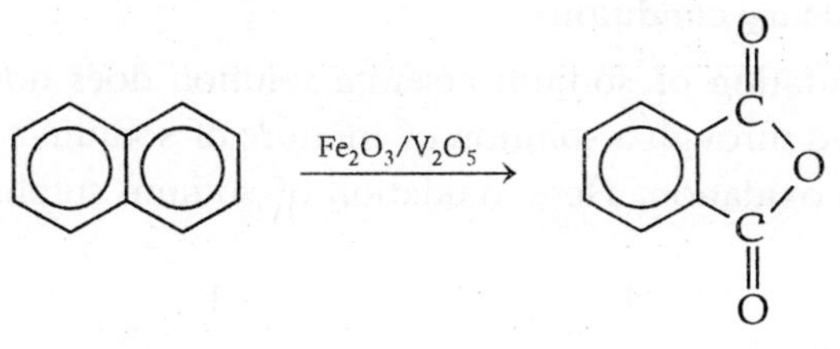

Here V_2O_5 acts as an inhibitor.

Sulphur and its compounds are important inhibitor.

In fact, the inhibitors get adsorbed on the active centres on the surface of the catalyst.

Poisoners: A substance which destroys the activity of catalyst by his presence is called catalytic poisoner. A small amount of that substance is quite enough to stop the reaction. Catalytic poisons are not inhibitors as they are added for a specific purpose.

Examples:

(*i*) AS_2O_3 acts as a poisoner during the oxidation of SO_2 using Pt catalyst.

(*ii*) H_2S acts as a poisoner in the combination of nitrogen and hydrogen using iron as a catalayst.

(*iii*) Carbon monoxide acts as a poisoner during the synthesis of water using Pt catalyst.

It is believed that the action of the catalytic poisoner is due to the selective adsorption of it on the surface of the catalyst. Hence the catalyst become inactive. In some extreme cases, it may combine with the catalyst. On the basis of the extent and strength of adsorption, catalytic poisoning is of two types:

1. **Temporary poisoning:** If the activity of catalyst is poisoned for sometimes then it is called temporary poisoning. When the catalytic poisoner is removed from the reaction mixture, the activity of catalyst is restored. For example, Aluminium silicate is used as a catalyst during the catalytic cracking of hydrocarbons. The decomposition of carbon act as temporary poisoner because they occupied the active sites on the surface of the catalyst. The catalyst is again activated after burning the deposited carbon.
2. **Permanent poisoning:** If the activity of catalyst is poisoned to such an extent, that its activity is virtually changed then it is called permanent poisoning.

2.4.1 Auto Catalysis

There are many reactions where one of the products of the reaction acts as a catalyst. Such type of catalyst is called auto catalyst and such reactions are known as auto catalysed reactions. In such type of reactions the reaction is slow in the initial stage. But after sometime the rate of reaction increases due to the formation of product acting as auto catalyst.

For example, during the hydrolysis of ethyl acetate, acetic acid is formed which acts as auto catalyst. Hence the rate of reaction is initially slow but accelerated after sometimes.

$$CH_3COOC_2H_5 + H_2O \longrightarrow CH_3COOH + C_2H_5OH$$

2.4.2 Induced Catalysis

The phenomenon by which a chemical reaction increases the rate of some other chemical reaction which does not occur at ordinary conditions.

For example, the oxidation of sodium arsenite solution does not take place under ordinary condition. But if air is passed through a solution of mixture of sodium arsenite and sodium sulphite, then both of them undergo oxidation. Here, oxidation of sodium sulphite catalyse the oxidation of sodium arsenite.

2.5 IMPORTANCE

Enzymes play very important role in many ways:

1. **Biological importance:** Enzymes have a great biological importance. Some biological importance of enzymes are following:

   (*i*) *Biochemical reaction:* The enzymes make the biochemical reactions very easily at ordinary temperature. For example, the enzyme urease break down 30,000 molecules of Urea into carbon dioxide and ammonia in just one second. In the absence of Urease the reaction would take 3 million years and more.

$$\underbrace{\underset{\text{Urea}}{NH_2 . CO . NH_2} + \underset{\text{Water}}{H_2O}}_{\text{Urine (Substrate)}} \xrightarrow{\text{Urease}} CO_2 + NH_3$$

(*ii*) *Physiology:* The enzymes are essential for respiration, digestion, blood clotting, muscle contraction, nerve impulse transmission etc. A missing or defective enzymes can be very harmful. For example, lack of sucrose enzyme causes abdominal pain and diarrhoea in persons on eating sucrose.

(*iii*) *Medical diagnosis:* An enzyme ELISA (Enzyme-linked immunosorbent assay) is used for detecting the diseases AIDS (Acquired immunodeficiency syndrome).

(*iv*) *Medical treatment:* Enzyme streptokinase is used in dissolving blood clot formed inside blood vessels.

(*v*) *Genetic engineering:* Enzymes like endonuclease, ligases etc. are used in genetic engineering.

2. **Industrial importance:** Enzymes have a great importance in industries. Some industrial importance are following:

(*i*) *Alcoholic drinks:* Enzyme zymase obtained from yeast is used in making alcoholic beverages by fermentation of sugar or like sugar substances.

(*ii*) *Cheese:* Enzyme renin obtained from calfs stomach is used in making cheese by coagulating milk protein caseinogen.

(*iii*) *Detergents:* Enzyme protease is used in making the detergents for washing the clothes.

EXERCISE

1. Define catalysis. Give general characteristic of catalysis.
2. What is catalysis? Discuss homogeneous and hetereogenous catalysis with examples.
3. Define promoter. How promoter increases the activity of catalyst?
4. What do you understand by Acid-Base catalysis?
5. Explain the kinetics (mechanism) of acid base catalysis.
6. What is Enzyme catalysis? Explain the characteristics of enzyme catalysis.
7. Explain the kinetics (mechanism) of Enzyme catalysed reactions.
8. Derive Michaelis-Menton equation for enzyme catalysis.
9. Why the reaction rate of an enzyme catalysed reaction changes from first order to zero order as the substrate concentration is increased?
10. Define the turn over of enzyme. Explain the effect of temperature and pH on the rate of enzyme catalysis.
11. Explain with examples of following:

    (*a*) Homogeneous catalysis

    (*b*) Heterogeneous catalysis.

12. What are active sites (centres) of enzymes?
13. Explain the following:
    (*a*) Inhibitors (*b*) Catalytic poisoners.
14. Define following:
    (*a*) Auto catalyst (*b*) Induced catalyst.
15. Explain the role of promoters in the catalytic reactions.
16. "Enzymes make life possible" Comment on this statement.
17. Explain the lock and key hypothesis of enzyme action.
18. Define following:
    Turnover number, Active centres, Denaturation, Autocatalyst.
19. How enzymes speed up the reaction?
20. Write a short note on following:
    (*a*) Proximity and orientation (*b*) Induced-fit hypothesis.
21. Explain the various factors affecting the activity of enzyme.
22. What are enzymes? Why they are known as biocatalyst?
23. Define enzymes. Write the names of five enzymes.
24. Write some importance of enzymes.

Chapter 3

WATER AND ITS TREATMENT (PART I)

3.1 INTRODUCTION

After air, water is a natural wonderful and is the most common important and useful compound of hydrogen and oxygen for surviving of all living organisms. Without food human being can survive for a number of days but without water no body can survive. Our body contains 70–75% of water which regulates life process such as digestion of food, transportation of nutrients, excreation of body wastes. It is important participant of photosynthesis. It regulates body temperature by the process of sweating and evaporation. It provides a medium for all biochemical reactions take place inside our body. It is a universal solvent and the versatile nature of water is due to some of its interesting properties. That's why it is widely used in laboratories, industrial purposes, irrigation, air-conditioning, steam generator, fire-fighting etc. besides drinking, bathing, sanitary, etc.

3.2 SOURCES OF WATER

Rainwater is the purest form of the natural water because it is obtained as a result of evaporation from the surface water. But due to air pollution created by human beings, some harmful gases make the rainwater impure. Water is found in free state as ice, snow, water, vapour as well as in the combined state. Pure water has pH 7 *i.e.*, neutral and is colourless, odourless and tasteless. It is poor conductor of heat and electricity.

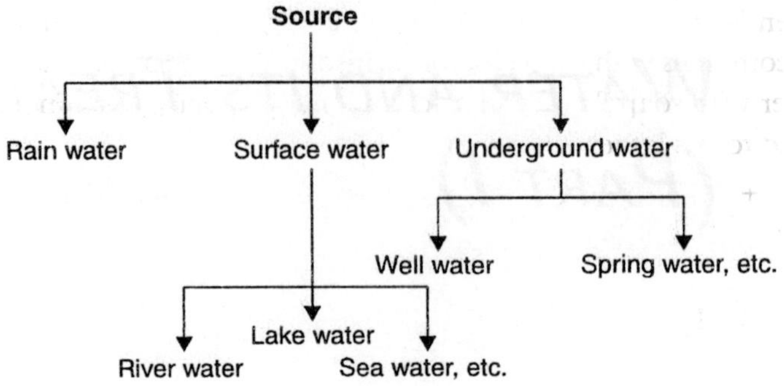

FIGURE 3.1

3.3 IMPURITIES IN WATER

These are the following impurities, which have been found as common, present in water.

3.3.1 Dissolved Impurities

Water may contain soluble inorganic or organic salts, gases etc.

(*i*) *Inorganic salts.* These are usually bicarbonates, chlorides, nitrates, sulphates of sodium, magnesium, potassium, calcium, aluminium, iron, zinc etc.

(*ii*) *Organic salts.* These are generally soluble organic compounds obtained by the domestic sewage, industrial wastes or decay of dead animals or plants etc.

(*iii*) *Gases.* From industries several gases like O_2, CO_2, H_2S, ammonia, oxides of nitrogen (N_xO_y), SO_2 etc. evolve which are soluble in water and make it impure. These gases are responsible for corrosion.

3.3.2 Suspended or Colloidal Impurities

Sand, clay, lime etc. present in water get suspended. The heavy particles settle down after standing sometimes while smaller particles remains in the water as colloidal particles.

3.3.3 Harmful Micro-organism

Various pathogenic micro-organisms like bacteria, viruses etc. also enter in water through waste and sewage. These are causes of various diseases.

3.4 HARDNESS OF WATER

Hardness is the characteristics of water by which water does not form lather with soap. Hard water is not fit for cooking, drinking, loundry purposes etc. it also does not suitable in industries. Actually hardness measures the capacity of precipitation of soap. Hard water contains soluble salts of calcium and magnesium. When soap comes in contact with hard water the calcium ion and magnesium ion present in hard water combines with soap and forms insoluble sticky precipitate. That's why hard water does not produce lather with soap. When all calcium ion and magnesium ion get consumed then hard water becomes soft and forms lather with soap.

$$\underset{\text{(Present in hard water)}}{M^{2+}} + \underset{\text{Sod. stearate (Soap)}}{2C_{17}H_{35}COONa} \longrightarrow (C_{17}H_{35}COO)_2M \downarrow + 2Na^+$$

where $M^{2+} = Ca^{2+}$ or Mg^{2+}.

3.4.1 Causes of Hardness

It is due to the presence of sulphate, chloride and bicarbonate of calcium or magnesium.

3.4.2 Type of Hardness

Hardness are of two types:

(*i*) *Temporary hardness.* It is due to the presence of bicarbonate of calcium or magnesium. It may be removed by boiling.

(*ii*) *Permanent hardness.* It is due to the presence of sulphate and chloride of calcium or magnesium. It cannot be removed by boiling.

Now-a-days temporary and permanent hardness is replaced by alkaline and non-alkaline hardness.

Alkaline Hardness (Carbonate Hardness)

When hardness of water is due to the presence of bicarbonate, carbonates and hydroxide of calcium and magnesium is called alkaline hardness or carbonate hardness. Generally such type of hardness is due to the presence of bicarbonates of calcium and magnesium.

Non-alkaline Hardness (Non Carbonate Hardness)

The non-alkaline hardness is determined by the subtracting of alkaline hardness from the total hardness. Such type of hardness is due to the presence of sulphate, chloride etc. of calcium and magnesium.

It has been noticed that the hardness in water is due to the presence of either calcium ion or magnesium ion. Hence hardness is also considered in terms of calcium hardness and magnesium hardness. If the water contains calcium salts then it is called calcium hardness and if magnesium salts are present then it is called magnesium hardness.

3.4.3 Hardness is Expressed in Terms of $CaCO_3$

Since a number of salts can cause hardness such as bicarbonate, sulphate and chloride of calcium and magnesium. For comparing the hardness of different sample of water it would be easier, if the hardness caused by different salts may be expressed in terms of single salt like $CaCO_3$. These are following reasons for choosing $CaCO_3$ as a standard (reference) compound for expressing the hardness.

(*i*) $CaCO_3$ is completely insoluble salt thus it can be easily precipitated completely during water treatment. The amount of hardness is calculated by weighing the dried precipitate. It is the main reason.

(*ii*) Its molecular weight is exactly 100 so the mathematical calculation becomes easy. It is very old convention and the fact is not so.

The equivalent of $CaCO_3$ for a hardness causing substance is given by

$$\text{Equivalent of } CaCO_3 = \frac{\begin{bmatrix}\text{Amount of substance causing} \\ \text{hardness (in mg / L)}\end{bmatrix} \times [\text{Equivalent mass of } CaCO_3]}{\text{Equivalent mass of hardness producing substance}}$$

If W = mass of substance causing hardness

and E = equivalent mass of hardness producing substance

$$\text{Then equivalent of } CaCO_3 = \frac{W \times 50}{E} \qquad [\because \text{ equivalent mass of } CaCO_3 = 50]$$

3.4.4 Units

1. *Parts per million (ppm).* It is the parts of calcium carbonate equivalent hardness per 10^6 part of water.

2. *Milligram per litre (mg/L).* It is the number of milligram of $CaCO_3$ equivalent hardness present in 1 litre water.

1 mg/L = 1mg of $CaCO_3$ equivalent hardness per litre of water

1 mg/L = 1 ppm.

3. *Clarke's (°Cl).* It is the part of $CaCO_3$ equivalent hardness per 70,000 parts of water.

4. *Degree French (°Fr).* It is the part of $CaCO_3$ equivalent hardness per 10^5 parts of water.

Relationship. 1 ppm = 1mg/L = 0.07°Cl = 0.1°Fr.

3.5 DETERMINATION OF HARDNESS

The hardness of water is determined by the following methods:

3.5.1 Hehner's Method

1. Temporary hardness

It is based on the following principle:

Principle. Temporary hardness is removed by boiling of water because calcium and magnesium bicarbonates get precipitated as calcium carbonate and magnesium carbonate.

$$Ca(HCO_3)_2 \xrightarrow{\text{boiling}} CaCO_3 \downarrow + H_2O + CO_2 \uparrow$$

$$Mg(HCO_3)_2 \xrightarrow{\text{boiling}} Mg(OH)_2 \downarrow + 2CO_2$$

A known volume of water is titrated before and after boiling against standard HCl. The difference gives the value of temporary hardness.

Procedure. It involves in following two steps:

Step I: A known volume of water is titrated against standard HCl using methyl orange indicator and the volume of consumed acid is noted (say V_1 ml).

Step II: A known quantity of water sample is evaporated to dryness and the residue is dissolved into hot distilled water. It is filtered and the volume of filtrate is made upto the original volume (starting in step II) by adding distilled water. Now a known volume (must be same as in step I) is titrated against standard HCl using methyl orange indicator and the volume of consumed acid (HCl) is noted (say V_2 ml).

Calculations

Let volume of acid used before boiling	$= V_1$ ml
Volume of acid used after boiling	$= V_2$ ml
Volume of water used in titration	$= x$ ml
Strength of acid (HCl) used	$= N/50$

∴ Volume of N/50 HCl used by temporary hardness present in x ml water $= (V_1 - V_2)$ ml

∴ Alkalinity due to temporary hardness $\equiv (V_1 - V_2)$ ml of N/50 HCl

$$\therefore \quad x \text{ ml } H_2O \equiv (V_1 - V_2) \text{ ml of N/50 HCl}$$

$$\therefore \quad 1 \text{ ml} \equiv \frac{[V_1 - V_2]}{x} \times \frac{1}{50}$$

$$\therefore \quad 1000 \text{ ml} = \frac{[V_1 - V_2]}{x \times 50} \times 1000$$

$$\therefore \quad \text{In terms of } CaCO_3 \text{ equivalent} = \frac{[V_1 - V_2]}{x \times 50} \times 50 \times 1000$$

where the eq. wt. of $CaCO_3 = 50$.

2. **Permanent hardness:** *Principle.* Permanent hardness is removed by boiling of water with a known excess amount of standard Na_2CO_3. The chloride and sulphates of calcium and magnesium gets precipitated as their carbonates. The residual Na_2CO_3 is then determined by titrating it against a standard HCl acid.

The following reactions take place:

$$CaCl_2 + Na_2CO_3 \longrightarrow CaCO_3 \downarrow + 2NaCl$$

$$MgCl_2 + Na_2CO_3 \longrightarrow MgCO_3 \downarrow + 2NaCl$$

$$CaSO_4 + Na_2CO_3 \longrightarrow CaCO_3 \downarrow + Na_2SO_4$$

$$MgSO_4 + Na_2CO_3 \longrightarrow MgCO_3 \downarrow + Na_2SO_4$$

Procedure. A known volume of standard Na_2CO_3 is titrated against standard acid (HCl or H_2SO_4, etc.) using methyl orange indicator and the volume of consumed acid is noted (say V_1 ml). It is called blank titration.

Now a known volume of water sample is boiled for sometimes with the sufficient excess known amount of standard Na_2CO_3 solution. All chlorides and sulphates of calcium and magnesium get precipitated as their carbonates. After cooling it is filtered and washed the residue with distilled water. All the washing is collected in a conical flask. The unused Na_2CO_3 present in the filtrate is calculated by titrating it against standard acid using methyl orange indicator. The volume of consumed acid is noted (say V_2 ml). The difference between titre value (*i.e.*, $V_1 - V_2$) corresponds to permanent hardness.

Calculation. Let volume of water sample = x ml

Strength of acid = $\frac{N}{50}$ (say)

Volume of consumed acid in blank titration = V_1 ml

Volume of consumed acid for unused Na_2CO_3 titration = V_2 ml

$\therefore$ Volume of acid corresponds to permanent hardness = $(V_1 - V_2)$ ml

Now, applying normality equation

$$\underset{\text{water}}{N_1V_1} = \underset{\text{acid}}{N_2V_2}$$

$$N_1 \times x = \frac{1}{50}(V_1 - V_2)$$

$$N_1 = \frac{1}{50}\frac{(V_1 - V_2)}{x}$$

$\therefore$ Strength = Normality × eq.wt.

$$= \frac{1}{50} \times \frac{(V_1 - V_2)}{x} \times 50\ g/l \qquad (\because \text{ eq. wt. of } CaCO_3 = 50)$$

$\therefore$ Permanent hardness in terms of $CaCO_3$ equivalent

$$= \frac{1}{50} \times \frac{(V_1 - V_2)}{x} \times 50 \times 1000 \text{ ppm}.$$

3.5.2 Soap Titration Method

Principle. When soap is added to hard water, it does not give lather because hardness producing metal ions (Ca^{2+} and Mg^{2+}) get precipitated. After complete precipitation of Ca^{2+} and Mg^{2+}, further addition of soap gives lather.

$$\underset{\text{Sodium Stearate(soap)}}{2C_{17}H_{35}COONa} + Ca(HCO_3)_2 \longrightarrow (C_{17}H_{35}COO)_2Ca \downarrow + 2NaHCO_3$$

$$2C_{17}H_{35}COONa + Mg(HCO_3)_2 \longrightarrow (C_{17}H_{35}COO)_2Mg \downarrow + 2NaHCO_3$$

$$2C_{17}H_{35}COONa + CaCl_2 \longrightarrow (C_{17}H_{35}COO)_2Ca \downarrow + 2NaCl$$

$$2C_{17}H_{35}COONa + MgCl_2 \longrightarrow (C_{17}H_{35}COO)_2Mg \downarrow + 2NaCl$$

$$2C_{17}H_{35}COONa + CaSO_4 \longrightarrow (C_{17}H_{35}COO)_2Ca \downarrow + Na_2SO_4$$

$$2C_{17}H_{35}COONa + MgSO_4 \longrightarrow (C_{17}H_{35}COO)_2Mg \downarrow + Na_2SO_4$$

The total hardness can be determined by titrating the water sample against the standard soap solution. The end point is detected as the appearance of a stable lather.

Procedure. A known volume of water sample is titrated against a standard solution of soap in alcohol. The end point is detected by the appearance of stable lather. The titre value corresponds to the total hardness of the sample. For estimating permanent and temporary hardness first of all water sample is boiled for about 20–30 minutes. Temporary hardness is removed by boiling the water. Now it is cooled. It contains only permanent hardness causing ions. Then it is again titrated against standard soap solution till the stable lather appearance (formation). The titrated value corresponds to the permanent hardness of the sample. The temporary hardness is determined by subtracting the value of permanent hardness from the value of total hardness. A blank titration of distilled water in which soap solution is prepared is also done against the prepared soap solution till the appearance of stable lather. The titre value is known as lather factor. For pure distilled water the lather factor should be zero. The value of lather factor is always subtracted from each titre value.

Calculations

Standardization of soap solution

Let volume of standard hard water (SHW) $= x$ ml

Volume of soap solution consumed $= V_1$ ml

V_1 ml of soap solution $\equiv x$ ml of SHW

$\equiv x$ mg of $CaCO_3$ ($\because$ 1 ml SHW = 1 mg $CaCO_3$)

1 ml of soap solution $= \frac{x}{V_1}$ mg of $CaCO_3$

Calculations of total hardness

Let volume of water sample for titration $= y$ ml

And volume of soap solution consumed $= V_2$ ml

Suppose the lather factor = zero

1 ml soap solution $= \frac{x}{V_1}$ mg of $CaCO_3$

V_2 ml soap solution $= \frac{x}{V_1} \times V_2$ mg of $CaCO_3$

y ml of water sample $= \frac{x}{V_1} \times V_2$ mg of $CaCO_3$

1000 ml water sample $= \frac{x}{V_1} \times \frac{V_2}{y} \times 1000$ mg $CaCO_3$

Total Hardness $= \frac{x}{V_1} \times \frac{V_2}{y} \times 1000$ ppm

Permanent hardness

Let volume of water sample after boiling $= z$ ml

And volume of soap solution used $= V_3$ ml

1 ml of standard soap solution $= \frac{x}{V_1}$ mg of $CaCO_3$

V_3 ml of standard soap solution $= \frac{x}{V_1} \times V_3$ mg of $CaCO_3$

$$i.e., z \text{ ml of water sample} = \frac{x}{V_1} \times V_3 \text{ mg of } CaCO_3$$

$$1000 \text{ ml water sample} = \frac{x}{V_1} \times \frac{V_3}{z} \times 1000 \text{ mg of } CaCO_3$$

$$i.e., \text{ Permanent Hardness} = \frac{x}{V_1} \times \frac{V_3}{z} \times 1000 \text{ ppm}$$

$$\text{Temporary Hardness} = \text{Total hardness} - \text{Permanent hardness}$$

3.5.3 EDTA Method

It is the most important and more accurate method to determine the hardness of water.

Principle. The total hardness (permanent as well as temporary) in water is determined by titrating the water sample against standard ethylenediaminetetraacetic acid (EDTA) solution in ammonia buffer solution having pH = 10 using eriochrome black-T as an indicator. The calcium and magnesium ion present in hard water forms stable compound with EDTA and less stable complex with eriochrome black-T solution.

The structure of EDTA is as follow:

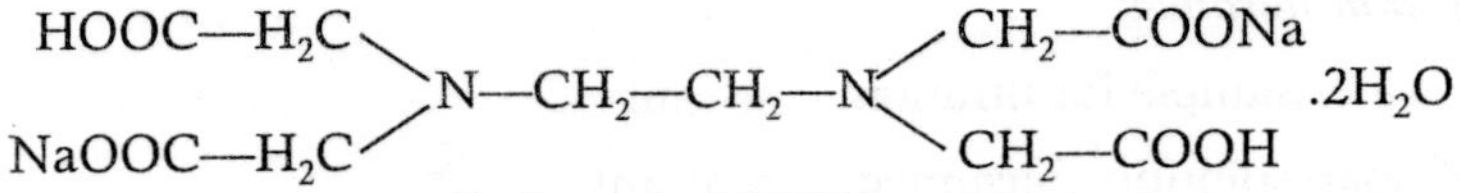

(Sodium salt of EDTA)

It ionises in water to give two Na^+ and a strong chelating agent. It is a hexadentale ligand and represented by H_2Y^{2-}. It forms complexes with bivalent cations (Ca^{2+}, Mg^{2+} etc.) or higher valent cations. The complexes with hardness causing diavalent ions are stable in alkaline medium (pH 8–10).

Eriochrome black-T may be represented as:

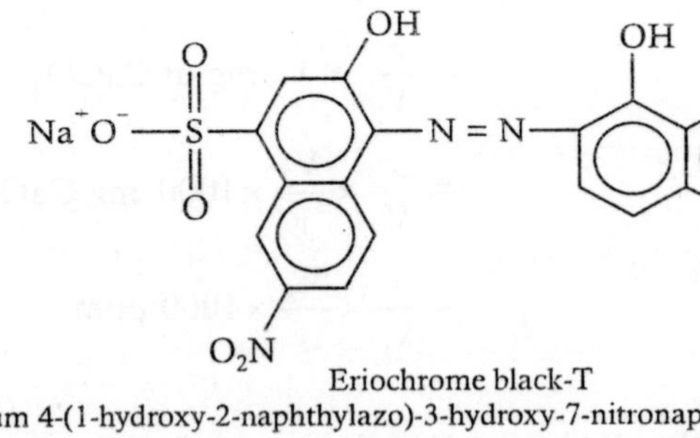

Eriochrome black-T
{Sodium 4-(1-hydroxy-2-naphthylazo)-3-hydroxy-7-nitronaphthalene-1-sulphonate}

The eriochrome black-T has two ionisable phenolic hydrogen atoms and for simplicity is represented by $Na^+H_2I_n^-$

$$\underset{\text{Red}}{H_2I_n^-} \underset{\text{pH } 5.5}{\overset{\text{pH } 7.0}{\rightleftharpoons}} \underset{\text{Blue}}{HI_n^{2-}} \underset{\text{pH } 11.0}{\overset{\text{pH } 11.5}{\rightleftharpoons}} \underset{\text{Yellowish orange}}{I_n^{3-}}$$

The calcium and magnesium ion present in hard water combine with the indicator eriochrome black-T at pH 9–10 to form less stable wine red complex.

$$M^{2+} + HI_n^{2-} \longrightarrow MI_n^- + H^+ \quad \text{(Where M = Ca or Mg)}$$
(EBT, blue)

Magnesium ion produces wine red colour with the indicator.

When EDTA is added, the free Ca^{2+} and Mg^{2+} forms a stable complex of metal-EDTA.

$$Ca^{2+} + H_2Y^{2-} \longrightarrow CaY^{2-} + 2H^+$$
$$Mg^{2+} + H_2Y^{2-} \longrightarrow MgY^{2-} + 2H^+$$

Mg-EDTA complex is less stable than Ca-EDTA complex but more stable than Mg-Indicator complex.

When all Ca^{2+} and Mg^{2+} get complexed with EDTA then further addition of EDTA sets free the metal ion from metal indicator complex and form more stable complex.

$$CaI_n^- + H_2Y^{2-} \longrightarrow CaY^{2-} + HI_n^{2-} + H^+$$
$$MgI_n^- + H_2Y^{2-} \longrightarrow MgY^{2-} + HI_n^{2-} + H^+$$
(Wine red) (EDTA) (Metal-EDTA complex, more stable) (Free indicator Blue)

The metal-EDTA complex may be represented as:

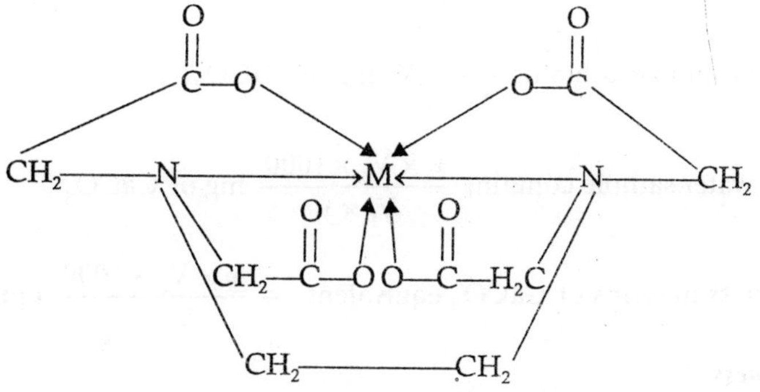

Procedure. A known volume of water sample is treated with ammonium buffer (NH_4Cl+NH_4OH) solution having pH 9–10. Now 4-5 drops of eriochrome black-T indicator is added to get wine red colour. After that it (mixture) is titrated against standard EDTA solution till the appearance of permanent blue colour. The titre value corresponds to the total hardness of the water sample. For permanent hardness water sample is boiled at first for sometimes and then filtered. (For better result the water sample is evaporated to dryness and the residue is dissolved in distilled water and then filtered). Now filtrate is titrated against standard EDTA solution in the presence of ammonium buffer solution using eriochrome black-T indicator till the appearance of blue colour. The titre value corresponds to the permanent hardness. Temporary hardness is determined by subtracting the value of permanent hardness from the value of total hardness.

1. **Role of ammonium buffer solution:** The main role of ammonium buffer solution is to maintain the pH of the medium during titration. The metal-EDTA complexes are stable only in alkaline medium (pH range 8–10). During the titration of hard water sample by EDTA method, large number of acid molecules are formed. It may alter the pH of the medium. Hence it is necessary to control the pH of the medium. The optimum pH for the experiment is 10.0 ± 0.1 and is adjusted by using ammonium buffer solution.

Calculation ***(Standardization of EDTA)***

Let the volume of standard water (SHW) = x ml

Volume of EDTA used $= V_1$ ml

V_1 ml of EDTA $\equiv x$ ml of SHW

$\equiv x$ mg of $CaCO_3$ (let 1 ml of SHW $\equiv$ 1 mg of $CaCO_3$)

$\therefore$ 1 ml of EDTA $= \dfrac{x}{V_1}$ mg of $CaCO_3$

Total hardness

Let the volume of water sample $= y$ ml

Volume of EDTA used $= V_2$ ml

$\because$ 1 ml of EDTA $\equiv \dfrac{x}{V_1}$ mg of $CaCO_3$

$\therefore$ V_2 ml of EDTA $\equiv \dfrac{x}{V_1} \times V_2$ mg of $CaCO_3$

i.e., y ml of water sample contains $\dfrac{x}{V_1} \times V_2$ mg of $CaCO_3$

$\therefore$ 1000 ml of water sample contains $\dfrac{x \times V_2 \times 1000}{V_1 \times y}$ mg of $CaCO_3$

$\therefore$ Total Hardness in terms of $CaCO_3$ equivalent $= \dfrac{x \times V_2 \times 1000}{V_1 \times y}$ ppm

Permanent hardness

Let the volume of water sample after boiling $= z$ ml

Volume of EDTA used $= V_3$ ml

$\because$ 1 ml of EDTA $\equiv \dfrac{x}{V_1}$ mg of $CaCO_3$

$\therefore$ V_3 ml of EDTA $\equiv \dfrac{x}{V_1} \times V_3$ mg of $CaCO_3$

i.e., z ml of water sample contains $\dfrac{x}{V_1} \times V_3$ mg of $CaCO_3$

$\therefore$ 1000 ml of water sample contains $\dfrac{x \times V_3 \times 1000}{V_1 \times z}$ mg of $CaCO_3$

$\therefore$ Permanent Hardness in terms of $CaCO_3$ equivalent $= \dfrac{x \times V_3 \times 1000}{V_1 \times z}$ ppm

Temporary Hardness = Total hardness – Permanent hardness

2. Determination of calcium hardness and magnesium hardness: *Theory.* Calcium ion present in water gets precipitated as calcium oxalate by adding calcium precipitating buffer $[NH_4OH + NH_4Cl + (NH_4)_2C_2O_4;\ pH \simeq 10]$ in water sample.

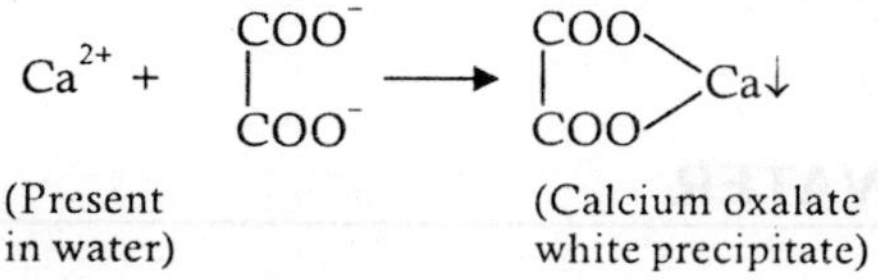

After filtering, the hard water contains magnesium salts. The filtrate containing magnesium salts is titrated against standard EDTA solution in ammonia buffer $[NH_4OH + NH_4Cl;\ pH \simeq 10]$ using EBT indicator as usual method. From the consumed volume of EDTA, the amount of magnesium hardness is calculated. The amount of calcium hardness is calculated by subtracting the magnesium hardness from the total hardness.

Reactions. (Kindly see EDTA method).

Procedure. First of all total hardness of water is determined by usual method. For determination of calcium hardness and magnesium hardness, a known amount of calcium precipitating buffer is added to a known volume of hard water sample. Now it is shaked well and left for about one hour. (for better result it is left for overnight). All calcium ions get precipitated as calcium oxalate. Now it is filtered and filtrate is collected carefully. Take some filtrate (few ml) and is treated with ammonium buffer solution. A few drops of EBT indicator is added to the filtrate to get wine red colour. The content is shaked well and then titrated against standard EDTA solution till the appearance of permanent blue colour. The magnesium hardness is calculated from the titre value.

Calcium hardness = Total hardness – Magnesium hardness

Calculation. Let volume of water sample = A ml

Volume of calcium precipitating buffer = B ml

Volume of filtrate for titration = C ml

Volume of EDTA consumed during titration = V_4 ml

$\because$ (A + B) ml of solution contains A ml H_2O

$\therefore$ C ml of solution contains $\dfrac{A.C}{A+B}$ ml H_2O

$= X$ ml H_2O (say)

$\because$ 1 ml of EDTA consumes $\dfrac{x}{V_1}$ mg of $CaCO_3$ (see EDTA method)

$\therefore$ V_4 ml of EDTA consumes $\dfrac{x}{V_1} \times V_4$ mg $CaCO_3$

$\because$ X ml water contains $\frac{x}{V_1} \times V_4$ mg $CaCO_3$

$\therefore$ 1000 ml water contains $\frac{x \times V_4}{V_1 X} \times 1000$ mg $CaCO_3$

Hence Magnesium hardness in terms of $CaCO_3$ equivalent $= \frac{x \times V_4}{V_1 \times X} \times 1000$ ppm

and Calcium hardness = Total hardness – Mg-hardness.

3.6 ALKALINITY OF WATER

The alkalinity of water is due to the presence of those type of substances in water which have tendency to increase the concentration of OH^- ions either by hydrolysis or dissociation in water. The alkalinity of water is due to the following factors :

(*i*) The presence of salts of weak organic acids which undergo hydrolysis and consume H^+ of water. As a result concentration of OH^- increases in water and water becomes alkaline.

(*ii*) The presence of HCO_3^-, $HSiO_3^-$, SiO_3^{2-}, etc. in water which make the water alkaline because they have tendency to take up H^+ from water. Hence concentration of OH^- in water increases.

Classification

Alkalinity of water is mainly classified as : (*i*) Hydroxide alkalinity (*ii*) Carbonate alkalinity and (*iii*) Bicarbonate alkalinity.

Both alkalinity and hardness are expressed in terms of $CaCO_3$ equivalent. Alkalinity is also expressed generally in ppm or mg/l.

When alkalinity < total hardness

Carbonate hardness in ppm = Alkalinity in ppm

When alkalinity ≥ total hardness

Carbonate hardness in ppm = Total hardness in ppm and

Non-Carbonate hardness = Total hardness – Carbonate hardness.

Determination

The alkalinity of water is determined by titrimetric method. A known volume of water sample is titrated against a standard acid (H_2SO_4) using phenolphthalein and methyl orange indicator. The reactions take place are following:

$$\left.\begin{array}{l} \left.\begin{array}{l} OH^- + H^+ \longrightarrow H_2O \\ CO_3^{2-} + H^+ \longrightarrow HCO_3^- \end{array}\right] P \\ HCO_3^- + H^+ \longrightarrow H_2O + CO_2 \end{array}\right] M$$

[**Note:** Here HCO_3^- represents the amount of HCO_3^- obtained from CO_3^{2-} and extra amount of HCO_3^-, if any, present in water sample.]

Here 'P' and 'M' represents the volume of consumed acid by using phenolphthalein indicator and methyl orange indicator respectively.

From the above relations the following conclusions may be drawn:

(*i*) When P = 0; it means that the alkalinity is due to the presence of only HCO_3^-. Both OH^- and CO_3^{2-} are absent. Thus, the alkalinity due to HCO_3^- = M.

(*ii*) When P = M; it means that the alkalinity is due to the presence of only OH^-. Thus, the alkalinity due to OH^- = P = M.

(*iii*) When $P = \frac{1}{2}M$; it means that the alkalinity is due to the presence of only CO_3^{2-}. In this case phenolphthalein end point indicates the half neutralization of CO_3^{2-} *i.e.*, neutralization of CO_3^{2-} upto HCO_3^-.

$$CO_3^{2-} + H^+ \longrightarrow HCO_3^-$$

Whereas methyl orange end point indicates the complete neutralization of CO_3^{2-} as :

$$CO_3^{2-} + H^+ \longrightarrow HCO_3^-$$

$$HCO_3^- + H^+ \longrightarrow H_2CO_3 \longrightarrow H_2O + CO_2$$

It is obviously that the alkalinity due to CO_3^{2-} = 2P or M.

(*iv*) When $P > \frac{1}{2}M$; it means that the alkalinity is due to the presence of OH^- in addition to CO_3^{2-}. In this case phenolphthalein end point indicates the complete neutralization of OH^- ions and half neutralization of CO_3^{2-} *i.e.*, neutralization of CO_3^{2-} upto HCO_3^-.

$$\left.\begin{array}{l} OH^- + H^+ \longrightarrow H_2O \\ CO_3^{2-} + H^+ \longrightarrow HCO_3^- \end{array}\right] P$$

Whereas methyl orange indicator end point indicates that the complete neutralization of OH^- and CO_3^{2-} as :

$$\left.\begin{array}{l} OH^- + H^+ \longrightarrow H_2O \\ CO_3^{2-} + H^+ \longrightarrow HCO_3^- \\ HCO_3^- + H^+ \longrightarrow H_2CO_3 \longrightarrow H_2O + CO_2 \end{array}\right] M$$

Therefore, it is clear that the neutralization of HCO_3^- is equal to M – P and the complete neutralization of CO_3^{2-} is equal to 2(M – P). (Since complete neutralization of CO_3^{2-} is twice the neutralization of HCO_3^- ions).

Thus, the alkalinity due to OH^- = M – 2(M – P) = M – 2M + 2P = 2P – M.

(*v*) When $P < \frac{1}{2}M$; it means that the alkalinity is due to the presence of extra amount of HCO_3 in addition to CO_3^{2-}. In this case phenolphthalein end point indicates the neutralization of CO_3^{2-} up to HCO_3^- *i.e.*, half neutralization of CO_3^{2-}.

$$CO_3^{2-} + H^+ \longrightarrow HCO_3^-$$

Whereas methyl orange indicator end point indicates the complete neutralization of CO_3^{2-} and extra amount of HCO_3^- presents as:

$$CO_3^{2-} + H^+ \longrightarrow HCO_3^-$$

$$HCO_3^- + H^+ \longrightarrow H_2CO_3 \longrightarrow H_2O + CO_2$$

It is obviously that the alkalinity due to $CO_3^{2-} = 2P$

and alkalinity due to $HCO_3^- = M - 2P$.

[**Note :** HCO_3^- and OH^- both do not exist simultaneously because they react to form CO_3^{2-} and H_2O as

$$HCO_3^- + OH^- \longrightarrow CO_3^{2-} + H_2O$$

So this combination is not possible.]

The conclusions may be summarised in Table 3.1.

Table 3.1

Result upto Phenolphthalein End Point (P) and Methyl Orange End Point (M)	OH^-	CO_3^{2-}	HCO_3^-
$P = 0$	0	0	M
$P = M$	P or M	0	0
$P = \frac{1}{2}M$	0	2P or M	0
$P > \frac{1}{2}M$	2P – M	2(M – P)	0
$P < \frac{1}{2}M$	0	2P	M – 2P

Procedure

A known volume of water sample is titrated against standard acid (H_2SO_4) by using phenolphthalein indicator. The end point is detected by disappearence of pink colour. The volume of acid consumed is noted. Now the titrated water sample is again titrated against the same standard acid (H_2SO_4) by using methyl orange indicator. The end point is detected by appearance of red colour. The volume of acid consumed is noted.

Calculations

Let volume of water sample for titration = V ml

Strength of acid = N/50

Volume of acid used in case of phenolphthalein indicator = A

Volume of acid used in case of methyl orange indicator = B

For phenolphthalein indicator

Applying normality equation,

$$\underset{\text{(Water)}}{N_1V_1} = \underset{\text{(Acid)}}{N_2V_2}$$

$$N_1 \times V = \frac{1}{50} \times A$$

$$N_1 = \frac{1}{50} \times \frac{A}{V}$$

$$\text{Strength} = N_1 \times \text{Eq. wt.}$$

$$= \frac{1 \times A}{50 \times V} \times 50\ g/L \qquad (\because \text{ Eq. wt. of } CaCO_3 = 50)$$

$\therefore$ Phenolphthalein alkalinity, in terms of $CaCO_3$ equivalent (P) $= \frac{1 \times A \times 50 \times 1000}{50 \times V}$ ppm

$$= x \text{ ppm}$$

Similarly, for methyl orange indicator (Total alkalinity)

The volume used of standard acid = (A + B) ml

Applying normality equation

$$\underset{\text{(Water)}}{N_1V_1} = \underset{\text{(Acid)}}{N_2V_2}$$

$$N_1 \times V = \frac{1}{50} \times (A + B)$$

$$N_1 = \frac{1}{50} \times \frac{(A + B)}{V}$$

$$\text{Strength} = N_1 \times \text{Eq. wt.}$$

$$= \frac{1}{50} \times \frac{(A+B)}{V} \times 50 \text{ g/L} \qquad (\because \text{ Eq. wt. of } CaCO_3 = 50)$$

$\therefore$ Methyl Orange (total) alkalinity, in terms of $CaCO_3$ equivalent

$$(M) = \frac{1}{50} \times \frac{(A+B)}{V} \times 50 \times 1000 \text{ ppm}$$

$$= Y \text{ ppm}$$

The type and amount of alkalinity is determined by comparing the value of P and M using table 3.1.

Disadvantages of Hard Water

In domestic use hard water is not suitable due to following reasons:

1. *Cooking.* Vegetables and other food like pulses etc. do not cook well in hard water.
2. *Wastage of soap.* Since hard water does not form lather with soap readily. Hence soap is wasted in removing the Ca^{2+} and Mg^{2+} ions present in water.
3. *Spoiling clothes.* The Ca^{2+} and Mg^{2+} present in hard water combine with soap to form insoluble compound, which sticks to the clothes. It is not easily removed and spoiled the clothes.
4. *Consumption of more fuels.* Ca^{2+} and Mg^{2+} salts are deposited inside the boiler. These form a layer as a scale, which are not removed easily. Hence consumption of fuel is increased and the cooking utensiles are spoiled.

1. *In Industries*

1. *Textile industry.* The water used in textile and dyeing industries should be free from hardness (Ca^{++} and Mg^{++} ions), organic matter, suspended particles etc. hard water precipitates basic dyes. Iron and manganese salt produce coloured spots on fabrics.

2. *Sugar industry.* Water should be free from hardness, suspended particles as well as pathogenic microorganisms. Hard water causes difficulties in the crystallization of sugar from molasses.
3. *Paper and Pulp industry.* The water should be free from hardness, suspended particles, iron, manganese etc. hardness increases the ash contents of paper. Dissolved silica produce cracking tendency of paper. Iron and manganese salts decrease the brightness and colour of paper.
4. *Aluminium industry.* The water should be of high quality. Any type of impurity and hardness make it low grade.
5. *Carbonated beverage industry.* The water should be free from colouring matter, pathogenic micro-organisms, hardness, iron and manganese salts, suspended particles etc. It also should low alkalinity. Hardness producing salts make it of low grade.
6. *Starch industry.* Water should be free from any type of hardness impurities (organic or inorganic), suspended particles etc. Iron and manganese salts produce yellow colour, which is undesirable. Hardness producing salts causes the precipitation of salts and accumulates in the starch.
7. *Iron industry.* Hard water makes the iron of low quality. It corrodes the iron and alloy. Water pipes are blocked due to precipitation of salts during boiling of hard water. Hence water should be free from hardness and impurities.
8. *Fibre industry.* The water used in fibre industry should be of high quality. Silica, salts of iron and manganese, dissolved organic and inorganic materials make the fibre of very low grade. Alkalinity and hardness causing ions are very much harmful for fibres.

2. *In Boilers*

A large quantity of water is used in boilers for generating the steam. Boiler plays very important role for generation of steam in industries. The water used in boiler is called boiler feed water. Hard water creates a number of problems like corrosion, scale and sludge formation, caustic embrittlement, priming and foaming etc. It is very dangerous because it causes the explosions of boilers due to high pressure.

Natural water is not directly used in boiler because it contains hardness producing salts. Hence water should be properly softened and sufficiently pure before feeding into the boiler.

Generally a boiler feed water should satisfy the following requirements:

Hardness < 0.5 ppm

Caustic alkalinity 0.15–0.45 ppm

Soda alkalinity < 1 ppm

Excess soda ash < 0.55 ppm

3.7 SCALE AND SLUDGE FORMATION

When hard water is boiled in the boiler to generate the steam, the dissolved salt starts separating in order to their solubilities after the saturation point. The least soluble salts separate out first and the precipitate forms as a layer inside the boiler. When precipitation of dissolved salts take place inside the boiler in the form of soft, slimy and non-adherent layer, it is called 'sludge' and if the precipitation takes place in the form of hard adhering coating inside the boiler walls, it is called 'scales'.

3.7.1 Sludge

Sludge is formed inside the boiler walls. These are formed by substances, which have greater solubility in hot water than in cold water. It is formed at comparatively colder portions of the boiler and collects in the area where flow rate is slow.

Composition. The main compositions of sludge are $MgCO_3$, $MgCl_2$, $MgSO_4$, $CaCl_2$, etc.

Disadvantages

1. They are poor conductor of heat so they waste lot of amount of heat.
2. The formation of sludges decrease the efficiency of the boiler.
3. Excess formation of sludges may cause choking of water pipes in the industry.

Preventions

1. It may be prevented by using soft water.
2. It is minimized by frequently cleaning operations, *i.e.*, the removal of a portion of concentrated water and replacing by fresh water.
3. It may be removed by scrapping off with a hard brush.

3.7.2 Scale

These are very hard and adhere firmly inside the boiler. They are very difficult to remove even by hammer and chisel.

Compositions. The main compositions of scales are $CaCO_3$, $Ca(OH)_2$, $CaSO_4$, $Mg(OH)_2$, $CaSiO_3$, $MgSiO_3$, etc.

Formation

The formation of scale are due to following:

1. *Decomposition of calcium bicarbonate.* Calcium bicarbonate is decomposed into calcium carbonate on boiling

$$Ca(HCO_3)_2 \xrightarrow{\Delta} \underset{\text{(Scale)}}{CaCO_3 \downarrow} + H_2O + CO_2 \uparrow$$

In low pressure boilers $CaCO_3$ causes scale formation. In high-pressure boilers $CaCO_3$ becomes soluble.

$$CaCO_3 \xrightarrow{\Delta} CaO + CO_2 \uparrow$$

$$CaO + H_2O \longrightarrow Ca(OH)_2$$

2. *Presence of calcium sulphate.* Calcium sulphate is soluble in cold water but the solubility decreases with rise in temperature. At high temperature in boilers $CaSO_4$ gets precipitated as hard scale on the inner portions of boiler. It is the main cause of scale formation in high pressure boilers.
3. *Hydrolysis of magnesium salts.* Magnesium salts dissolved in water undergo hydrolysis and forms precipitates of $Mg(OH)_2$ as a scale inside the boiler walls.

$$MgCl_2 + 2H_2O \longrightarrow \underset{\text{(Soft scale)}}{Mg(OH)_2 \downarrow} + 2HCl$$

$Mg(OH)_2$ is a soft type of scale.

4. *Presence of silica.* Silica present in small quantities deposits as calcium silicate ($CaSiO_3$) or magnesium silicate ($MgSiO_3$). These silicates are sparingly soluble in cold water and almost insoluble in hot water. These are very difficult to remove, as they are very hard and sticky on the side of boilers.

Disadvantages

1. *Wastage of fuel.* They are bad conductor of heat so a lot amount of heat will be useless.
2. *Danger of explosion.* Since scales are not uniformly formed so these may crack due to expansion and water suddenly comes in contact with overheated iron plates. This causes in the formation of large amount of steam suddenly. This results high pressure and explosion may occur.
3. *Bagging.* The distortion of boiler material is called bagging. Due to overheating the boiler rapid reaction may take place between water and iron at high temperature.

$$3Fe + 4H_2O \longrightarrow Fe_3O_4 + 4H_2 \uparrow$$

It causes the distortion of boiler materials. At high temperature corrosion may takes place.

4. *Decrease in efficiency.* Due to deposition of scales in boilers valve and pipes, it may be choked and hence efficiency of boiler is decreased.

Removal of Scales

1. These are removed by scrapper or hard brush.
2. These can be removed by thermal shocks *i.e.*, heating the boilers and suddenly cooling with cold water.
3. These can be removed by dissolving in some suitable chemicals. $CaCO_3$ scales can be dissolved in 5–10% diluted HCl. Similarly, calcium ions can be removed by EDTA solution.

Preventions

The scale formation may be minimized by the following treatments :

1. External treatment, and
2. Internal treatment.

1. **External treatment.** It includes the water softening techniques like lime soda process, zeolite process or demineralized (ion exchange process). (See Art. 4.2 of chapter 4)
2. **Internal treatment.** In this treatment suitable chemicals are added to the boiler containing hard water. The scales are either precipitated or converted into soluble complexes. This process is also known as "sequestration" and the chemical which is added is called 'sequestering agent'. These are important methods for internal treatment of water.

(*a*) *Colloidal conditioning.* To avoid scale formation in low pressure boilers we add organic substances like kerosene, tannin, agar-agar etc., these substances act as protective coatings. Hence salt remains as non-sticky deposits in the boiler as a sludge and be removed very easily.

(*b*) *Carbonate conditioning.* Carbonate conditioning is applicable for avoiding the scale formation in low pressure boilers. Sodium carbonate is added to the boiler containing hard water. Calcium sulphate ($CaSO_4$) present in hard water gets converted into calcium carbonate ($CaCO_3$) and it

$$CaSO_4 + Na_2CO_3 \longrightarrow CaCO_3 + Na_2SO_4$$

can be removed by blow down operation. Hence deposition of $CaSO_4$ as a scale does not takes place. Carbonate conditioning is not suitable for high pressure boiler because in high pressure boiler, excess amount of Na_2CO_3 may undergoes in hydrolysis and NaOH is formed as a product which causes caustic embrittlement.

(*c*) *Phosphate conditioning.* Phosphate conditioning is generally applicable for avoiding the scale formation in high-pressure boiler. Sodium phosphate is added to the hard water in boiler. It reacts with the hardness producing substances or ions to form soft sludges of calcium phosphate and magnesium phosphate.

$$3CaSO_4 + 2Na_3PO_4 \longrightarrow \underset{\text{(Sludge)}}{Ca_3(PO_4)_2} \downarrow + 3Na_2SO_4$$

$$3MgCl_2 + 2Na_3PO_4 \longrightarrow \underset{\text{(Sludge)}}{Mg_3(PO_4)_2} \downarrow + 6NaCl$$

These soft sludges can be removed by blow-down operation. For this conditioning we use generally sodium dihydrogenphosphate NaH_2PO_4 (acidic), disodiumhydrogenphosphate Na_2HPO_4 (weakly alkaline), trisodiumphosphate Naq_3PO_4 (alkaline) and sodium pyrophosphate $Na_4P_2O_7$.

(*d*) *Calgon conditioning.* Calgon is sodium hexametaphosphate $Na_2[Na_4(PO_3)_6]$. Calgon is added to the boiler water. It forms soluble complex compound with calcium salt like $CaSO_4$, $CaCl_2$ etc., present in water and hence prevents from the formation of scale and sludge.

$$Na_2[Na_4(PO_3)_6] \rightleftharpoons 2Na^+ + [Na_4P_6O_{18}]^{2-}$$

$$2CaSO_4 + [Na_4P_6O_{18}]^{2-} \longrightarrow \underset{\text{(Soluble complex ion)}}{[Ca_2P_6O_{18}]^{2-}} + 2Na_2SO_4$$

(*e*) *Treatment with sodium aluminate.* Sodium aluminate ($NaAlO_2$) is added to hard water in boiler. It gets hydrolysed to form NaOH and gelatinous precipitate of $Al(OH)_3$. The product NaOH reacts with $MgCl_2$ present in water to form $Mg(OH)_2$ as a precipitate.

$$NaAlO_2 + 2H_2O \longrightarrow NaOH + Al(OH)_3 \downarrow$$

$$MgCl_2 + 2NaOH \longrightarrow 2NaCl + Mg(OH)_2 \downarrow$$

The flocculent precipitated of $Mg(OH)_2$ and $Al(OH)_3$ produced inside the boiler. These precipitates entraps finely suspended and colloidal impurities including silica and oil drops. These can be removed by blow-down operation.

(*f*) *Radioactive conditioning.* Some tablets containing radioactive salts are placed in boiler containing hard water. The energy radiation emitted by radioactive salts prevent scale formation.

(*g*) *Complexometric conditioning (EDTA conditioning).* About 1.5% alkaline solution of ethylenediaminetetraceticacid (EDTA) having pH 8.5 solution is added to the boiler water. The EDTA binds the scale-forming cations to form stable and soluble complex. Hence scale and sludge formation in boiler is prevented. It is better than phosphate conditioning because phosphate conditioning fails to prevent the formation of iron oxide, corrosion etc. whereas this method prevents from corrosion, formation of iron oxide, etc.

(*h*) *Electrical conditioning.* Sealed glass tubs, containing mercury connected to a battery are set rotating in the boiler. When water boils, mercury bulbs emit electrical discharges and prevents scale formation.

3.8 PRIMING AND FOAMING

When water is boiled in a boiler the formation of persist foam or bubbles at the surface of water takes place. It is called foaming. But when water is boiled rapidly the bubbles and drops of liquid water along with steam is entered into engine. It is called priming. Both process usually occur together.

3.8.1 Causes of Priming

It is caused due to the presence of:

1. Large amounts of dissolved solids like alkali sulphates and chloride
2. Improper designing of water
3. High water level
4. High velocity of steam.

Prevention

It is prevented by:

1. Proper designing of boiler
2. Using soft and filtered water
3. Keeping the low water level
4. Using mechanical steam purifier.

3.8.2 Causes of Foaming

It is caused due to the presence of:

1. Oil or grease in water
2. Finely divided sludge in water
3. Some chemicals which produces foams by reducing the surface tension of water.

Prevention

It is prevented by:

1. Addition of antifoaming chemicals like castor·oil etc.
2. Addition of some suitable coagulants like ferrous sulphate etc., so that the oil or grease present in water may be removed.
3. Using soft and filtered water.

Disadvantages of Priming and Foaming

Following are the disadvantages of priming and foaming:

1. Due to the presence of foaming boiling point of water is increased hence wastage of fuel occurs.
2. It is very difficult to maintain the constant pressure of steam.

3. Due to excess foaming the bubbles entered into the engine along with steam. It lowers the efficiency of engine and engine part is spoiled.
4. Corrosion take place in the part of engine.
5. Due to the presence of foam water, level is not identified.

3.9 CAUSTIC EMBRITTLEMENT

It is actually a special type of boiler corrosion and is caused by the use of highly alkaline water in the boiler. During softening of hard water by soda-lime process free sodium carbonate (Na_2CO_3) is usually present in small proportion in the softened water. In high-pressure boilers, Na_2CO_3 decomposes to give sodium hydroxide and CO_2.

$$Na_2CO_3 + H_2O \longrightarrow 2NaOH + CO_2$$

The formation of NaOH makes the boiler water caustic. The caustic water flows into the minute hair crack parts (like bends, joints, rivets etc.) inside the boiler by the capillary action. Water evaporate and the concentration of NaOH increases slightly. This caustic soda attack the surrounding area and iron of boiler is dissolved into NaOH to form sodium ferroate which decomposes and forms rust magnetite (Fe_3O_4).

$$3Na_2FeO_2 + 4H_2O \longrightarrow 6NaOH + Fe_3O_4 + H_2$$

or

$$6Na_2FeO_2 + 6H_2O + O_2 \longrightarrow 12NaOH + 2Fe_3O_4$$

This causes the formation of irregular intergranular cracks on the boiler metal, which is known as embrittlement. It always occurs particularly at highly stressed parts like joints, rivets etc. It is actually electrochemical phenomenon and may be explained as :

Iron at rivets, joint etc. (anode) $\oplus$	Concentrated NaOH solution	Dilute NaOH solution	$\ominus$ Iron at plane surface (cathode)

By the electrochemical action the point of high local stress gets corroded. Here the iron surface surrounded by dilute NaOH area acts as cathode whereas iron cracks surrounded by concentrated NaOH acts as anode.

Preventions

It is prevented by the following:

1. Adding tannin or lignin to boiler water because these block the hair cracking inside to boiler.
2. Using Na_2SO_4 to the boiler water. Na_2SO_4 also blocks the hair cracking inside the boiler. It has been observed that in boiler if the ratio of Na_2SO_4 : NaOH concentration is 3 : 1 then the boiler works smoothly at 20 atmospheric pressure and above.
3. Using sodium phosphate as softening reagent instead of Na_2CO_3 during softening of hard water in lime-soda process.

3.10 BOILER CORROSION

It is the decay of boiler material by attack of some chemicals or electrochemical attack by its environment.

3.10.1 Factors Causing Boiler Corrosion

1. *Presence of dissolved oxygen.* When water is heated, the dissolved oxygen gets liberated and attacks the boiler material

$$4Fe + 3O_2 + 2xH_2O \longrightarrow \underset{\text{(Rust)}}{2[Fe_2O_3 \, . \, xH_2O]}$$

The reaction may also be represented as follows:

$$2Fe + 2H_2O + O_2 \longrightarrow 2Fe(OH)_2$$

$$4Fe(OH)_2 + O_2 \longrightarrow \underset{\text{(Rust)}}{2[Fe_2O_3 \, . \, 2H_2O]}$$

2. *Presence of dissolved CO_2.* Small quantity of CO_2 is also dissolved in water. If water contains bicarbonate then CO_2 is released on boiling the water inside the boiler.

$$Ca(HCO_3)_2 \xrightarrow{\Delta} CaCO_3 \downarrow + H_2O + CO_2 \uparrow$$

$$Mg(HCO_3)_2 \xrightarrow{\Delta} MgCO_3 \downarrow + H_2O + CO_2 \uparrow$$

Carbon dioxide dissolves in water to form a carbonic acid. Carbonic acid is a weak acid.

$$CO_2 + H_2O \longrightarrow H_2CO_3$$

3. *Presence of acid forming salts.* Some inorganic salts like $MgCl_2$, $CaCl_2$ etc. present in water gets hydrolyse to produce acids which corrodes the boiler material.

$$MgCl_2 + 2H_2O \longrightarrow Mg(OH)_2 \downarrow + 2HCl$$

$$CaCl_2 + 2H_2O \longrightarrow Ca(OH)_2 \downarrow + 2HCl$$

Calcium chloride undergoes hydrolysis very slowly but in the presence of silicic acid the hydrolysis take place rapidly at lower temperature. Corrosion takes place is as follows :

$$Fe + 2HCl \longrightarrow FeCl_2 + H_2 \uparrow$$

Preventions

Boiler corrosion may be prevented by following methods:

1. *By removal of dissolved oxygen.* Dissolved oxygen is removed by adding a calculated amount of sodium sulphite or hydrazine or sodium sulphide. The reactions take place are as follows :

$$2Na_2SO_3 + O_2 \longrightarrow 2Na_2SO_4$$

$$\underset{\text{(Hydrazine)}}{N_2H_4} + O_2 \longrightarrow N_2 + 2H_2O$$

$$Na_2S + 2O_2 \longrightarrow Na_2SO_4$$

Dissolved oxygen and dissolved CO_2 is removed by mechanical deaeration method as shown in Fig. 3.2. Water is sprayed in a perforated plate-fitted tower, heated from sides and connected to a vacuum pump. The water passes through the perforated plates undergoes deaeration at high temperature and low pressure. Dissolved O_2 and CO_2 escapes. This is because the solubility of a gas in water is directly proportional to pressure and inversely proportional to temperature. Hence the water gets deaerated.

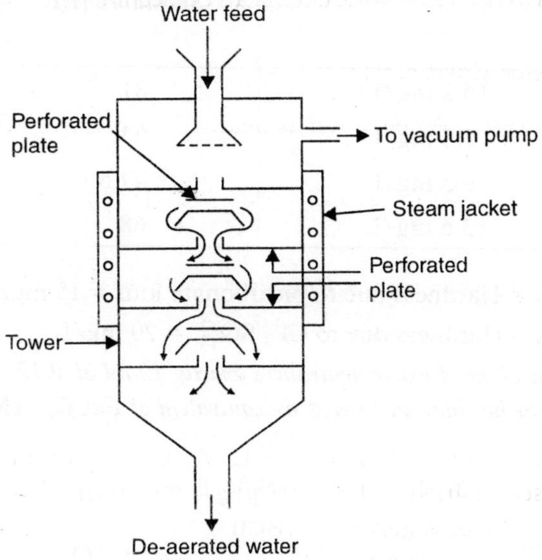

FIGURE 3.2 **Mechanical deaeration of water**

2. *By removal of dissolved CO_2.* The dissolved CO_2 is removed by mechanical deaeration method. It is also removed by adding a calculated amount of ammonia.

$$2NH_4OH + CO_2 \longrightarrow (NH_4)_2 CO_3 + H_2O$$

3. *By addition of alkali.* Corrosion by acids may be prevented by adding some alkalies from outside so that the product acid may be neutralized.

SOLVED NUMERICALS

Based on Hardness of Water

1. *A water sample contains 180 mg of $MgSO_4$ per litre. Calculate the hardness in term of $CaCO_3$ equivalent.*

Solution.

Mass of $MgSO_4$ (W) $= 180$ mg/L

Equivalent mass of $MgSO_4$ (E) $= 60$

Equivalent of $CaCO_3$ $= \dfrac{50 \times W}{E} = \dfrac{50 \times 180}{60}$

$= 150$ mg/L $= 150$ ppm.

2. *Calculate temporary hardness and permanent hardness of a sample of water which on analysis is found to contain the following:*

$Ca(HCO_3)_2 = 16.2\ mg/L$, $Mg(HCO_3)_2 = 7.3\ mg/L$

$MgCl_2 = 9.5\ mg/L$, $CaSO_4 = 13.6\ mg/L$

Solution.

Substance	*Mass of Substance*	*Equivalent Mass of Substance (E)*	*$CaCO_3$ Equivalent* = $\left(\frac{W \times 50}{E}\right)$
$Ca(HCO_3)_2$	16.2 mg/L	81	10 mg/L
$Mg(HCO_3)_2$	7.3 mg/L	73	5 mg/L
$MgCl_2$	9.5 mg/L	47.5	10 mg/L
$CaSO_4$	13.6 mg/L	68	10 mg/L

Temporary Hardness = Hardness due to bicarbonate ions = 15 mg/L

Permanent Hardness = Hardness due to Cl^-, SO_4^{2-} = 20 mg/L.

3. *100 ml of a sample of hard water neutralizes exactly 12 ml of 0.12 N HCl using methyl orange as indicator. Express the hardness in term of an equivalent of $CaCO_3$.* **(M.D.U. Rohtak May 2005)**

Solution.

$$\underset{\text{(Hard Water)}}{N_1V_1} = \underset{\text{(HCl)}}{N_2V_2}$$

$$N_1 \times 100 = \frac{12}{100} \times 12$$

$$N_1 = \frac{12 \times 12}{100 \times 100}$$

Strength of hardness in terms of an equivalent of

$$CaCO_3 = \frac{12 \times 12}{100 \times 100} \times 50 \times 1000 = 720 \text{ ppm}$$

4. *Braham Sarover water of Kurukshetra on analysis gives the following results : CO_2 = 22 ppm, HCO_3^- = 305 ppm, Ca^{2+} = 80 ppm, Mg^{2+} = 48 ppm, total solid = 5000 ppm. Calculate carbonate and non-carbonate hardness of water sample.*

Solution.

Constituent	*Quantity (W)*	*Equivalent wt. (E)*	*$CaCO_3$ Equivalent* = $\left(\frac{W \times 50}{E}\right)$
CO_2	22 ppm	22	50 ppm
HCO_3^-	305 ppm	61	$\left(\frac{305 \times 50}{61}\right)$ = 250 ppm
Ca^{2+}	80 ppm	20	$\left(\frac{80 \times 50}{20}\right)$ = 200 ppm
Mg^{2+}	48 ppm	12	$\left(\frac{48 \times 50}{12}\right)$ = 200 ppm

Carbonate hardness (due to bicarbonates of Ca and Mg) = 250 ppm

Non-carbonate hardness = 200 ppm Ca^{2+} + 200 ppm Mg^{2+} – 250 ppm HCO_3^- = 150 ppm.

Based on Hardness Determination by Soap Solution Method

5. *100 ml of hard water is titrated with soap solution gave the following datas*

total hardness = 9.4 ml soap solutions

permanent hardness = 3.4 ml soap solutions

Standard hard water (100 mg/L of $CaCO_3$*) = 18.4 ml soap solutions*

lather factor = 0.4 ml soap solution.

Calculate the amount of temporary and permanant hardness.

Solution.

1000 ml water contains 100 mg $CaCO_3$ (Given)

1 ml water contains $\frac{100}{1000}$ mg $CaCO_3$

100 ml water contains $\frac{100 \times 100}{1000} = 10$ mg $CaCO_3$.

The lather factor is given 0.4 ml soap solution.

Now,

100 ml hard water contains 10 mg $CaCO_3$ required (18.4 – 0.4) ml soap solution

= 18 ml soap solution

$\because$ 18 ml soap solution consumes 10 mg $CaCO_3$

$\therefore$ 1 ml soap solution consumes $\frac{10}{18} = 0.55$ mg $CaCO_3$

For Total Hardness

100 ml hard water requires 9.4 – 0.4 = 9.0 ml soap solution

$\because$ 1 ml soap solution consumes 0.55 mg $CaCO_3$

$\therefore$ 9.0 ml soap solution consumes 0.55 × 9.0 = 4.95 mg $CaCO_3$

100 ml hard water contains 4.95 mg $CaCO_3$

1 ml hard water contains $\frac{4.05}{100}$ mg $CaCO_3$

1000 ml hard water contains $\frac{4.95 \times 1000}{100} = 49.5$ ppm.

$\therefore$ Total hardness in terms of $CaCO_3$ equivalent = 49.5 ppm

For Permanent Hardness

100 ml hard water requires 3.4 – 0.4 3.0 ml soap solution

$\because$ 1 ml soap solution consumes 0.55 mg $CaCO_3$

$\therefore$ 3.0 ml soap solution consumes 0.55 × 3 = 1.65 mg $CaCO_3$

100 ml hard water contains $1.65 \text{ mg } CaCO_3$

1 ml hard water contains $\frac{1.65}{100} \text{ mg } CaCO_3$

1000 ml hard water contains $\frac{1.65 \times 1000}{100} = 16.5 \text{ ppm}$

$\therefore$ Permanent hardness in terms of $CaCO_3$ equivalent = 16.5 ppm

Temporary hardness in terms of $CaCO_3$ equivalent

$$= \text{Total hardness} - \text{Permanent hardness}$$
$$= (49.5 - 16.5) \text{ ppm}$$
$$= 33 \text{ ppm.}$$

Based on Hardness Determination by EDTA Method

6. *50 ml of water sample requires 10 ml 0.01 N EDTA when titrated using buffer solution (pH = 10) to attain the end point. Calculate the total hardness of a sample in terms of ppm equivalent of $CaCO_3$ per litre.*

Solution.

$$\underset{\text{(Water)}}{N_1V_1} = \underset{\text{(EDTA)}}{N_2V_2}$$

$$N_1 \times 50 = \frac{1}{100} \times 10$$

$$N_1 = \frac{10}{100 \times 50}$$

$$\text{Strength} = \text{Normality} \times \text{Equivalent weight}$$

$$= \frac{10}{100 \times 50} \times 50 \times 1000 \text{ mg/L}$$

($\because$ equivalent weight of $CaCO_3$ = 50)

$\therefore$ Total hardnness in terms of $CaCO_3$ equivalent = 100 ppm

7. *200 ml of hard water sample require 30 ml of 0.02 M EDTA with $NH_4Cl - NH_4OH$ buffer and EBT indicator. Another 200 ml of the sample is boiled for about half an hour and after filtering the precipitate, the volume of the filtrate is made 200 ml again by the addition of distilled water. 20 ml of this boiled sample requires 5 ml 0.01 M EDTA following the same procedure. Calculate the temporary and permanent hardness of the sample.*

Solution.

$$\underset{\text{(Water)}}{M_1V_1} = \underset{\text{(EDTA)}}{M_2V_2}$$

$$M_1 \times 200 = \frac{2}{100} \times 30$$

$$M_1 = \frac{2 \times 30}{100 \times 200}$$

$\therefore$ Strength = Molarity × mol. wt.

$$= \frac{2 \times 30}{100 \times 200} \times 100 \text{ mg/L} \quad (\because \text{ mol. wt. of } CaCO_3 = 100)$$

$\therefore$ Total hardness in terms of $CaCO_3$ equivalent $= \frac{2 \times 30}{100 \times 200} \times 100 \times 1000 = 300$ ppm

Now after boiling sample contain only permanent hardness.

$$\underset{\text{(Water)}}{M_1V_1} = \underset{\text{(EDTA)}}{M_2V_2}$$

$$M_1 \times 20 = \frac{1}{100} \times 5$$

$$M_1 = \frac{5}{100 \times 20}$$

$\therefore$ Strength = Molarity × mol. wt.

$$= \frac{5}{100 \times 20} \times 100 \text{ g/L}$$

$\therefore$ Permanent hardness in terms of $CaCO_3$ equivalent $= \frac{5}{100 \times 20} \times 100 \times 1000 = 250$ ppm

Temporary Hardness in terms of $CaCO_3$ equivalent = Total hardness – Permanent hardness

= 300 – 250 = 50 ppm.

8. *20 ml of standard hard water [containing 1.5 g $CaCO_3$ per litre] required 25 ml EDTA solution for end point. 100 ml of water sample required 18 ml EDTA solution, while same water after boiling required 12 ml EDTA solution. Calculate carbonate and non-carbonate hardness of water.*

Solution.

1000 ml H_2O contains 1.5 g $CaCO_3$ (Given)

20 ml H_2O contains $\frac{1.5 \times 1000 \times 20}{1000} = 30$ mg $CaCO_3$

For total hardness

$\because$ 25 ml EDTA required for 20 ml SHW = 30 mg $CaCO_3$

$\therefore$ 1 ml EDTA required for $\frac{30}{25}$ mg $CaCO_3$

$\therefore$ 18 ml EDTA required for $\frac{30}{25} \times 18$ mg $CaCO_3$

100 ml H_2O contains $\frac{30}{25} \times 18$ mg $CaCO_3$

$\therefore$ 1000 ml H_2O contains $\frac{30 \times 18 \times 1000}{25 \times 100} = 216$ mg $CaCO_3$

$\therefore$ Total hardness in terms of $CaCO_3$ equivalent = 216 ppm

For (non-carbonate) permanent hardness

$\because$ 1 ml EDTA required for $\frac{30}{25}$ mg $CaCO_3$

$\therefore$ 12 ml EDTA required for $\frac{30}{25} \times 12$ mg $CaCO_3$

100 ml H_2O contains $\frac{30}{25} \times 12$ mg EDTA

$\therefore$ 1000 ml H_2O contains $\frac{30 \times 12 \times 1000}{25 \times 100} = 144$ mg $CaCO_3$

$\therefore$ Non Carbonate hardness in terms of $CaCO_3$ equivalent = 144 ppm.

Temporary (Carbonate) Hardness in terms of $CaCO_3$ equivalent

= Total hardness – Non-carbonate hardness = 72 ppm

9. *10 ml of standard hard water (containing 1 g $CaCO_3$ per litre) required 12.5 ml of EDTA solution using EBT indicator in ammonia buffer solution. 50 ml of hard water sample required 10 ml of same EDTA solution in the same manner. On the other hand 80 ml of calcium precipitating buffer was added to 400 ml of water and left for 2 hours. After filtering, 60 ml of filtrate required 3 ml of same EDTA solution under similar conditions. Calculate total hardness, magnesium hardness and calcium hardness of water. [Assumed all calcium ions were precipitated]*

Solution.

1000 ml H_2O contains 1 g $CaCO_3$ (Given)

10 ml H_2O contains 10 mg $CaCO_3$

For total hardness

12.5 ml EDTA required for 10 ml SHW ≡ 10 mg $CaCO_3$

$\therefore$ 1 ml EDTA required for $\frac{10}{12.5}$ mg $CaCO_3$

$\therefore$ 10 ml EDTA required for $\frac{10}{12.5} \times 10$ mg $CaCO_3$

50 ml water sample contains $\frac{10}{12.5} \times 10$ mg $CaCO_3$

$\therefore$ 1000 ml water sample contains $\frac{10}{12.5} \times \frac{10}{50} \times 1000$ mg $CaCO_3$

$\therefore$ Total hardness in terms of $CaCO_3$ equivalent = 160 ppm.

For magnesium hardness

Volume of water sample = 400 ml

Volume of calcium precipitating buffer solution = 80 ml

$\therefore$ Total volume of solution = (400 + 80) ml

= 480 ml

480 ml solution contains 400 ml water

$\therefore$ 60 ml solution (after filtration) contains $\frac{400}{480} \times 60$ ml water

= 50 ml water

Volume of EDTA consumed = 3 ml (Given)

$\because$ 1 ml EDTA required for $\frac{10}{12.5}$ mg $CaCO_3$

$\therefore$ 3 ml EDTA required for $\frac{10}{12.5} \times 3$ mg $CaCO_3$

50 ml water contains $\frac{10}{12.5} \times 3$ mg $CaCO_3$

$\therefore$ 1000 ml water contains $\frac{10}{12.5} \times \frac{3}{50}$ mg $CaCO_3$

$\therefore$ Magnesium hardness in terms of $CaCO_3$ equivalent

= 48 ppm

Calcium hardness in terms of $CaCO_3$ equivalent

= Total hardness – Magnesium hardness

= (160 – 48) ppm

= 112 ppm.

Based on Alkalinity of Water

10. *200 ml of water sample required 25 ml of $\frac{N}{50}$ H_2SO_4 during titration by using phenolphthalein indicator and 26 ml of the same acid by using methyl orange indicator. Calculate the alkalinity of each type in terms of $CaCO_3$ equivalent.*

Solution.

P = 25 ml, M = 26 ml, $P > \frac{1}{2}M$

Volume of acid in OH^- hardness = 2P – M

Volume of acid in CO_3^{2-} hardness = 2(M – P)

For alkalinity due to OH^-

$$\underset{\text{(Water)}}{N_1V_1} = \underset{\text{(Acid)}}{N_2V_2}$$

$$N_1 \times 200 = \frac{1}{50}\ (2P - M)$$

$$N_1 = \frac{1 \times 24}{50 \times 200}$$

Alkalinity due to OH^-, in terms of $CaCO_3$ equivalent $= \frac{1 \times 24 \times 50 \times 1000}{50 \times 200} = 120$ ppm

($\because$ Alkalinity = normality × eq. wt. of $CaCO_3$ (50) × 1000 ppm.)

For alkalinity due to CO_3^{2-}

$$\underset{\text{(Water)}}{N_1V_1} = \underset{\text{(Acid)}}{N_2V_2}$$

$$N_1 \times 200 = \frac{1}{50}\ 2(M - P)$$

$$N_1 = \frac{2}{50 \times 200}$$

Alkalinity due to CO_3^{2-}, in terms of $CaCO_3$ equivalent $= \frac{1 \times 2 \times 1000 \times 50}{50 \times 200} = 10$ ppm.

11. *100 ml of a water sample on titration with 0.03 N HCl using phenolphthalein as an indicator gave end point with 7.5 ml of acid. Another sample of same volume of water require 15 ml of acid (0.03 N) using methyl orange indicator to obtain complete neutralization. Calculate the alkalinity of water in terms of $CaCO_3$.*

Solution.

Volume of acid used in case of phenolphthalein indicator = 7.5 ml

Volume of acid used in case of methyl orange indicator = 15 ml

So $P = \frac{1}{2}M$

So alkalinity in water is due to CO_3^{2-}

Volume of acid = 2P

$$\underset{\text{(Water)}}{N_1V_1} = \underset{\text{(Acid)}}{N_2V_2}$$

$$N_1 \times 100 = \frac{3}{100} \times 2P$$

$$N_1 = \frac{3 \times 15}{100 \times 100}$$

Alkalinity due to CO_3^{2-}, in terms of $CaCO_3$ equivalent $= \frac{3 \times 15 \times 50 \times 1000}{100 \times 100} = 225$ ppm.

12. *100 ml of a water sample required 4 ml of $\frac{N}{50}$ H_2SO_4 for neutralization to phenolphthalein end-point. Another 16 ml of same acid was needed for further titration to methyl orange end point. Determine the type and amount of alkalinity.* **(K.U.K. Jan 2004, June 2008)**

Solution.

$$P = 4 \text{ ml}, \qquad M = 4 + 16 = 20 \text{ ml}, \qquad P < \tfrac{1}{2}M$$

Alkalinity is due to presence of CO_3^{2-} and HCO_3^-

$\therefore$ Volume of acid in CO_3^{2-} type $=$ 2P

and Volume of acid in HCO_3^- type $=$ M – 2P

For CO_3^{2-} alkalinity

$$\underset{\text{(Water)}}{N_1V_1} = \underset{\text{(Acid)}}{N_2V_2}$$

$$N_1 \times 100 = \frac{1}{50} \times 2P$$

$$N_1 = \frac{1 \times 8}{50 \times 100}$$

Alkalinity due to CO_3^{2-}, in terms of $CaCO_3$ equivalent $= \dfrac{1 \times 8 \times 50 \times 1000}{50 \times 100} = 80$ ppm

For HCO_3^- alkalinity

$$\underset{\text{(Water)}}{N_1V_1} = \underset{\text{(Acid)}}{N_2V_2}$$

$$N_1 \times 100 = \frac{1}{50} \times (M - 2P)$$

$$N_1 = \frac{1 \times 12}{50 \times 100}$$

Alkalinity due to HCO_3^-, in terms of $CaCO_3$ equivalent $= \dfrac{1 \times 12 \times 1000 \times 50}{50 \times 100} = 120$ ppm

$\therefore$ Total alkalinity in terms of $CaCO_3$ equivalent = 120 + 80 = 200 ppm.

EXERCISE

1. Define hardness of water. Differentiate between hard water and soft water.
2. What are the type of hardness? Explain its causes. Write the units of hardness. How are they related each other? ***(K.U.K. June 2006, June 2009, Jan. 2005, Jan. 2006)***
3. What do you understand by alkaline and non-alkaline hardness?
4. Write the principle of soap solution method and EDTA methods for determining the hardness of water.
5. Name the important sources of water. Why is rain water considered as the purest water? Define priming and foaming in boilers.
6. What do you understand by scale and sludge? How are they formed? What are their disadvantages and how are they prevented? ***(K.U.K. June 2006, June 2008, K.U.K. Jan. 2004, Jan. 2006)***
7. Write a short note on 'caustic embrittlement' and 'boiler corrosion'. ***(K.U.K. June 2004)***
8. How priming and foaming are caused? What are their disadvantages? How can they be prevented in boilers?

9. Describe the soap solution method for the estimation of hardness of water.
10. What is the principle of EDTA titration? Briefly discuss the estimation of hardness of water by EDTA method and role of ammonia buffer solution used in it. ***(K.U.K. Jan. 2005, Jan. 2007)***
11. What are the factors which cause alkalinity in water? How is alkalinity of water determined by titrimetric method?
12. What is boiler feed water? What are the qualities of water used in sugar industry, starch industry, aluminium industry, fibre industry and pulp industry?
13. Write short notes on the following :
    (*a*) Phosphate conditioning ***(K.U.K. June 2004)***
    (*b*) Calgon conditioning ***(K.U.K. June 2008)***
    (*c*) Colloidal conditioning
    (*d*) Carbonate conditioning
    (*e*) EDTA conditioning
14. What do you understand by boiler feed water? What are its requisites?
15. Why hardness or alkalinity of water is expressed in terms of calcium carbonate only? Write the structures of EDTA and the complex formed by it with a bivalent metal ion. ***(K.U.K. Jan. 2006)***
16. How is the permanent hardness of water determined? ***(K.U.K. Jan. 2004)***
17. Define alkalinity. How is it determined? Write the names of three sludge-forming and three scale forming compounds. ***(K.U.K. June 2004)***
18. Name a chemical substance which by dissolving in distilled water makes it alkaline as well as hard? Explain why is the combination of OH^- and HCO_3^- not possible in alkalinity ? ***(K.U.K. Jan. 2005, Jan. 2007)***
19. Name the three gases dissolved in water responsible for causing corrosion. ***(K.U. K. Jan. 2005)***
20. Name the various internal methods used for treating boiler-feed water. Discuss in detail the phosphate conditioning. ***(K.U. K. June 2005, June 2006, Jan. 2005)***
21. Comment on the statement "Distilled water containing sodium bicarbonate exhibits only alkalinity whereas aqueous solution of calcium bicarbonate imparts both alkalinity and the hardness." ***(K.U.K. June 2005)***
22. Define lather factor. Why its numerical value is always subtracted from each titre value during the estimation of hardness by soap titration method?
23. What is meant by hardness of water? How is it determined by complexometric titration (EDTA method) method? ***(K.U.K. June 2006)***
24. What do you mean by calcium hardness and magnesium hardness of water? How will you determine the calcium hardness and magnesium hardness of water sample by complexometric method? Explain in details.
25. Define buffer solution. What is the role of ammonia buffer solution in the determination of hardness of water during EDTA method?
26. Differentiate between scales and sludges. ***(K.U.K. June 2009)***
27. Define hardness. How is it expressed? Give various units of hardness and relation between them. ***(K.U.K. June 2009)***
28. Give the name and structure of indicator used in EDTA titration method for hardness estimation. How is required pH obtained? ***(K.U.K. June 2009)***
29. What is internal conditioning of water? Explain calgon or phosphate conditioning of water. ***(K.U.K. June 2009)***
30. What is meant by 'Alkalinity of water sample'? How is it determined volumetrically? Explain with the help of suitable reactions involved in it. ***(K.U.K Jan 2009)***

31. Describe the factors which cause 'Alkalinity' in water. How is it determined by titrimetric method? ***(K.U.K. Jan 2009)***
32. Define alkaline (carbonate) and non alkaline (non carbonate) hardness of water. How is it determined by titrimetric method? Explain the methods with chemical reactions involved in it. ***(K.U.K. June 2008)***
33. How alkalinity is differ from hardness? Write the units of alkalinity and give relation between them.
34. What do you mean by alkaline and non alkaline water? How alkalinity is determined experimentally?
35. Write the name and structure of indicators used in the determination of alkalinity of water. Explain the role of indicators in the determination of alkalinity.
36. What is boiler corrosion? Explain factors causing boiler corrosion. How is it prevented?
37. What do you understand by caustic embrittlement? Explain its mechanism. How is it prevented?
38. Define priming and foaming? What are its causes and how they are prevented? What are disadvantages of priming and foaming?
39. Define bagging. Explain the disadvantages of scale and sludges. What is meant by temporary hardness of water? How is it removed? Write the structural formula of ***(K.U.K. Jan 2006)***
    (*i*) Disodium salt of EDTA
    (*ii*) Eriochrome black-T.
40. What are boiler scales? Give reasons for their formation and their ill effects.
41. Why is calgon conditioning better than phosphate conditioning?
42. What is degree of hardness of water? Why is rain water a purest form of natural water?
43. Why should natural water not be feed to boiler? Explain the Hehner's method for determination of hardness of water. Write the chemical reactions involved in it.

PROBLEMS FOR PRACTICE

1. 100 ml of water sample requires 20 ml N/100 EDTA in titration using $NH_4Cl—NH_4OH$ buffer and EBT indicator. Calculate its hardness. [**Ans.** 100 ppm]
2. Calculate temporary and permanent hardness of water during soap titration method, when 50 ml of water is titrated with soap solution gave the following datas:
   Lather factor = 0.3 ml soap solution; total hardness = 9.3 ml soap solution; permanent hardness = 3.1 ml soap solution; 50 ml standard hard water (200 mg/L of $CaCO_3$) = 18.3 ml of soap solution.
   [**Ans.** Temporary hardness 69 ppm
   Permanent hardness 31 ppm]
3. 50 ml of water sample was titrated against soap solution gave the following data:
   (*i*) Lather factor = 0.5 ml soap solution.
   (*ii*) Total hardness = 15.5 ml soap solution.
   (*iii*) Permanent hardness = 12.5 ml soap solution.
   (*iv*) 100 ml standard hard water (0.2 g of $CaCO_3$ per litre) = 40.5 ml soap solution.
   Calculate total hardness, permanent hardness and temporary hardness of water.
   [**Ans.** Total hardness 150 ppm
   Permanent hardness = 120 ppm
   Temporary hardness = 30 ppm]

4. 40 ml of standard hard water (containing 1.5g $CaCO_3$ per litre) required 50 ml of soap solution for permanent lather formation. 100 ml of water sample required 20 ml of soap solution while same volume of water after boiling required 15 ml of soap solution. If the lather factor is zero then calculate the temporary and permanent hardness of water. [**Ans.** Temporary hardness = 60 ppm
Permanent hardness = 180 ppm]

5. 50 ml of standard hard water (containing 1.0 g $CaCO_3$ per litre) required 60.5 ml of soap solution for end point whereas 50 ml of pure distilled water (in which soap solution is prepared) required only 0.5 ml of soap solution for end point. 100 ml of water sample required 30.5 ml of soap solution while same volume of water after boiling required 10.5 ml of soap solution. Calculate temporary and permanent hardness of water. [**Ans.** Temporary hardness = 166.7 ppm
Permanent hardness = 83.3 ppm]

6. 70 ml of a standard hard water (containing 0.28 g $CaCO_3$ per litre) required 20.1 ml of a given soap solution for persisting lather formation for 5 minutes. 70 ml of hard water sample under the same condition required 14.5 ml of soap solution. The same volume of water after boiling required 5.4 ml of the soap solution. If the lather factor is 0.5 ml of soap solution then calculate total permanent and temporary hardness in ppm, degree clark and degree french. [**Ans.** Total hardness = 200 ppm, 14°Cl, 20° Fr
Permanent hardness = 70 ppm, 4.9°Cl, 7°Fr
Temporary hardness-130 ppm 9.1°Cl, 13°Fr]

7. 100 ml of water sample require 30 ml of $\frac{N}{50}$ HCl using methyl orange indicator for end point while the same volume after boiling requires 20 ml of the same acid. Calculate the temporary hardness of water.
[**Ans.** 100 ppm]

8. 100 ml of water sample is boiled for half an hour with 20 ml of $\frac{N}{50}$ Na_2CO_3. After cooling, filtering the filtrate required 12 ml of $\frac{N}{50}$ H_2SO_4. Calculate the permanent hardness of water. [**Ans.** 80 ppm]

9. 100 ml of hard water sample requires 20 ml of $\frac{N}{50}$ EDTA solution during the titration using EBT indicator in ammonia buffer solution. Calculate hardness of water. [**Ans.** 200 ppm]

10. 20 ml of standard hard water (containing 1.5 g $CaCO_3$ per litre) required 25 ml EDTA solution for end point. 100 ml of water sample required 18 ml EDTA solution; while same water after boiling required 12 ml EDTA solution. Calculate carbonate and non-carbonate hardness of water.
[**Ans.** 72 ppm, 144 ppm]

11. 50 ml of alkaline water sample required 20 ml N/50 H_2SO_4 for phenolphthalein end point and another 5 ml for methyl orange indicator *i.e.*, complete neutralisation. Describe the type of alkalinity and calculate the amount and type of alkalinity.
[**Ans.** OH^- = 300 ppm, CO_3^{2-} = 200 ppm] ***(K.U. K. Jan. 2007, June 2009)***

12. 100 ml of alkaline water sample required 7.3 ml N/50 H_2SO_4 using phenolphthalein indicator and another sample of 100 ml water also required 7.3 ml of N/50 H_2SO_4 using methyl orange indicator. Calculate the amount and type of alkalinity present in water sample. [**Ans.** OH^- = 73 ppm]

13. A water sample is alkaline to both phenolphthalein and methyl orange. 50 ml of this sample required 15 ml of N/50 H_2SO_4 for phenolphthalein indicator and another 10 ml of complete neutralisation. Calculate the amount and type of alkalinity in terms of $CaCO_3$ equivalent.
[**Ans.** OH^- = 100 ppm, CO_3^{2-} = 400 ppm]

14. 100 ml of a water sample has hardness equivalent to 12.5 ml of 0.08 N $MgSO_4$. Calculate its hardness in ppm? [**Ans.** 500 ppm]

15. A water sample contains 180 mg of $MgSO_4$ per litre. Calculate the hardness in terms of $CaCO_3$ equivalent. [**Ans.** 150 ppm]

16. A sample of irrigation water sample contains 104 mg of Ca^{2+} per litre. Calculate the hardness of water in terms of $CaCO_3$ equivalent. [**Ans.** 260 ppm]

17. A water sample having 204 mg of calcium sulphate per litre. Calculate the hardness of sample in terms of $CaCO_3$ equivalents. [**Ans.** 150 ppm]

18. A sample of irrigation water contains 16.8 mg/L of $Mg(HCO_3)_2$, 20 mg/L of $MgCl_2$, 25 mg/L of $CaCl_2$. Calculate the total hardness in terms of $CaCO_3$ equivalent. [**Ans.** 53.41 mg/L]

19. A water sample contains 140 mg of $CaSO_4$ per litre. Calculate the hardness in terms of ppm. [**Ans.** 102.9 ppm]

20. A sample of water on analysis has been found to contain following in ppm :

$Ca(HCO_3)_2 = 10.5$, $Mg(HCO_3)_2 = 12.5$,

$CaCl_2 = 8.2$, $CaSO_4 = 7.5$, $MgSO_4 = 2.6$

Calculate the temporary and permanent hardness in degree French. [**Ans.** Temporary hardness 1.504°Fr
Permanent hardness 1.509°Fr]

21. 400 ml of a water sample contains 3.01 mg of $CaSO_4$, 2.0 mg of $MgCl_2$ and 3.5 mg of $Ca(HCO_3)_2$. Calculate the total hardness in terms of Clark's degree. [**Ans.** 1.13°Cl]

22. Calculate the hardness of water containing the following dissolved salts:

$CaSO_4 = 28$ mg/L, $Mg(HCO_3)_2 = 22$ mg/L,

$MgCl_2 = 30$ mg/L, $CaCl_2 = 85$ mg/L.

[**Ans.** Temporary hardness = 15.07 ppm, Permanent hardness = 128.7 ppm]

23. A water sample contain $Ca(HCO_3)_2 = 32.4$ mg/L, $Mg(HCO_3)_2 = 29.2$ mg/L and $CaSO_4 = 13.5$ mg/L. Calculate temporary and permanent hardness. [**Ans.** 40 ppm, 10 ppm]

24. 100 ml of water sample requires 20 ml $\frac{N}{100}$ EDTA when titrated using NH_4Cl-NH_4OH buffer and EBT indicator. Calculate the hardness of the sample. [**Ans.** 100 ppm]

25. 1 g of $CaCO_3$ was dissolved in dil HCl and the solution diluted to 1 litre. 50 ml of this solution required 45 ml of EDTA solution. 50 ml of hard water required 18 ml of EDTA solution during titration in ammonia buffer using EBT indicator. On the other hand, 50 ml of boiled water sample required 9 ml of EDTA solution under same condition. Calculate each type of hardness in ppm.

[**Ans.** Total hardness = 400 ppm, Permanent hardness = 200 ppm,
Temporary hardness = 200 ppm]

26. 0.30 g of $CaCO_3$ was dissolved in HCl and the solution made upto 1 litre with distilled water. 100 ml of the above solution required 30 ml of EDTA solution on titration. 100 ml of hard water sample required 35 ml of same EDTA solution on titration. After boiling 100 ml of this water, cooling, filtering and then titration required 20 ml of EDTA solution. Calculate the temporary and permanent hardness of water.

[**Ans.** Total hardness = 350 ppm, Permanent hardness = 200 ppm,
Temporary hardness = 150 ppm]

27. 20 ml of SHW (containing 1.5 g $CaCO_3$ per litre) required 25 ml EDTA solution for end point. 100 ml of water sample required 18 ml EDTA solution while same water after boiling required 12.5 ml EDTA solution. Calculate total, permanent and temporary hardness. [**Ans.** 216 ppm, 150 ppm, 66 ppm]

28. 0.28 g of $CaCO_3$ was dissolved in HCl and the solution made upto 1 litre with distilled water. 100 ml of the above solution required 28 ml of EDTA solution on titration. 100 ml of a hard water sample required

33 ml of the same EDTA solution on titration. After boiling 100 ml of this water, cooling and filtering and then titrated 10 ml of EDTA solution. Calculate the temporary and permanent hardness.
[**Ans.** 230 mg/L, 100 mg/L] ***(M.D.U. Rohtak 2004)***

29. 50 ml of SHW containing 1 mg of pure $CaCO_3$ per ml consumed 20 ml of EDTA. 50 ml of water sample consumed 25 ml of same EDTA solution using Eriochrome black-T indicator. Calculate the total hardness of water sample. [**Ans.** 1250 ppm]

30. 100 ml of water sample on tiration with phenolphthalein indicator require 5 ml of 0.02 N H_2SO_4. Another 100 ml of the water sample when titrated with methyl orange end point require 5 ml of 0.02 N H_2SO_4. What type of alkalinity is present in sample and what is its magnitude? [**Ans.** OH^- = 50 ppm]

31. A sample of water was alkaline both to phenolphthalein end methyl orange. 50 ml of this water sample required 15 ml of $\frac{N}{50}$ sulphuric acid for phenolphthalein end point and another 10 ml for complete neutralisation. Calculate the type and amount of alkalinity in ppm.

[**Ans.** OH^- = 100 ppm, CO_3^{2-} = 400 ppm]

32. 100 ml of a water sample require 20 ml of $\frac{N}{50}$ sulphuric acid for neutralization to phenolphthalein end point. After this methyl orange indicator was added to this and further acid required was again 20 ml. Calculate the alkalinity of water as $CaCO_3$ in ppm. [**Ans.** CO_3^{2-} = 400 ppm]

33. 200 ml of water sample require 25 ml of $\frac{N}{50}$ sulphuric acid for neutralization to phenolphthalein end point. After that methyl orange was added to this and further acid requires was 35 ml $\frac{N}{50}$. Calculate the type and amount of alkalinity of water as $CaCO_3$ in ppm. [**Ans.** CO_3^{2-} = 250 ppm ; HCO_3^- = 50 ppm]

34. 100 ml of a hard water sample were taken in a 250 ml beaker and boiled to precipitate temporary hardness. 20 ml of an alkali mixture solution (NaOH + Na_2CO_3), whose 10 ml required 8 ml of N/10 HCl for neutralization to methyl orange end-point, were then added. The reaction mixture was heated to boiling, cooled and filtered. The filtrate required 13 ml of N/10 HCl for complete neutralization. Calculate permanent hardness of the water sample. [**Ans.** 150 ppm]

35. 100 ml of water sample required 25 ml of M/100 EDTA with NH_4Cl-NH_4OH buffer and EBT indicator. Another 100 ml of the sample is boiled for about half an hour and after filtering the precipitate, the volume of the filtrate is made up to 100 ml again by adding distilled water. 20 ml of the boiled sample require only 4 ml of M/100 EDTA following the same procedure. Calculate the temporary and permanent hardness. [**Ans.** 50 ppm, 200 ppm]

36. 200 ml of alkaline water sample was titrated against a mixture of acids (30 ml of N/40 H_2SO_4 and 20 ml of N/80 HCl) and the reading was found 5 ml for phenolphthalein indicator and 15 ml for methyl orange indicator. Calculate the amount and type of alkalinity of water in ppm.

[**Ans.** CO_3^{2-} = 50 ppm ; HCO_3^- = 25 ppm]

37. 100 ml of water sample required 4 ml of $\frac{N}{50}$ H_2SO_4 for neutralization to phenalphthalein end-point. Another 16 ml of the same acid was needed for further titration to methyl orange end point. Determine the type and amount of alkalinity. [**Ans.** CO_3^{2-} 80 ppm, HCO_3^- 120 ppm] ***(K.U.K. Jan. 2004, June 2008)***

38. 10 ml of standard hard water (1 ml of it = 1 mg of $CaCO_3$) in presence of ammonical buffer solution/Eriochrome black–T indicator consumes 13.8 ml of EDTA solution. 50 ml of a given hard water sample under the similar experimental conditions, consumes 25 ml of the same EDIT solution. 50 ml of the given hard water sample upon boiling, cooling, filtering etc. consumes 10 ml of same EDTA solution. Calculate the various types of hardness in terms of ppm as $CaCO_3$ equivalents.

[**Ans.** Total hardness = 362.31 ppm Permanent hardness = 144.92 ppm
Temporary hardness = 217.39 ppm] ***(K.U.K. Jan. 2005)***

39. 100 ml of an alkaline water in presence of phenolphthalein indicator consumes 50 ml of $\frac{N}{100}$ HCl. The resulting mixture in presence of methyl orange indicator consumes 35 ml of same HCl. Predict the alkalinities and calculate their amounts in ppm as $CaCO_3$ equivalents. ***(K.U.K. Jan. 2005)***

[**Ans.** OH^- = 75 ppm, CO_3^{2-} = 350 ppm]

40. 400 ml of alkaline water sample was titrated against a mixture of acids (50 ml of $\frac{N}{100}$ H_2SO_4 and 20 ml of $\frac{N}{40}$ HCl) and the reading was found 9 ml for phenolphathalein indicator and further 13 ml for methyl orange indicator. Calculate the amount and type of alkalinity of water in ppm.

[**Ans.** CO_3^{2-} = 32.14 ppm, HCO_3^- = 7.14 ppm]

41. 250 ml of alkaline water sample was titrated against a mixture of acids (40 ml of $\frac{N}{100}$ H_2SO_4 and 30 ml of $\frac{N}{20}$ HCl) and the reading was found 5 ml for phenolphthalein indicator. The same volume of water required 5 ml of the same acid mixture during the titration using methyl orange indicator only. Calculate the amount and type of alkalinity of water. [**Ans.** OH^- = 27.14 ppm]

42. A water sample responds to methyl orange but not to phenolphtalein. 50 ml of the above water sample completely neutralizes 20 ml of $\frac{N}{50}$ HCl. Calculate the amount in ppm as $CaCO_3$ equivalents of the alkalinity present in the sample. [**Ans.** HCO_3^- = 400 ppm) ***(K.U.K. June 2005)***

43. A water sample is alkaline both to phenolphthalein (HPh) and methyl orange indicators. 100 ml of an alkaline water sample in presence of HPh consumes exactly 30 ml of $\frac{N}{5}$ HCl. However, the resulting mixture in presence of methyl orange consumes only 10 ml of the same acid. Predict the types of alkalinities and their amounts in ppm as $CaCO_3$ equivalents. [**Ans.** OH^- = 2000 ppm, CO_3^{2-} = 2000 ppm]
(K.U.K. Jan. 2006)

44. 100 ml of standard $\left(\frac{N}{50}\right)$ hard water sample requires 20 ml of EDTA solution whereas 50 ml hard water sample requires 12.5 ml of EDTA solution. Calculate total hardness of water. [**Ans.** 1250 ppm]

45. 20 ml of standard $\left(\frac{N}{40}\right)$ hard water requires 25 ml of EDTA solution whereas 100 ml of hard water requires 18 ml of EDTA solution. On the other hand same amount of hard water after boiling required 12.5 ml of same EDTA solution under similar condition. Calculate total hardness, permanent hardness and temporary hardness of water.

[**Ans.** Total hardness = 180 ppm; Permanent hardness = 125 ppm; Temporary hardness = 55 ppm]

46. 10 ml of standard $\left(\frac{N}{40}\right)$ hard water consumes 25 ml of EDTA solution in ammonia buffer using EBT indicator. 100 ml of hard water sample requires 18.5 ml of same EDTA solution whereas the same amount of hard water after boiling consumes 10.0 ml of same EDTA solution under similar condition. Calculate temporary hardness of water sample. [**Ans.** 42.5 ppm]

47. 20 ml of standard hard water (containing 1.5 g $CaCO_3$ per litre) required 25 ml of EDTA solution in ammonia buffer solution using EBT indicator. 100 ml of water sample required 20 ml of same EDTA solution. On the other hand 40 ml of calcium precipitating buffer solution was added to 200 ml of water sample and left for overnight for complete precipitation of calcium ions. After that it was filtered and 60 ml of filtrate required 4.5 ml of same EDTA solution under similar conditions. Calculate magnesium hardness and calcium hardness of water. [**Ans.** Mg hardness 108 ppm Ca hardness 132 ppm]

48. 50 ml of hard water sample required 10 ml of $\frac{N}{40}$ EDTA solution in ammonia buffer solution using EBT indicator. 100 ml of calcium precipitating buffer solution was added to 500 ml of water sample and left for 1 hour for complete precipitation of calcium ions. After filtering, 60 ml of filtrate required 4 ml of same EDTA solution under similar conditions. Calculate calcium hardness of water. [**Ans.** 150 ppm]

49. 100 ml of hard water sample required 20 ml of $\frac{N}{40}$ EDTA solution in $NH_4Cl - NH_4OH$ buffer using EBT indicator. 60 ml of calcium precipitating buffer solution was mixed with 300 ml of water and left for 1 hour. Assumed that all calcium ions were precipitated. After filtration 60 ml of filtrate required 3.5 ml of EDTA solution under same conditions. Calculate magnesium hardness as well as calcium hardness of water. [**Ans.** Mg. hardness = 87.5 ppm; Ca. hardness = 162.5 ppm]

50. 100 ml of water sample was boiled with 20 ml of $\frac{N}{10}$ sodium carbonate solution. After filtration the filtrate required 30 ml of $\frac{N}{20}$ H_2SO_4 for complete neutralization. Calculate the permanent hardness of the water sample. [**Ans.** 250 ppm]

51. 25 ml of $\frac{N}{50}$ sodium carbonate solution was added to 100 ml of water sample. Now the mixture was boiled for 1 hour for complete precipitation of insoluble carbonate. After cooling, it was filtered. The filtrate required 10 ml of $\frac{N}{50}$ H_2SO_4 for complete neutralization. Calculate the permanent hardness of the water sample. [**Ans.** 150 ppm]

52. A water sample contains 50 mg of Mg^{2+} per litre. Calculate the hardness of water in terms of $CaCO_3$ equivalent. [**Ans.** 20.83 ppm]

53. 100 ml of water sample contains 5 mg of $MgCO_3$. Calculate the hardness of water in terms of $CaCO_3$ equivalent. [**Ans.** 59.59 ppm]

54. 350 mg of $CaSO_4$ is found in 5 litres of water. Calculate the hardness of water in terms of $CaCO_3$ equivalent. [**Ans.** 51.45 ppm]

55. How many grams of $MgCO_3$ dissolved per litre water to produce 84 ppm of hardness ? [**Ans.** 0.0706 ppm]

56. 20 ml of $CaCl_2$ solution, whose strength is equivalent to 1.5 g of $CaCO_3$ per litre required 30 ml of EDTA solution. Whereas 10 ml of hard water sample required 10 ml of same EDTA solution. Calculate the hardness of water in terms of $CaCO_3$ equivalent. [**Ans.** 1000 ppm]

57. How many grams of $FeSO_4$ dissolved per litre water to produce 210.5 ppm of hardness ? [**Ans.** 0.319 g]

58. 100 ml of hard water sample required 20 ml $\frac{N}{40}$ H_2SO_4 for complete neutralisation. Calculate the hardness of water. [**Ans.** 250 ppm]

59. 200 ml of water sample contains following $CaSO_4$ = 2.64 mg $MgCl_2$ = 1.06 mg and $Ca(HCO_3)_2$ = 3.74 mg. Calculate the total hardness of the water sample in terms of $CaCO_3$ equivalent.

[**Ans.** 26.75 ppm]

60. 25 ml of N/50 Na_2CO_3 was boiled with 100 ml of water sample for complete precipitation of insoluble carbonate. After cooling and filtration the filtrate required 5 ml of N/50 H_2SO_4 for complete neturalisation. Calculation the amount of permanent hardness of the water sample. [**Ans.** 200 ppm]

61. 10 ml of N/10 Na_2CO_3 was added to 100 ml of hard water sample and boiled for some times for complete precipitation of insoluble carbonate. After filtration the filtrate required 30 ml of N/40 H_2SO_4 for complete neutralisation. Calculate the amount of permanent hardness of the water sample. [**Ans.** 125 ppm]

62. 100 ml of hard water sample required 10 ml of 0.02 M EDTA in ammonical medium using EBT indicator. On the other hand the same amount of water sample is boiled for half an hour and after cooling and filtering, the volume of the filtrate is made up to 100 ml again by adding distilled water. It required 6 ml of the same EDTA solution under similar condition. Calculate the amount of temporary and permanent hardness of water. [**Ans.** 40 ppm, 60 ppm]

63. 20 ml of standard hard water (1g $CaCO_3$ per litre) required 40 ml of EDTA solution in ammonical buffer solution using EBT indicator, whereas 100 ml of hard water sample required 25 ml of the same EDTA solution under similar conditions. Calculate the amount of hardness of water. [**Ans.** 125 ppm]

64. 50 ml of water sample required 5 ml of $\frac{N}{50}$ H_2SO_4 using methyl orange indicator but did not give any colouration with phenolphthalein. Calculate the type and amount of alkalinity in terms of $CaCO_3$ equivalent. [**Ans.** HCO_3^- = 100 ppm] ***(M.D.U. Rohtak Dec. 2003)***

65. 100 ml of water sample required 16 ml of $\frac{N}{50}$ H_2SO_4 for neutralisation to phenolphthalein end point. After this methyl orange indicator was added and further 20 ml of the same acid was required for neutralisation. Calculate the type and amount of alkalinity in terms of $CaCO_3$ equivalent.

[**Ans.** CO_3^{2-} = 320 ppm, HCO_3^- = 40 ppm] ***(M.D.U. Rohtak Dec. 2004)***

66. A water sample is alkaline to both phenolphthalein and methyl orange indicator. 100 ml of this water sample required 18 ml of 0.02 N H_2SO_4 upto phenolphthalein end point and another 6 ml upto methyl orange end point *i.e.*, complete neutralisation. Calculate the type and amount of alkalinty present in that water sample. [**Ans.** OH^- = 120 ppm, CO_3^{2-} = 120 ppm]

67. A water sample is not alkaline to phenolphthalein. 100 ml of the water on titration with N/50 HCl, required 16.9 ml to obtain the end point, using methyl orange indicator. Calculate the type and amount of alkalintiy present in the sample. [**Ans.** HCO_3^- = 169 ppm]

68. 100 ml of alkaline water sample, on titration with N/50 H_2SO_4 gave a titre value of 5.8 ml to phenolphtalein end point and another 100 ml of same sample on titration with same acid gave a titre value of 11.6 ml to methyl orange end point. Calculate the type and amount of alkalinity of water in terms of $CaCO_3$ equivalent.

[**Ans.** CO_3^{2-} = 116 ppm]

69. 100 ml of an alkaline water sample, on titration against N/50 H_2SO_4 using phenolphthalein indicator, gave an end point when 5 ml of acid were run down. Another 100 ml of the same sample also required 5 ml of the same acid to obtain methyl orange end point. Calculate they type and magnitude of alkalinity.

[**Ans.** OH^- = 50 ppm]

70. 100 ml of a raw water sample on titration with H_2SO_4 required 12.4 ml of the acid to phenolphthalein end-point and 15.2 ml of the acid to methyl orange end-point. Determine the type and extent of alkalinity present in the water sample. [**Ans.** CO_3^{2-} = 56 ppm; OH^- = 96 ppm]

71. 50 ml of alkaline water sample required 18 ml of N/50 H_2SO_4 to the phenolphthalein end-point and 24 ml to the methyl orange end-oint. Calculate the type and extent of alkalinity present in the alkaline water sample. [**Ans.** OH^- = 240 ppm CO_3^{2-} = 240 ppm]

Chapter **4**

WATER AND ITS TREATMENT (PART II)

4.1 TREATMENT OF WATER FOR DOMESTIC USE

Water, which is safe to drink, is known as potable water. Drinking (potable) water has following characteristics:

1. It should be colourless, odourless, tasteless and transparent.
2. It should be free from hardness, suspended particles and pathogenic bacterias (micro-organisms).
3. It should neither be very hard nor be very soft.
4. The turbidity should not exceed 10 ppm.
5. Its pH value should be about 7.0–7.5.
6. It should be free from harmful dissolved solids like compounds of arsenic or lead etc. and harmful dissolved gases like H_2S, SO_2, etc.
7. The total dissolved solids (TDS) should be less than 500 ppm.

Pure soft water is plumbo-solvent *i.e.,* it attacks lead used in plumbing. For health point of view some minerals and ions must be present in water. Drinking water possesses about 60–70 ppm hardness. It is firm and suitable for health.

The natural water obtained from river, canals etc. does not satisfy the characteristics of portable water hence it is necessary to treatment of water for domestic use. These are some following important methods, which have been employed for treatment of water.

4.1.1 Screening

The raw water is allow to passed through screens having large number of holes. The floating matters are retained by them and only water is passed through the holes.

4.1.2 Sedimentation

It is the process by which water is allow to stand in a big tank for sometimes. Most of the suspended particles get settled down at the bottom due to the force of gravity. The supernatant water is drawn continuously from the tank. It is a very slow process and may take several hours or a few days. About 60–70% of suspended matter is removed by this process. When water contains fine clay particles and colloidal matter so it is necessary to add some suitable chemicals before sedimentation for maximum coagulation and sedimentation.

4.1.3 Coagulation

Coagulation is the process by which the fine suspended and colloidal particles are removed from the water by the addition of suitable chemicals (coagulants). When fine clay particles or colloidal particles are present in water, they do not get settle down by ordinary sedimentation because colloidal particles have some charges which repel each other. Due to mutual repulsion they do not combine to form a large size. But when coagulants are added in water they neutralize the charges on colloidal particles. The chargeless particles now come close to one another and combines to form larger molecules, which get settled down, under the gravitational force. The coagulants are generally used in solution form. Some useful chemical coagulants are following:

Potash Alum ($K_2SO_4Al_2(SO_4)_3.24H_2O$)

Potash alum is very important coagulant and is the most widely used in water treatment plants. It reacts in the presence of alkalinity of water hence if natural alkalinity is not present in water then it is necessary to add some sufficient lime. Actually aluminium sulphate reacts with calcium bicarbonate or magnesium bicarbonate present in water to form a flocculant precipitate of $Al(OH)_3$.

$$Al_2(SO_4)_3 + 3Ca(HCO_3)_2 \longrightarrow 2Al(OH)_3 \downarrow + 3CaSO_4 + 6CO_2$$

$$Al_2(SO_4)_3 + 3Mg(HCO_3)_2 \longrightarrow \underset{\text{(White Ppt.)}}{2Al(OH)_3} \downarrow + 3MgSO_4 + 6CO_2$$

Sodium Aluminate ($NaAlO_2$)

It is used for treating water having no alkalinity (*i.e.*, pH <7). It is also used along with $Al_2(SO_4)_3$ for the treatment of water in water plants.

$$NaAlO_2 + 2H_2O \longrightarrow Al(OH)_3 \downarrow + NaOH$$

$$6NaAlO_2 + Al_2(SO_4)_3 + 12H_2O \longrightarrow \underset{\text{(White Ppt.)}}{8Al(OH)_3} \downarrow + 3Na_2SO_4$$

Ferrous Sulphate ($FeSO_4.7H_2O$)

Ferrous sulphate is widely used as a coagulant for alkaline water (*i.e.*, pH >8.0). Water is gently agitated with ferrous sulphate. As a result $Fe(OH)_3$ gets precipitated. The reactions take place as follow.

$$FeSO_4 + Ca(HCO_3)_2 + H_2O \longrightarrow Fe(OH)_2 + CaCO_3 + H_2SO_4 + CO_2 \uparrow$$

$$FeSO_4 + Mg(HCO_3)_2 + H_2O \longrightarrow Fe(OH)_2 + MgCO_3 + H_2SO_4 + CO_2 \uparrow$$

$$4Fe(OH)_2 + 2H_2O + O_2 \longrightarrow 4Fe(OH)_3 \downarrow$$

The precipitate is removed mechanically or blow down process. During this process H_2SO_4 is produced which is neutralised by the alkalinity of water. If water is not alkaline then sufficient amount of lime is added to make it alkaline, then it is treated with ferrous sulphate.

4.1.4 Filtration

It is the process by which the colloidal particles, suspended matter etc. and some micro-organisms are removed by passing water through a porous material consisting of a fine bed of sand and other granular materials. It is known as filter. There are two types of filter, *i.e.*,

1. Gravity type filters and 2. Pressure type filters.

1. **Gravity type filters**—It is based on the filtration due to gravitation force.

Gravity type filter is also classified on the basis of their efficiency as slow sand filters and rapid sand filters.

Sand Filter. The sand filter may be represented as in Fig. 4.1.

It consists of three layers. The top layer consists of fine sand and is thick layer. The middle layer consists of coarse sand and the bottom layer consists of gravels. It is provided with an inlet for sedimented water and an under drain channel at the bottom for exist of filtered water. The sedimented water enter into the filter and is uniformly distributed over the fine sand bed. The filtration get start and the impurities are retained by the filtering medium. Filtered water comes out from the outlet.

During filtration the sand pores get clogged, due to retention of impurities in the pores. As a result the rate of filtration becomes slow. For solving this problem, about 2–3 cm of the top fine sand layer is scrapped off and replaced with clean and fresh sand. The filter is ready to reuse. The scrapped sand is washed with water and dried for reuse at the time of next scrapping operation.

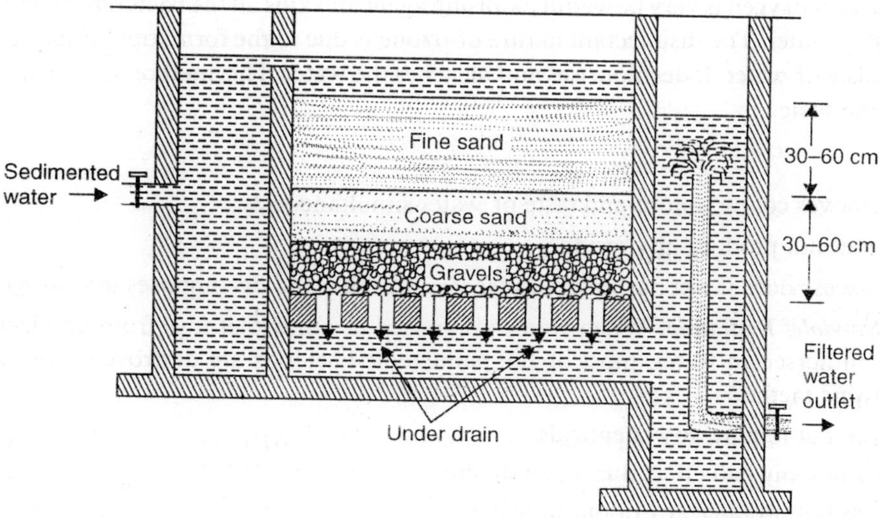

FIGURE 4.1 **Sand filter**

2. **Pressure type filters**—It is very similar to gravity type filters. The pressure filters are located in air tight vessels. The impure water is pumped into the vessels by using pumps or motor. Impure water

is passed through pressure filters at high pressure (the pressure should be greater than the atmospheric pressure). The rate of filtration is greater than gravity filtration.

4.1.5 Disinfection (Sterilization) of Water

Sterilization of water means the removal of micro-organisms (bacteria, virus etc.) from the water. The filtered water may contain pathogenic bacteria and other micro-organisms. These are very harmful and make the water unfit for domestic use. Hence it is necessary to remove the micro-organisms and pathogenic bacteria from water. The chemicals or substances, which are added, to water for killing the bacteria etc. are known as disinfectants. Several methods have been adopted for sterilization of water. Some of them are given below.

Boiling. It is a simple sterilization method. Water is boiled about 20–30 minutes. Harmful disease causing bacteria are killed by this process. This method is not widely used due to the following limitations:

1. It is very much expensive for municipal supply of water. Hence this method is only useful for individuals.
2. The taste of water may change.
3. A large quantity of fuel is required to boil of water on a large scale.
4. The process kills pathogenic bacterias at the time of boiling. It does not protect the water against any future infection.

By Ozone. The process is called ozonization of water. In this process water is treated with ozone for half an hour. Ozone is highly unstable. It undergoes decomposition and liberates nascent oxygen.

$$O_3 \longrightarrow O_2 + [O]$$

The nascent oxygen is very powerful oxidising agent and kills all the bacterias. It oxidises organic matter present in water. The disinfectant nature of ozone is due to the formation of nascent oxygen. It improves the taste of water. It decolourises and bleaches the water. This method is expensive so it is not useful on a large scale.

Advantages

(*i*) It removes colour, odour and taste of water simultaneously.

(*ii*) It does not produce any residue.

(*iii*) Its excess dose is not harmful because it is unstable and decomposes into oxygen.

By Ultra-violet Radiation. When water is exposed to ultraviolet rays from an electric mercury vapour lamp immersed in water, most of the pathogenic bacterias are destroyed. Following are the advantages of this method:

1. It does not require any chemicals.
2. It has not any bad effect during treatment.
3. It does not produce any odour in water.
4. It takes very small time.

Although it has great advantage but due to very expensive it is not widely used on a large scale.

By Chlorination. It is the most important method for sterilizing the water. Chlorination is done by the following methods:

1. *As a gas or concentrated aqueous solution.* Chlorine either in gas form or in concentrated aqueous solution form produces nascent oxygen which is a powerful germicide. When water is treated with chlorine, the nascent oxygen destroys the harmful bacteria and germs by oxidation.

$$Cl_2 + H_2O \longrightarrow HCl + \underset{\text{Hypochlorous Acid}}{HClO}$$

$$HClO \longrightarrow HCl + [O]$$

The hypochlorite ion (OCl^-) produced by the ionization of hypochlorous acid (HOCl) destroy the cell membranes of pathogenic micro-organisms and hence harmful bacterias are destroyed, so they also act as disinfectants.

2. *As bleaching powder.* Bleaching powder ($CaOCl_2$) contains about 30% available chlorine, which is a strong oxidising agent. When water is treated with a calculated amount of bleaching powder, it produces hypochlorous acid (HClO), which is a powerful disinfectant.

$$CaOCl_2 + H_2O \longrightarrow Ca(OH)_2 + Cl_2$$

$$Cl_2 + H_2O \longrightarrow HCl + HClO$$

$$HOCl \longrightarrow HCl + [O]$$

About 1 kg of bleaching powder is sufficient of 1000 kiloliters of water and this mixture is allowed to stand for several hours. It has following drawbacks.

(*i*) It introduces calcium in water so water may become hard.

(*ii*) Its excess amount produces bad smell in water.

(*iii*) Its storage is difficult because it is unstable.

3. *As chloramine.* Chloramine is obtained by treating chlorine with ammonia in the ratio 2: 1 by volume

$$Cl_2 + NH_3 \longrightarrow \underset{\text{(Chloramine)}}{NH_2Cl} + HCl$$

When chlorine or bleaching powder is used in excess for disinfection of water then it produces unpleasant taste or disagreeable odour in water. But chloramine does not produce such type of bad smell or unpleasant taste. Hence it is considered as better germicidal than chlorine. Chloramine gives hypochlorous acid (HOCl) in water and this HOCl act as a powerful disinfectant. Hypochlorous acid also liberates nascent oxygen which destroy the harmful bacterias by oxidation

$$NH_2Cl + H_2O \longrightarrow HOCl + NH_3$$

$$HOCl \longrightarrow HCl + [O]$$

Advantages

It has great advantages.

1. It gives good taste to water.
2. It does not produce bad smell in water.
3. It removes irritating smell due to excess of chlorine.
4. It provides a greater lasting effect than chlorine.

Advantages of Chlorine used as Disinfectant

1. It is economically cheap and very effective.
2. It can be stored very easily in a small space.
3. It is a stable and does not deteriorate on keeping.
4. It is the most ideal disinfectant.
5. It may be used at low or high temperature.
6. It does not produce any salt impurities in water.

Although it has a great advantage but following are some disadvantages of chlorine used as disinfectant in water.

Disadvantage of Chlorine used as Disinfectant

1. Its excess amount produces unpleasant odour and taste in water.
2. Its excess amount produces an irritation on muscus membrane.
3. It is more effective below the pH 6.5 but less effective at higher pH values.

The quantity of free chlorine in treated water should not excess (0.1 to 0.2 ppm.).

Super chlorination. When excess amount of chlorination is added for disinfection of water, it is called superchlorination. Superchlorination destroy the pathogenic micro-organism as well as organic impurities by oxidation process. The excess amount of chlorine after treatment may be removed by using calculated amount of ammonia etc.

Prechlorination and Postchlorination. When water sample is treated with chlorine before filtration, is known as prechlorination. A higher chlorine does is required to satisfy the chlorine demand of filterable matter. Due to this the cost is increased but the quality of water obtained is superior as the chlorinated products with unpleasant tastes and odours may be adsorbed during filtration.

When water sample is treated with chlorine after filtration, is known as postchlorination. It is cheaper than prechlorination due to lower chlorine demand but the treated water may have unpleasent taste and odour.

Break-point chlorination. If chlorination of water is done carefully and in a controlled way, then a point comes where free residual chlorine begins to appear is called break-point chlorination. During break-point chlorination water is treated with such an amount of chlorine which is just sufficient to

1. Destroy the bacteria.
2. Oxidize the organic matter present in water.
3. Oxidize the ammonia if present in water
4. Leave behind slight excess of free chlorine which could act as disinfectant during storage of water. When the dosage of applied chlorine is gradually increased to the treating water rich in organic compound or ammonia and the results obtained is plotted graphically, a typical relationship is obtained between the amount of chlorine added to water and the experimentally determined free residual chlorine as shown in Fig. 4.2.

The break point chlorination of pure distilled water and impure water are represented by curve I and curve II respectively. From the graph it is clear that in case of pure distilled water, a straight line 'OL' is obtained as represented by curve I. It indicates that due to purity when the dose of applied chlorine is increased the amount of residual chlorine increases. On the other hand in case of impure water a curve ABCD is obtained as represented by curve II.

When the dose of applied chlorine is low, all the added chlorine gets consumed in complete oxidation of reducing substances present in water hence no free residual chlorine occurs. It is represented by 'OA'. On increasing the amount of added chlorine, the amount of residual chlorine (experimentally determined) is also increased slowly. This stage is correspond to the formation of chloro-organic compounds without oxidizing them and is represented by 'AB'.

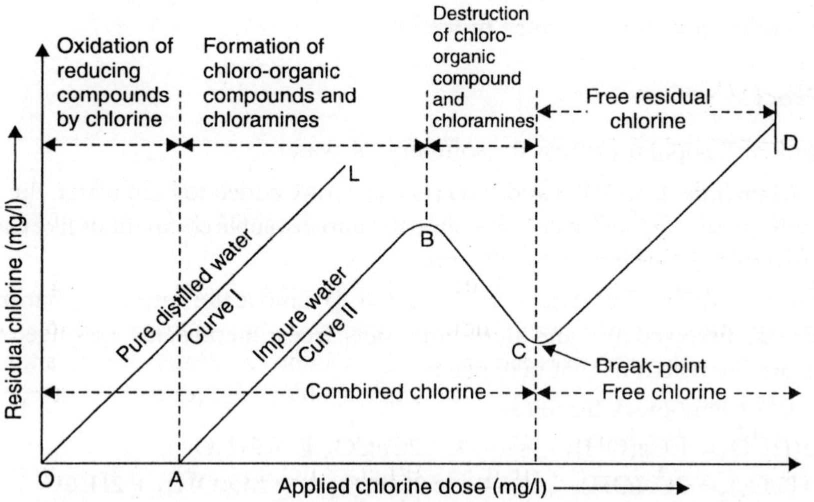

FIGURE 4.2

At high concentration of applied chlorine, oxidation of chloro-organic compounds or micro-organisms takes place. Hence the amount of free residual chlorine is decreased. The destruction of chloro-organic compounds and chloramines takes place continuously and reaches a minima and is represented by curve BC. The reactions may take place as follow:

$$4NH_2Cl + 3Cl_2 + H_2O \longrightarrow N_2 + N_2O + 10HCl$$

$$2NHCl_2 + 3Cl_2 + 4H_2O \longrightarrow 2NO_2 + 10HCl$$

After reaching the minima (*i.e.*, the point C) the added chlorine does not take part in chemical reaction. As a result the amount of residual chlorine is increased by adding the amount of chlorine and is represented by curve CD. The point 'C' is known as 'break-point' chlorination. The point at which free residual chlorine begins to appear is called the 'break-point' chlorination. Hence for effectively killing the micro-organisms, sufficient amount of chlorine (minima of the curve *i.e.*, point 'C') must be added. Thus, break-point chlorination helps in eliminating disagreeable odour and bad taste in water.

Advantages and Significance

1. It completely destroys all the pathogenic bacteria.
2. It indicates the complete destruction of organic compounds, which are present in water.
3. It helps to calculate the just sufficient amount of chlorine for adding in water.
4. It prevents the growth of any weeds in water.
5. It also signifies complete decomposition of NH_3, removal of colouring materials and improvement of taste and odour of the water sample.

4.2 WATER SOFTENING

Water plays very important role in industry and domestic purpose. It has a variety of applications in industries. The natural water contains several types of dissolved solid so it is necessary to remove the dissolved solid, which causes hardness of water. Softening of water is the process of removing the hardness of water. The most common methods for softening of water are given below:

1. Lime soda process
2. Zeolite process (Base exchange process)
3. Demineralization (Ion exchange process)

1. Lime Soda Process

It is very important and popular process for softening of water.

Principle. When lime [$Ca(OH)_2$] and soda (Na_2CO_3) are added to hard water, the soluble calcium and magnesium salts present in hard water are converted into insoluble compounds like $CaCO_3$, $Mg(OH)_2$ etc. which are separated as precipitate by filtration.

Functions of lime. Lime [$Ca(OH)_2$] is very effective to remove the temporary hardness, permanent magnesium hardness, dissolved iron and aluminium salts, free mineral acids, dissolved CO_2 and H_2S in water. Following are the reactions that take place.

1. Removal of temporary hardness

$$Ca(HCO_3)_2 + Ca(OH)_2 \longrightarrow 2CaCO_3 \downarrow + 2H_2O$$
$$Mg(HCO_3)_2 + 2Ca(OH)_2 \longrightarrow 2CaCO_3 \downarrow + Mg(OH)_2 + 2H_2O$$

2. Removal of permanent magnesium hardness

$$MgCl_2 + Ca(OH)_2 \longrightarrow Mg(OH)_2 \downarrow + CaCl_2$$
$$MgSO_4 + Ca(OH)_2 \longrightarrow Mg(OH)_2 \downarrow + CaSO_4$$

3. Removal of dissolved iron and aluminium salts

$$FeSO_4 + Ca(OH)_2 \longrightarrow Fe(OH)_2 \downarrow + CaSO_4$$
$$4Fe(OH)_2 + 2H_2O + O_2 \longrightarrow 4Fe(OH)_3 \downarrow$$
$$Al_2(SO_4)_3 + 3Ca(OH)_2 \longrightarrow 2Al(OH)_3 \downarrow + 3CaSO_4$$

4. Removal of free mineral acids

$$H_2SO_4 + Ca(OH)_2 \longrightarrow CaSO_4 + 2H_2O$$
$$2HCl + Ca(OH)_2 \longrightarrow CaCl_2 + 2H_2O$$

5. Removal of dissolved CO_2 and H_2S

$$CO_2 + Ca(OH)_2 \longrightarrow CaCO_3 \downarrow + H_2O$$
$$H_2S + Ca(OH)_2 \longrightarrow CaS \downarrow + 2H_2O$$

Functions of soda. When lime is used to remove the hardness producing substances like permanent magnesium hardness or ions, Fe^{2+}, Al^{3+}, mineral acids etc, it has been found that permanent calcium hardness ($CaCl_2$ and $CaSO_4$) is introduced in water. Soda is very effective to remove the permanent calcium hardness as follows:

$$CaCl_2 + Na_2CO_3 \longrightarrow CaCO_3 \downarrow + 2NaCl$$
$$CaSO_4 + Na_2CO_3 \longrightarrow CaCO_3 \downarrow + Na_2SO_4$$

At room temperature the precipitates of $CaCO_3$ and $Mg(OH)_2$ are fine and do not settle down easily, so the filtration becomes difficult. These fine particles may take part in clogging and corrosion in the pipes and boiler tubes. These drawbacks may be improved by adding some chemicals as a coagulants like alum, sodium aluminate or ferrous sulphate etc. The amount of lime and soda is also consumed in the removal of coagulants.

Process. Lime-soda process is carried out at room temperature as well as high temperature (90–100°C) and is known as cold lime-soda process and hot soda-lime process respectively.

Cold lime-soda process. Cold lime-soda process provides water containing a residual hardness of 50–60 ppm. A calculated amount of lime and soda are mixed with water at room temperature. Small amount of coagulant like alum or $NaAlO_2$ etc. are also added. The coagulant helps in the entraption of the fine precipitates. $NaAlO_2$ also helps in the removal of silica and oil, if present in water.

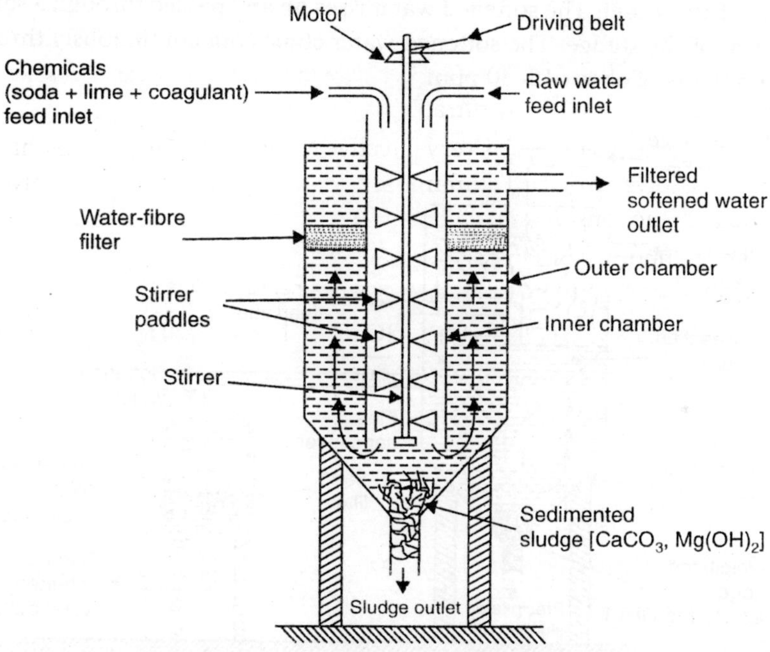

***FIGURE 4.3* Cold lime-soda softener**

Method. The apparatus used in cold lime-soda process is represented as in Fig. 4.3. A calculated amount of lime and soda are mixed with raw water alongwith a small amount of a coagulant. This mixture is fed from the top of the inner vertical circular chamber. The inner chamber is fitted with a vertical rotating shaft carrying a number of paddles. When the raw water and chemicals flow down, there is a vigorous stirring and continuous mixing takes place. Here chemical reactions take place and hard water is converted into soft water. The insoluble precipitate forms a heavy sludge which comes out from the outlet from time to time at the bottom of apparatus. The softened water comes into the outer co-axial chamber and rises upwards. It passes through a filtering media (usually made of wood fibres) to ensure complete removal of the sludge. Filtered softer water finally comes out continuously through the outlet at the top. The softened water contains residual hardness of about 50–60 ppm.

Hot lime-soda process. It is the fast process of softening the water at high temperature (90–100°C). Since the temperature is about the boiling of water hence the reactions becomes fast, the precipitation and filtration take place very easily and rapidly.

Method. The apparatus used in hot lime-soda process is represented as in Fig. 4.4. It consists of three parts:

1. A reaction tank,
2. A conical sedimentation vessel and
3. A sand filter.

Raw water is fed from the top of the reaction tank where it is thoroughly mixed with the calculated amount of lime and soda along with some coagulants. The mixture is heated in the superheated steam. The reactions take place and water gets softened. Softened water along with sludge reaches into a conical sedimentation vessel where sludge settles down and comes out from time to time through the outlet at the bottom of the vessel. The softened water rises up and passed through a sand filter to ensure the complete removal of the sludge. The softened water comes out continuously through the outlet. It contains residual hardness of about 15–30 ppm.

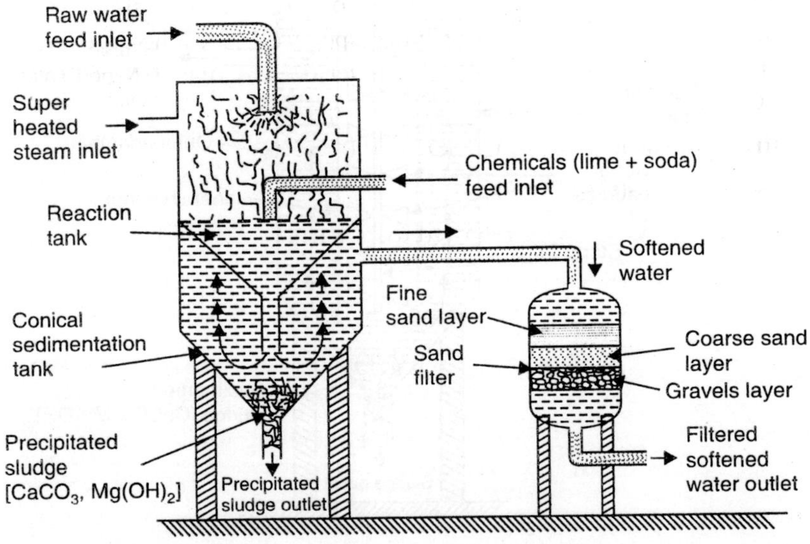

FIGURE 4.4 **Hot lime-soda softener**

Advantages of Lime-Soda Process

1. It is economical.
2. Hot soda-lime process is much faster than the cold lime-soda process.
3. During this process pH value of water is increased hence the corrosion of pipe is reduced.
4. Besides the removal of hardness, the quantity of minerals in water are also reduced.
5. Due to alkaline nature of water, amount of pathogenic bacterias in water get reduced.
6. It requires less amount of coagulants.

Disadvantages

1. The softened water is not completely free from hardness. It contains 15–30 ppm of hardness.
2. Disposal of large amount of sludge is a problem.
3. Careful operation and skilled supervision is required for efficient treatment of water.

Some Important Tips for Solving Numerical Problems

1. First of all calculate the amount of all substances present in the water sample in terms of $CaCO_3$ equivalent

$$CaCO_3 \text{ equivalent} = \frac{\text{Mass of the impurity} \times 50}{\text{Equivalent mass of the impurity}} \text{ mg/L}$$

2. Add (sum up) all $CaCO_3$ equivalent of substances to get total hardness.
3. Substances like NaCl, KCl, Na_2SO_4, SiO_2, Fe_2O_3 etc. do not impart any hardness, therefore, these do not consume any soda or lime. Hence these should not be taken into consideration for calculating the lime and soda requirements.
4. When the impurities are given as $CaCO_3$ or $MgCO_3$ these should be considered due to bicarbonates of calcium and magnesium respectively. The amount expressed as $CaCO_3$ does not require any further conversion. However the amount of $MgCO_3$ should be converted into $CaCO_3$ equivalent.

Lime requirement. Mol. mass of $CaCO_3$ = 100

Molar mass of lime $Ca(OH)_2$ = 74

100 parts by mass of $CaCO_3$ are required to 74 parts by mass of lime.

The amount of lime required for softening of water = $\frac{74}{100}$ [Temporary calcium hardness + 2 × Temporary magnesium hardness + Permanent magnesium hardness + CO_2 + HCl + H_2SO_4 + Fe^{2+} + Al^{3+} + HCO_3^- – $NaAlO_2$; All expressed in term of $CaCO_3$ equivalents]

Here 1 equivalent of temporary magnesium hardness *i.e.*, $Mg(HCO_3)_2$ consumes two equivalent of lime for complete neutralization.

Following are the reactions take place

$$Mg(HCO_3)_2 + Ca(OH)_2 \longrightarrow Ca(HCO_3)_2 + Mg(OH)_2$$

$$Ca(HCO_3)_2 + Ca(OH)_2 \longrightarrow 2CaCO_3 + 2H_2O$$

That's why $Mg(HCO_3)_2$ require double amount of lime for softening of hard water.

Soda requirement. Molar mass of soda Na_2CO_3 = 106

100 parts by mass of $CaCO_3$ are required to 106 parts by mass of soda.

∴ The amount of soda required for softening of water = $\frac{106}{100}$ [Permanent calcium hardness + Permanent magnesium hardness + Fe^{2+} + Al^{3+} + HCl + H_2SO_4 – HCO_3^- – $NaAlO_2$; All expressed in term of $CaCO_3$ equivalents]

5. If the quantities of Ca^{2+} and Mg^{2+} are given in the analytical report then 1 equivalent soda is required for Ca^{2+} whereas 1 equivalent of lime and 1 equivalent of soda (both) is required for Mg^{2+}.
6. 1 equivalent of HCO_3^- requires 1 equivalent of lime during lime-soda process to produce 1 equivalent of CO_3^{2-}. 1 equivalent of CO_3^{2-} is regarded as 1 equivalent of soda. That is why in case of soda requirement the quantity of HCO_3^- in equivalent is subtracted.
7. Sodium aluminate ($NaAlO_2$) requires neither soda nor lime. However, 1 equivalent of $NaAlO_2$ undergoes hydrolysis to produce 1 equivalent of OH^- which may be regarded as 1 equivalent of lime. That's why the quantity of $NaAlO_2$ in equivalents is always subtracted in both the cases (soda as well as lime process)

$$NaAlO_2 + 2H_2O \longrightarrow NaOH + Al(OH)_3$$

8. If purity factor is given (*e.g.*, 85% pure lime, 70% pure soda etc.) then use the formula.

   Total lime requirement = Lime requirement × Purity factor × Volume of water

   Total soda requirement = Soda requirement × Purity factor × Volume of water
9. If the analytical data represents the presence of calcium nitrate and magnesium nitrate then 1 equivalent of soda is required for calcium nitrate whereas 1 equivalent lime and 1 equivalent soda (both) are required for magnesium nitrate. Following are the reactions take place.

$$Na_2CO_3 + Ca(NO_3)_2 \longrightarrow CaCO_3 + 2NaNO_3$$

$$Ca(OH)_2 + Mg(NO_3)_2 \longrightarrow Ca(NO_3)_2 + Mg(OH)_2$$

10. If the analytical report shows interms of alkalinity and hardness then
    (*i*) Ca–alkalinity = Ca-hardness or total alkalinity, whichever is small

    (*ii*) (*a*) Mg-alkalinity = Mg-hardness (if total alkalinity is equal to or greater than total hardness)

    (*b*) Mg-alkalinity= Total alkalinity–Ca-hardness (if total alkalinity is greater than Ca-hardness but less than total hardness)

    (*iii*) Sodium alkalinity = Alkalinity–Total hardness

    (*iv*) Calcium non carbonate

    (permanent hardness = Ca-hardness–Ca-alkalinity

    (*v*) Magnesium non carbonate

    (permanent) hardness = Mg-hardness–Mg alkalinity

    (*vi*) Total non carbonate hardness = Total hardness–Total alkalinity.
11. If Alkalinity present in water then it first combines to Ca^{2+}, then to Mg^{2+} and only then to Na^+.

Table 4.1 Differences between Cold and Hot Lime-soda Process

Sr. No.	*Cold Lime-Soda Process*	*Hot Lime-Soda Process*
1.	It is a slow process.	It is a fast process.
2.	Filtration is not easy.	Filtration is easy and rapid.
3.	It is done at room temperature.	It is done at high temperature (90-100°C).
4.	It has low softening capacity.	It has high softening capacity.
5.	Softened water has residual hardness about 50–60 ppm.	Softened water has residual hardness about 15–30 ppm.

2. Zeolite Process (Permutit Process)

Zeolite is known as permutit *i.e.*, boiling stone. Zeolites are naturally occurring hydrated sodium alumino silicate minerals ($Na_2O.Al_2O_3.xSiO_2.yH_2O$) where x = 2–10 and y = 2–6). These are capable for exchanging the ions causing hardness in water by sodium ions. Zeolites are of two types:

1. *Natural zeolites.* These are non porous, amorphous and durable. *e.g.*, $Na_2O.Al_2O_3.3SiO_2.2H_2O$ (Natrolite), $CaO.Al_2O_3.4SiO_2.4H_2O$ (Laumontite).
2. *Synthetic zeolites.* Synthetic zeolites are generally porous in nature. They have gel like structures and are prepared by heating together Na_2CO_3, Al_2O_3 and SiO_2. Synthetic zeolite has higher exchange capacity than natural zelolite. For example, sodium zeolite $Na_2O.Al_2O_3.xSiO_2.yH_2O$ (x = 2–10 and y = 2–6). Sodium zeolite is represented by Na_2Z for simplicity, where 'Z' stands for the insoluble zeolite radical framework.

Process. The apparatus is shown as in Fig. 4.5. The hard water is introduced at a specified rate through a zeolite bed kept in a cylinder. The zeolite is loosely packed over a gravel. The hardness causing ions Ca^{2+}, Mg^{2+} etc. present in hard water are retained by the zeolite as calcium zeolite or magnesium zeolite while the outgoing water contains sodium salts. The following reaction takes place.

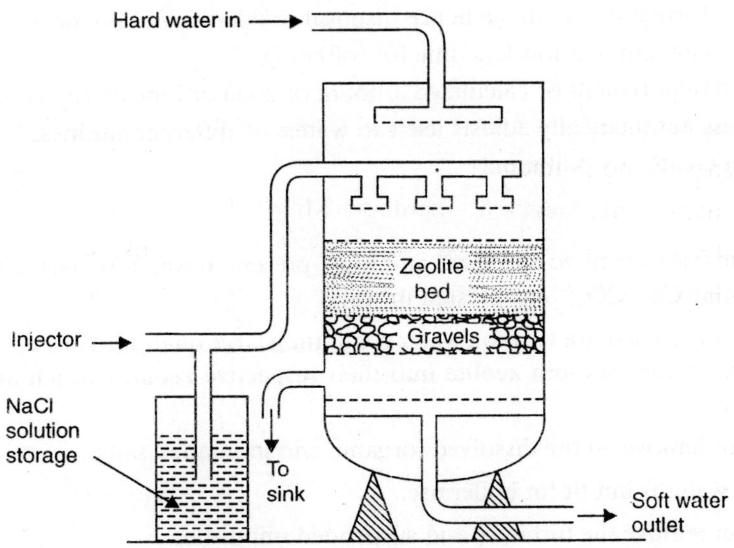

FIGURE 4.5 **Zeolite softener**

$$Ca(HCO_3)_2 + \underset{\text{(Sod. Zeolite)}}{Na_2Z} \longrightarrow \underset{\text{(Cal. Zeolite)}}{CaZ} + 2NaHCO_3$$

$$Mg(HCO_3)_2 + Na_2Z \longrightarrow \underset{\text{(Mag. Zeolite)}}{MgZ} + 2NaHCO_3$$

$$CaCl_2 + Na_2Z \longrightarrow CaZ + 2NaCl$$

$$MgCl_2 + Na_2Z \longrightarrow MgZ + 2NaCl$$

$$CaSO_4 + Na_2Z \longrightarrow CaZ + Na_2SO_4$$

$$MgSO_4 + Na_2Z \longrightarrow MgZ + Na_2SO_4$$

The dissolved sodium salt does not cause any hardness.

Regeneration. After working continuously about 10–12 hours the sodium zeolite gets completely exhausted due to the conversion of sodium zeolite into calcium zeolite and magnesium zeolite. Hence it is necessary to regenerate it. The regeneration of zeolite is carried by treating the exhausted zeolite with concentrated brine (10% NaCl) solution. Calcium zeolite and magnesium zeolite is converted into calcium chloride and magnesium chloride respectively which is washed away during washing with water. The sodium zeolite is thus regenerated and reused for softening the hard water. The following reactions take place:

$$\underset{\text{Calcium Zeolite}}{CaZ} + 2NaCl \longrightarrow \underset{\text{Sodium Zeolite}}{Na_2Z} + CaCl_2$$

$$\underset{\text{Magnesium Zeolite}}{MgZ} + 2NaCl \longrightarrow \underset{\text{Sodium Zeolite}}{Na_2Z} + MgCl_2$$

Advantages

1. It removes hardness more than lime-soda process. It contains the residual hardness about 10 ppm.
2. It is cheap and compact in size, so it occupying a small space.
3. There is no formation of sludge hence disposal problems does not occur.
4. It is quite clean and requires less time for softening.
5. There is no requirement of calculated amount of soda or lime or any coagulants.
6. The process automatically adjusts itself to waters of different hardness.
7. It does not create any pollution.

Disadvantages

1. It is applicable to remove only Ca^{2+} and Mg^{2+} present in water. It does not remove the acidic ions lime HCO_3^-, CO_3^{2-} etc. present in water.
2. It is not very efficient for treating water containing large quantities of Fe^{2+} and Mn^{2+} because these ions convert sodium zeolite into their respective zeolites which are very difficult to regenerate.
3. It does not remove all the dissolved (organic and inorganic) salts.
4. Softened water is not fit for boiler use.
5. It does not remove the turbidity and suspended impurities.

6. It does not destroy all pathogenic bacterias.
7. Anions are not removed by this process.

Comparison of Zeolite Process and Lime-Soda Process

Both processes are effective for softening of water but they are differ to each other by the following points:

Table 4.2

Sr. No.	*Zeolite Process*	*Lime-Soda Process*
1.	It dos not require lime or soda or coagulants.	It requires lime and soda along with some coagulants.
2.	Treated water contains sodium salts.	Treated water does not contain sodium salt.
3.	This process is not useful for removal of acidic impurities, turbidity etc.	There is not such type of limitations.
4.	Softened water contains residual hardness about 10 ppm.	Softened water contains 50–60 ppm in cold lime-soda process and 15–30 ppm in hot lime-soda process.
5.	The process is fast.	The process is slow in cold lime soda process and fast in hot lime soda process.
6.	The capital cost is higher.	The capital cost is low.
7.	It can operate under pressure and can be designed for fully automatic operation.	It cannot be operated under pressure.
8.	Operation expenses are lower as cheap NaCl is required for regeneration.	Operation expenses are higher as costly lime, soda and coagulants are required.
9.	The plant occupies less space.	The plant occupies large space.
10.	There is no problem of settling coagulation, filtration and removal of sludge etc.	There are difficulties in settling coagulation, filtration and removal of sludge etc.
11.	Treated water contains more $NaHCO_3$ which creates problem in boiler.	Treated water is free from $NaHCO_3$ so it is used in boiler.
12.	It is not fit for industrial purposes especially in boiler.	It is fit for industrial purpose especially in boiler.

3. Demineralization

It is also known as ion exchange process or deionization process. Demineralization is the process of complete removal of all ions (cations and anions) present in water by using ion-exchange resins.

Ion exchange resins consist of cross-linked, long chain organic polymer. Generally polystyerene-divinylbenzene (PS-DVB), polymethacrylic acid-divinylbenzene or phenol-formaldehyde polymer is used as ion exchange resin. The functional groups of resin may be acidic or basic. On the basis of functional group ion exchange resins are of two types.

1. Cation exchange resins and 2. Anion exchange resins.

1. *Cation exchange resins.* These contain acidic groups like —COOH or —SO_3H. These are usually co-polymer of styrene and divinylbenzene which on carboxylation or sulphonation become

cation exchanger *i.e.*, such type of resins exchange their H^+ ions with the cations in hard water. For example, a cation exchange may be represented as:

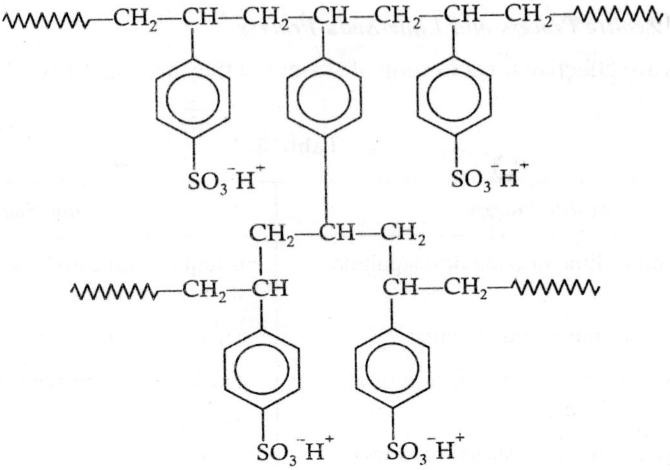

In general it is represented as RH where R represents the resin network for simplicity. Resins containing —SO_3H group are strongly acidic than those containing —COOH group.

2. *Anion exchange resins.* These resins contain basic functional groups. These are usually copolymer of styrene and divinylbenzene containing amine or substituted amine or quaternary ammonium groups as their hydroxide salts. For example, an anion exchange resins may be represented as :

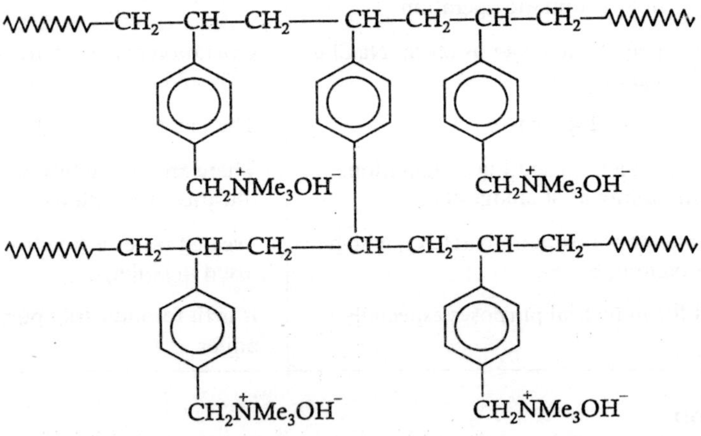

(Me = —CH_3 Methyl Group)

In general it is represented as R′OH for simplicity where R′ represents the resin network. These resins exchange their anion (OH^-) with the anion part of hard water. Resin containing quaternary ammonium salts are strongly basic than those containing —NH_2 or —NH—NH_2 group.

Working. Cation exchanger and anion exchanger both are interconnected with a pipe as shown in Fig. 4.6. The hard water is passed through first cation exchange resin. The Ca^{2+} and Mg^{2+} in water get exchange by H^+ from cation exchanger as follows:

$$2RH + Ca^{2+} \longrightarrow R_2Ca + 2H^+$$

$$2RH + Mg^{2+} \longrightarrow R_2Mg + 2H^+$$

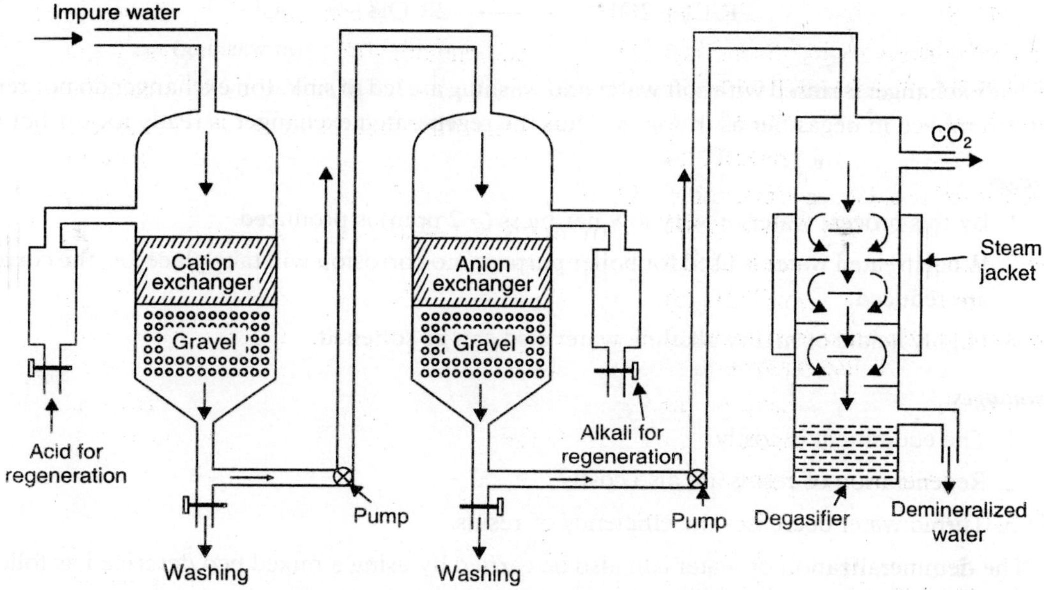

FIGURE 4.6 **Demineralization by ion exchangers**

Then the treated hard water is passed through anion exchanger resin columns where anions like SO_4^{2-}, Cl^- etc. present in hard water are replaced by OH^- ions from resins as follow.

$$R'OH + Cl^- \longrightarrow R'Cl + OH^-$$

$$2R'OH + SO_4^{2-} \longrightarrow R'_2SO_4 + 2OH^-$$

The released H^+ from cation exchanger treatment and OH^- from anion exchanger treatment get combined to produce water.

$$H^+ + OH^- \longrightarrow H_2O$$

It is called deionized water because it is free from cations and anions.

Regeneration of resins. When these exchangers are used for a long time then the columns get exhausted and it is necessary to regenerate. The exchanger is regenerated by first back washing and then by passing a solution of appropriate ion. The cation exchanger is regenerated by passing suitable acids (2% HCl or H_2SO_4) and anion exchanger is regenerated by passing an alkali (dil NaOH).

For cation exchanger

$$R_2Ca + 2H^+ \longrightarrow 2RH + \underset{\text{(in washing)}}{Ca^{2+}}$$

$$R_2Mg + 2H^+ \longrightarrow 2RH + \underset{\text{(in washing)}}{Mg^{2+}}$$

For anion exchanger

$$R'_2SO_4 + 2OH^- \longrightarrow 2R'OH + \underset{\text{(in washing)}}{SO_4^{2-}}$$

$$R'Cl + 2OH^- \longrightarrow 2R'OH + \underset{\text{(in washing)}}{Cl^-}$$

The exchanger is rinsed with soft water and washing are led to sink. Ion exchanger do not remove CO_2. It is removed in degasifier as in Fig. 6. Thus the regenerated exchanger is ready for further use.

Advantages

1. By this process water of very low hardness (≈ 2 ppm) is produced.
2. When treated water is used for boiler purpose, no corrosion will take place *i.e.*, the corrosion are reduced.
3. Highly acidic or highly alkaline water also can be softened.

Disadvantages

1. The equipment is costly.
2. Regeneration of resins are also costly.
3. Turbid water decreases the efficiency of resins.

The demineralization of water can also be carried by using a mixed bed described as follows:

Mixed bed deionizer. It consists of a single column containing a mixture of strongly cation exchanger and a strongly anion exchanger. The hard water is allowed to pass through it as in Fig. 4.7.

The hard water comes in contact with both exchanger several times, so it is highly efficient process than separate column of ion exchanger process. The purified water contains very less (about 1 ppm or less) hardness. It is more convenient and widely used.

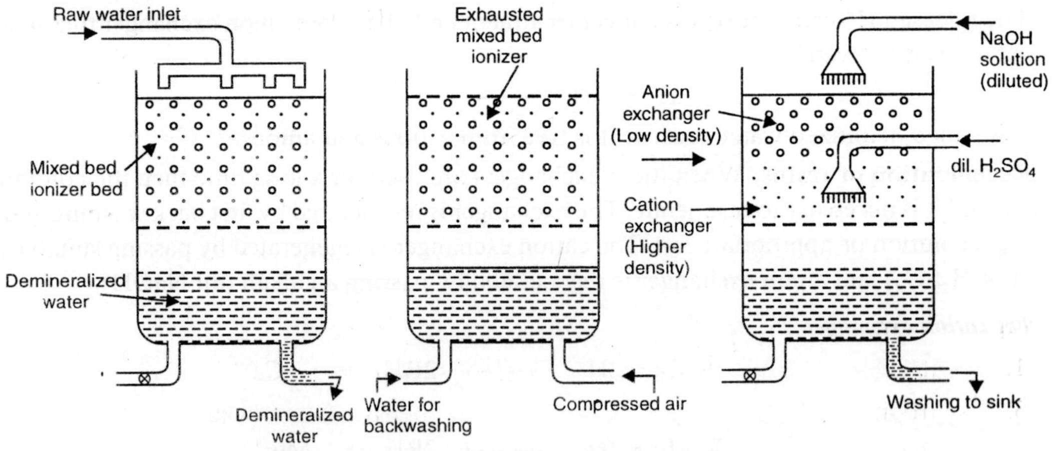

FIGURE 4.7 **Regeneration of mixed-bed ion exchanger**

Regeneration of resins. After long time working, the resins are exhausted. For regeneration the beds of resins are back washed. During this process anion exchanger resin gets displaced above the cation exchanger resin because anion exchanger resins are lighter than cation exchanger resin. It makes very easy to separate both the resins. The anion exchanger resin is regenerated by passing dil. NaOH solution and cation exchanger resin is regenerated by passing dil H_2SO_4. After that these beds are then rinsed with soft water. The beds are mixed again by passing a stream of compressed air from below. Now it is again ready to reuse.

Ion exchangers should possess following properties

1. They should not produce any colour or bad smell in treated water.
2. They should be non toxic, cheap and easily available.
3. They should possess a high ion-exchange capacity.
4. Since ion-exchange is a surface phenomenon, so they should have a large surface area.
5. They should be highly stable and inert for chemical attack.
6. Their regeneration and washing should be cheap and easy.

Now-a-days there are number of ion exchangers available in the market with different trade names. Generally polystyrene based cation–exchangers and melamines based anion-exchangers are used for market purpose. Such type of exchangers have been developed by National Chemical Laboratory, Pune.

4.3 DESALINATION

Desalination is the process of removing dissolve solids (NaCl etc.) from water. When water contains high concentration of dissolved solids then it is known as brackish water. It is unfit for drinking due to much salty taste. On the basis of dissolved amount of solids water is classified as:

	Amount of dissolved solids
1. Fresh water	<1000 ppm
2. Brackish water	1000 ppm–3500 ppm
3. Sea water	>3500 ppm

Although 'desalination' process is costly but it is very much important in these days due to scarcity of available fresh water. As we know that the demand of water is increasing day by day but fresh water is not available as per demand. That's why 'desalination' process has been adopted as much as possible in the world.

4.3.1 Methods of Desalination

Following are the methods for desalination of water.

1. Distillation,
2. Freezing,
3. Electrodialysis, and
4. Reverse osmosis.

1. *Distillation.* Brackish water may be purified by distillation of salty water into pure water on a large scale but it is very costly and not very common on a large scale.

2. *Freezing.* When salty water (brackish water) is allowed to freeze the ice crystals are separated free from salt. It is washed and melted into pure water. The yield is satisfactory but the cost of production is high on a large scale.

3. *By electrodialysis.* The electrodialysis is done by electrodialyzer. It is an important method for desalination of brackish water and may be applied at room temperature. It is based on the fact that when direct electric current is applied to saline (brackish) water by using electrodes and ion selective permeable membranes, the ions present in saline water migrate towards oppositely charged electrodes

and water gets purified. An ion selective permeable membrane is that which allow to pass either cation or anion not both. The systematic line diagram is shown as in Fig. 4.8.

The electrodialysis cell consists of electrodes and alternate cation and anion permeable membranes. Cathode is placed near cation permeable membrane and anode is placed near anion permeable membrane. A cation permeable membrane is permeable to cations only due to the presence of fixed charged functional groups like RSO_3^- or $RCOO^-$ inside the membrane, they reject the anions having same charge similarly anion-selective membrane is permeable to only anions due to the presence of fixed positively charged functional groups like $R_4N^+Cl^-$ inside the membrane, they reject the cations having same charge.

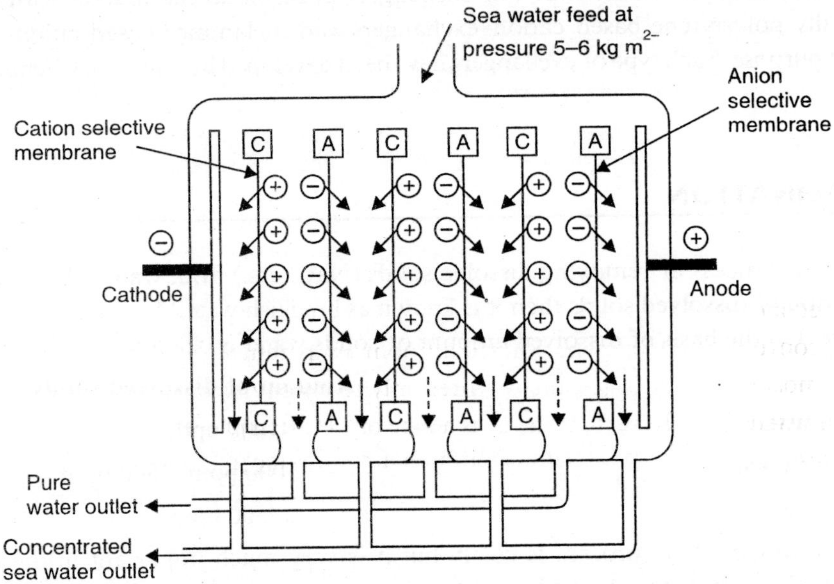

FIGURE 4.8 **Electrodialysis cell (Electrodialyzer)**

Brackish water is passed under a pressure about 5–6 kg m^{2-} between membrane pairs and an electric field is applied perpendicular to the direction of water flow. The cation selective permeable membrane only permits positively charged ions (Na^+) to pass through. The fixed negative charge inside the membrane do not allow the negatively charged Cl^- ions to pass through. Similarly, anion selective permeable membrane only permits negatively charged ions (Cl^-) to pass through. The fixed positive charge inside the membrane does not permit the positively charged (Na^+) ions to pass through. Hence salt concentration is increased in a separate compartment. Thus, water get purified. In general it may be represented as in Fig. 4.9.

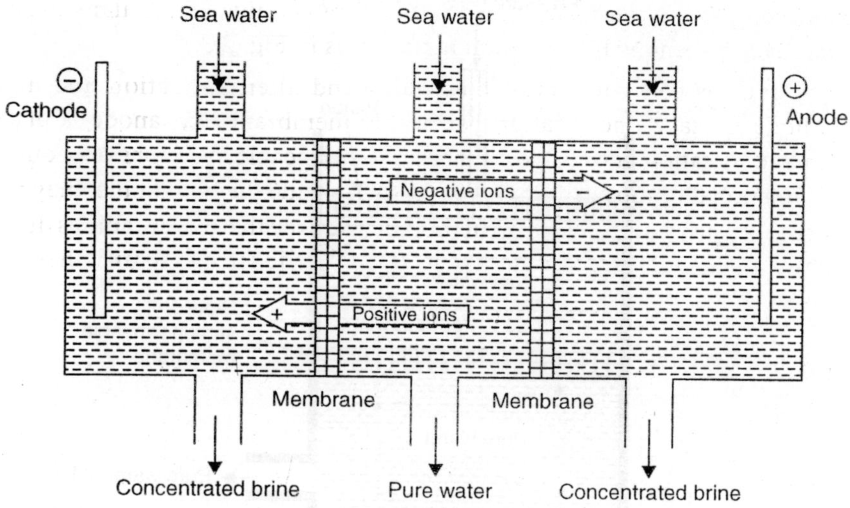

FIGURE 4.9

Advantages

(*i*) It is economical.

(*ii*) It is convenient and may be applied at room temperature.

(*iii*) It is most compact in size and requires only electricity for operating.

4. *By reverse osmosis.* Reverse osmosis is a type of filtration process that is often used for water. It works by using pressure to force a solution through a membrane, retaining the solute on one side and allowing the pure solvent to pass to other side. It is the reverse of the normal osmosis process.

If two solutions having different concentrations are separated by a semi permeable membrane then flow of solvent takes place from dilute side (lower concentration) to the higher concentration until the concentration becomes equal on the both sides. This phenomenon is called osmosis and the driving force is called the osmotic pressure, since it is analogous to flow caused by a pressure differential. If a hydrostatic pressure is applied in excess of osmotic pressure, on the highly concentrated side then the flow of solvent gets reversed *i.e.*, solvent passes from high concentration to low concentration areas through semi permeable membrane. It is called reverse osmosis. In this case, there are two forces influencing the movement of water: (*i*) the pressure caused by the difference in solute concentration between the components *i.e.*, osmotic pressure and (*ii*) the externally applied pressure. Due to high externally applied pressure water gets pure and is separated from its contaminates by reverse osmosis process. The membrane filtration is also known as super-filtration or hyper-filtration. This technique is very useful in sea-cost areas. The outsketching line diagram of this technique is as shown in Fig. 4.10.

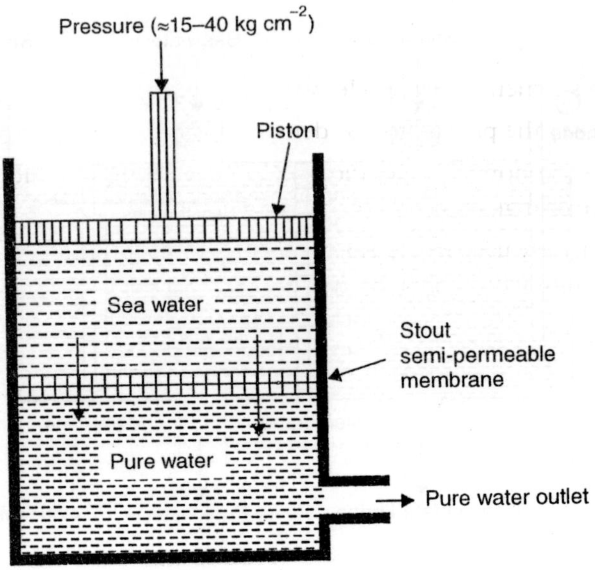

FIGURE 4.10

In this process, a high pressure (≈ 15–40 kg cm^{-2}) is applied to the sea water or brackish water which is to be treated. The semi-permeable membrane only allows the solvent molecule (pure water) to passes through. Thus dissolved ionic and non-ionic solutes left behind and water gets purified. Membrane pore sizes can vary from 0.1 to 5,000 nanometers (nm) depending on filter type. Particle filtration removes particles of 1,000 nm or larger. Microfiltration removes particles of 50 nm or larger. Ultrafiltration removes particles of roughly 3 nm or larger. Nano filtration removes particles of 1 nm or larger. The membrane used in the reverse osmosis process removes particles of 0.1 nm or larger.

The major problem is to find membranes strong enough to withstand the high pressure and impermeable to the dissolved solute particles. Hence synthetic membranes made of polymer like nylon. Cellulose triacetate (CTA) etc have been used. Recently membranes made of polymethaylmethacrylate, polyamide, thin film composites (TFC) are used.

The Reverse osmosis (R.O) system available in market, include a number of steps.

(*i*) a sediment filter to trap particles including rust and calcium carbonate.

(*ii*) optionally a second sediment filter with smaller pores.

(*iii*) an activated carbon filter to trap organic chemicals and chlorine, which will attack and degrade TFC reverse osmosis membranes.

(*iv*) a reverse osmosis (RO) filter which is a thin film composite membrance (TFM or TFC).

(*v*) optionally a second carbon filter to capture those chemicals not removed by the RO membrane.

(*vi*) optionally an ultraviolet lamp for disinfecting any microbes that may escape filtering by the reverse osmosis membrane.

In some systems, the carbon prefilter is omitted and cellulose triacetate (CTA) membrane is used. The CTA membrane is prone to rotting unless protected from chlorinated water, while the TFC membrane is prone to breaking down under the influence of chlorine. In CTA systems, a carbon post-filter is used to remove chlorine from the final product water.

Applications

1. It removes minerals from boiler water at power plants.
2. It is used for the production of deionised water.
3. It is more economical operation for concentrating food liquids like tomato juice and fruit juices like orange juice.
4. It is extensively used in the dairy industry for the production of whey (liquid remaining after cheese manufacture), protein powders and for the concentration of milk to reduce shipping costs.
5. It is widely used in the wine industry for the concentration of wine.
6. It is also used to clean effluent and brackish ground water.
7. The dialysis used by people with kidney failure is based upon this technique.

Advantages

1. It is economical, compact and very simple.
2. Ionic, non ionic, colloidal and high molecular weight organic matters are easily removed by this process. Colloidal silica is not removed by demineralisation process.
3. It has long life and easy to replace the membrane within few minutes.
4. The maintenance cost is also economical and depends upon membrane.

Disadvantages

1. It does not remove the odour of water.
2. It does not remove the bacteria, virus etc. present in water.
3. High percentage of chlorine present in water may damage the TFC membrane.

On keeping the above view the complete RO system may include additional water treatment stages like ultra violet light or ozone to prevent microbiological contramination (bacteria or viruses). Activited carbon filter is used for the removal of organic chemicals or chlorine. It enhances the life of TFC membrane. Carbon filter also removes the odour and those particles which is not removed by the RO membrene by any means.

Due to great advantages reverse osmosis is successful for converting sea water into drinking water.

SOLVED NUMERICALS

Based on Lime and Soda Process

1. *Calculate the amount of lime and soda required for softening of 15000 litres of water which analysed as follows: temporary hardness = 25 ppm, permanent Ca hardness = 20 ppm, permanent Mg hardness = 15 ppm.*

Solution.

$$\text{Lime requirement} = \frac{74}{100}\ [\text{Temp. hardness} + \text{Perm. Mg hardness}] \times \text{Volume of water}$$

$$= \frac{74}{100} [25 + 15] \times 15000 = 444000 \text{ mg} = 444 \text{ g.}$$

$$\text{Soda requirement} = \frac{106}{100} [\text{Perm. hardness}] \times \text{Volume of water}$$

$$= \frac{106}{100} \times 20 \times 15000 = 318 \text{ g.}$$

2. *Water having following composition has to be softened by lime-soda process:*

$Ca(HCO_3)_2$ = 220 ppm, *$Mg(HCO_3)_2$ = 76 ppm*

$MgCl_2$ = 150 ppm, *$CaSO_4$ = 100 ppm*

CO_2 = 44 ppm, *$MgSO_4$ = 60 ppm*

Calculate the amount of lime and soda required to soften 10^6 litres of water.

Solution.

Constituents	*Quantity (W)*	*Equivalent wt.(E)*	*$CaCO_3$ Equivalent = $\frac{W \times 50}{E}$*
$Ca(HCO_3)_2$	220 ppm	81	$220 \times \frac{50}{81} = 135.80$ ppm
$Mg(HCO_3)_2$	76 ppm	73	$76 \times \frac{50}{73} = 52.1$ ppm
$MgCl_2$	150 ppm	47.5	$150 \times \frac{50}{47.5} = 157.9$ ppm
$CaSO_4$	100 ppm	68	$100 \times \frac{50}{68} = 73.5$ ppm
CO_2	44 ppm	22	$50 \times \frac{44}{22} = 100$ ppm
$MgSO_4$	60 ppm	60	$60 \times \frac{50}{60} = 50$ ppm

$$\text{Lime required} = \frac{74}{100} [Ca(HCO_3)_2 + 2Mg(HCO_3)_2 + MgCl_2 + MgSO_4 + CO_2 \text{ as } CaCO_3 \text{ eq.}] \times \text{Vol. of water}$$

$$= \frac{74}{100} [135.80 + 2(52.1) + 157.9 + 50 + 100] \times 10^6$$

$$= 405.964 \times 10^6 = 405.96 \text{ kg}$$

$$\text{Soda required} = \frac{106}{100} [MgCl_2 + MgSO_4 + CaSO_4 \text{ as } CaCO_3 \text{ eq.}] \times \text{Vol. of water}$$

$$= \frac{106}{100} [157.9 + 50 + 73.5] \times 10^6 \text{ mg}$$

$$= 298.284 \times 10^6 = 298.28 \text{ kg}$$

3. *Calculate the amount of lime required for softening the 6500 litres of hard water which containing the 6500 litres of hard water which containing 75 ppm of $MgSO_4$.*

Solution.

Constituent	*Quantity (W)*	*Equivalent wt (E)*	*$CaCO_3$ Equivalent* $= \frac{W \times 50}{E}$
$MgSO_4$	75 ppm	60	$75 \times \frac{50}{60} = 62.5$ ppm

$$\text{Lime required} = \frac{74}{100} [MgSO_4 \text{ as } CaCO_3 \text{ eq.}] \times \text{Vol. of water}$$

$$= \frac{74}{100} [62.5] \times 6500 = 300.63 \text{ g}$$

4. *Calculate the amount of lime and soda for softening 10^5 litres which contain*

HCl 7.3 mg/l, *$Al_2(SO_4)_3$ 34.2 mg/l*

$MgCl_2$ 9.5 mg/l, *NaCl 34.2 mg/l*

The purity factor of lime is 90% and that of soda is 98%. 10% of chemicals are to be used in excess in order to complete the reaction quickly. **(Raj. Univ. May 2001)**

Solution.

Constituent	*Quantity (W)*	*Equivalent wt. (E)*	*$CaCO_3$ Equivalent* $= \frac{W \times 50}{E}$
HCl	7.3 mg/l	36.5	$7.3 \times \frac{50}{36.5} = 10$ mg/l
$Al_2(SO_4)_3$	34.2 mg/l	114	$34.2 \times \frac{50}{114} = 15$ mg/l
$MgCl_2$	9.5 mg/l	47.5	$9.5 \times \frac{50}{47.5} = 10$ mg/l

$$\text{Lime required} = \frac{74}{100} [HCl + Al_2(SO_4)_3 + MgCl_2] \times \text{Vol. of water} \times \text{Purity of water}$$

$$= \frac{74}{100} [10 + 15 + 10] \times 10^5 \times \frac{100}{90} = 2.87 \text{ kg}$$

$$\text{Total lime required [using 10\% excess]} = 2.87 + 2.87 \times \frac{10}{100} = 3.157 \text{ kg}$$

$$\text{Soda required} = \frac{106}{100} [HCl + Al_2(SO_4)_3 + MgCl_2] \times \text{Vol. of water} \times \text{Purity of water}$$

$$= \frac{106}{100} [10 + 15 + 10] \times 10^5 \times \frac{100}{98} = 37.8 \times 10^5 \text{ mg} = 3.78 \text{ kg}$$

$$\text{Total soda required [using 10\% excess]} = 3.78 + 3.78 \times \frac{10}{100} = 4.158 \text{ kg}$$

5. *A water sample, on analysis, shows the following results.*

Ca^{2+}	*160 ppm*	CO_2	*88 ppm*
Mg^{2+}	*72 ppm*	HCO_3^-	*488 ppm*

Calculate the line and soda required to softening 1000000 litres of water if $FeSO_4 \cdot 7H_2O$ is used as a coagulant at the rate of 139 ppm.

Solution.

Constituent	*Quantity (W)*	*Equivalent wt. (E)*	*$CaCO_3$ Equivalent* $= \frac{W \times 50}{E}$
Ca^{2+}	160 ppm	20	$\frac{160 \times 50}{20} = 400$ ppm
Mg^{2+}	72 ppm	12	$\frac{72 \times 50}{12} = 300$ ppm
CO_2	88 ppm	22	$\frac{88 \times 50}{22} = 200$ ppm
HCO_3^-	488 ppm	61	$\frac{488 \times 50}{61} = 400$ ppm
$FeSO_4.7H_2O$	139 ppm	139	$\frac{139 \times 50}{139} = 50$ ppm

$$\text{Lime required} = \frac{74}{100}[Mg^{2+} + CO_2 + HCO_3^- + FeSO_4.7H_2O] \times \text{vol. of water}$$

$$= \frac{74}{100}[300 + 200 + 400 + 50] \times 10^6 \text{ mg}$$

$$= \frac{703 \times 10^6}{10^6} = 703 \text{ kg.}$$

$$\text{Soda required} = \frac{106}{100}[Ca^{2+} + Mg^{2+} - HCO_3^- + FeSO_4.7H_2O] \times \text{vol. of water}$$

$$= \frac{106}{100}[400 + 300 - 400 + 50] \times 10^6 \text{ mg}$$

$$= \frac{371 \times 10^6}{10^6} = 371 \text{ kg.}$$

6. *A water sample, on analysis, gave the following results.*

Total alkalinity = *580 ppm as $CaCO_3$ equivalent*

Calcium hardness = *484 ppm as $CaCO_3$ equivalent*

Magnesium hardness = *126 ppm as $CaCO_3$ equivalent*

Calculate the amount of lime and soda required to soften 10^6 L of water.

Solution.

Here, Ca-alkalinity = Ca-hardness or Total alkalinity, whichever is small

= 484 ppm

∴ Ca-Temporary hardness = 484 ppm

Mg-alkalinity = Total alkalinity – Ca-hardness (because here total alkalinity is greater than Ca-hardness but lesser than total hardness)

= 580 – 484

= 96 ppm

∴ Mg Temporary hardness = 96 ppm

∴ Mg-permanent hardness = Mg-hardness – Mg-alkalinity

= 126 – 96

= 30 ppm

and Ca-permanent hardness = Ca-hardness–Ca-alkalinity

= 484 – 484

= zero ppm

∴ Lime required $= \frac{74}{100}$ [Ca-Temp. hardness + 2 × Mg-Temp. hardness + Mg-Perm. hardness] × Vol. of water

$= \frac{74}{100} [484 + 2 \times 96 + 30] \times 10^6$ mg

= 522.44 kg

and Soda required $= \frac{106}{100}$ [Ca-Perm. hardness + Mg-Perm. hardness] × Vol. of water

$= \frac{106}{100} [0 + 30] \times 10^6$ mg

= 31.8 kg.

7. *The analytical report shows the following results of a water sample.*

Ca^{2+} 80 ppm HCO_3^- = 36.6 ppm

Mg^{2+} 48 ppm SO_4^{2-} = 111.36 ppm

Na^+ 16.10 ppm Cl^- = 13.49 ppm

Express the results in terms of salts present as their $CaCO_3$ equivalents.

Solution.

Constitutents	*Quantity (W)*	*Equivalent wt. (E)*	$CaCO_3$ *Equivalent* $= \frac{W \times 50}{E}$
Ca^{2+}	80 ppm	20	$\frac{80 \times 50}{20} = 200$ ppm
Mg^{2+}	48 ppm	12	$\frac{48 \times 50}{12} = 200$ ppm
Na^{+}	16.10 ppm	23	$\frac{16.10 \times 50}{23} = 35$ ppm
HCO_3^-	36.6 ppm	61	$\frac{36.6 \times 50}{61} = 300$ ppm
SO_4^{2-}	111.36 ppm	48	$\frac{111.36 \times 50}{48} = 116$ ppm
Cl^-	13.49 ppm	35.5	$\frac{13.49 \times 50}{35.5} = 19$ ppm

Here,

Total alkalinity = 300 ppm (due to HCO_3^-)

Calcium alkalinity = 200 ppm

∴ Magnesium alkalinity = (300 – 200) ppm
= 100 ppm

Total hardness = 400 ppm (due to Ca^{2+} and Mg^{2+})

Calcium hardness = 200 ppm

∴ Magnesium hardness = (400 – 200) ppm
= 200 ppm

Calcium temporary hardness = Calcium alkalinity
= 200 ppm

Magnesium temporary hardness = Magnesium alkalinity
= 100 ppm

∴ Magnesium permanent hardness
= Magnesium hardness – Magnesium temporary hardness
= (200 – 100) ppm
= 100 ppm

Now, Since Cl^- = 19 ppm

∴ Sodium chloride = 19 ppm

and Sodium sulphate $= \text{Total sodium ion} - \text{Sodium chloride}$

$= (35 - 19)$ ppm

$= 16$ ppm.

Hence the salts present in terms of their $CaCO_3$ equivalent may be expressed as

$$Ca(HCO_3)_2 = 200 \text{ ppm}$$
$$Mg(HCO_3)_2 = 100 \text{ ppm}$$
$$MgSO_4 = 100 \text{ ppm}$$
$$NaCl = 19 \text{ ppm}$$
$$Na_2SO_4 = 16 \text{ ppm}$$

Based on Zeolite Process (Permutite Process)

8. *10,000 lt of hard water has to be soft by zeolite. If the zeolite require 5 lit of NaCl solution (100 g/L of NaCl) for regeneration after softening. Calculate the amount of hardness of water.*

Solution.

1 L NaCl contain = 100 g

5 L NaCl contain = $100 \times 5 = 500$ gm

the amount of substance (W) = 500 gm

equivalent mass of NaCl (E) = 58.5

$$\text{equivalent of } CaCO_3 = \frac{W \times 50}{E}$$

$$= \frac{500 \times 50}{58.5} = 427.35 \text{ g } CaCO_3$$

10,000 L water $= 427.35$ gm $CaCO_3$

$= 427.35 \times 1000$ mg $CaCO_3$

$$1 \text{ L water} = \frac{427.35 \times 1{,}000}{10{,}000} = 42.735 \text{ ppm}$$

9. *A zeolite softener was 90% exhausted by removing the hardness completely when 1,00,000 litres of hard water sample passed through it. The exhausted zeolite bed required 150 lt of 30% NaCl solution for its complete regeneration. Calculate the hardness of water.*

Solution.

150 lt of 30% NaCl solution required = $150 \times 30 = 4{,}500$ g NaCl

$W = 4{,}500$ g

$E = 58.5$

$$\text{equivalent of } CaCO_3 = \frac{W \times 50}{E} = \frac{4{,}500 \times 50}{58.5}$$

$= 3{,}846$ g $CaCO_3$

1,00,000 lit water = 3,846 gm $CaCO_3$ = $3{,}846 \times 1000$ mg $CaCO_3$

90% exhausted by removing the hardness

$$1,00,000 \text{ lit water} = 3,846 \times 1,000 \times \frac{100}{90} \text{ g } CaCO_3$$

$$1 \text{ lit water} = \frac{3,846 \times 1,000 \times 100}{90 \times 1,00,000} = \frac{3,846}{90} = 42.73 \text{ ppm.}$$

EXERCISE

1. Define hardness of water. Write lime-soda process for purification of water.
2. What are the functions of lime and soda in lime-soda process, give equations?
3. What is meant by 'desalination' of sea-water? Name the various methods for desalination of water. ***(K.U.K. Jan. 2006)***
4. Write a short note on break-point chlorination.
5. In what respect hot soda-lime process is superior to the cold lime-soda process?
6. What do you mean by softening of water? Explain zeolite process. How is zeolite process differ from lime soda process?
7. Define the following:

   Chlorination, Dechlorination, Superchlorination and Break-point chlorination.
8. What do you mean by demineralization and desalination? Discuss in details the electrodialysis process for desalination of sea water with the help of neat, cleaned and labelled diagram. What are its advantages and disadvantages? ***(K.U.K. Jan. 2006, June 2009, Jan. 2007, Jan. 2009)***
9. Explain the ion exchange method of purifying the water. Discuss their use and regeneration giving the reactions involved. ***(K.U.K. Jan. 2003, Jan. 2006, June 2006, Jan. 2007, June 2008)***
10. Explain the mixed bed demineralization process.
11. Describe the various processes used for domestic water treatment.
12. What are the characteristics of water? Write a short note on reverse osmosis. ***(K.U.K. 2003)***
13. What do you mean by ion selective membrane? How are exhausted resins regenerated in an ion-exchange process? Give chemical equations.
14. What do you mean by screening, sedimentation and coagulant sedimentations? How are colloidal impurities removed from water?
15. What are the advantages and disadvantages of lime-soda process? What are limitations of the use of bleaching powder as a sterilizing agent?
16. Write the principle of lime-soda process? Why should we use coagulants alongwith lime and soda? Why is water softened by zeolite process unfit for use in boilers?
17. Pure soft water is not fit for drinking purpose, why?
18. (*a*) Explain the electrodialysis process for desalination of brackish water.

    (*b*) Write a short note on Mixed-Bed Demineralization.

    (*c*) What do you understand by scale and sludge? What are their disadvantages?

    (*d*) What are the advantages of hot lime-soda softening over the cold process?

19. What is lime and soda? Why does $Mg(HCO_3)_2$ require double amount of lime for softening of hard water? *(K.U.K. Jan. 2006)*

20. What are requisites of water for drinking purposes? How water is made fit for drinking purposes? *(C.D.L.U. Dec. 2003)*

21. What do you understand by hard and soft water? Explain the zeolite or permutit process of removing the hardness of water. *(C.D.L.U. Dec. 2003)*

22. What do you mean by prechlorination, postchlorination and super chlorination? Write the significance of break point chlorination.

23. What is saline water? Explain the process of reverse osmosis for desalination of water. *(K.U.K. June 2009)*

24. What are various techniques for removing hardness from water? Explain the method which gives best results. *(K.U.K. June 2009)*

25. Give important characteristics of potable water. *(K.U.K. June 2009, Jan. 2004)*

26. Write a short note on break point chlorination. *(K.U.K. June 2009)*

27. Describe lime-soda process for softening of water. What are the limitations of this process. *(K.U.K. Jan. 2009)*

28. Define following

Break point Chlorination, Desalination, Hardness, Sedimentation, Disinfection, post chlorination, super chlorination. *(K.U.K. Jan. 2008)*

29. What are coagulants? Name any three coagulants frequently used in water treatment. *(K.U.K. Jan. 2006)*

30. What is meant by disinfection of water? Describe any one chemical method used for the disinfection of water. Name two microbes which are injurious to us. *(K.U.K. Jan. 2006)*

31. What are natural and synthetic zeolites? Explain the zeolite process for softening of hard water. *(K.U.K. June 2005)*

32. Distinguish between softening and demineralisation of water. *(K.U.K. June 2005)*

33. Differentiate among pure water, hard water, heavy water and boiler feed water. *(K.U.K. June 2005)*

34. Discuss in detail the demineralisation of water using the ion exchange resins. Explain your answer by mentioning the chemical reactions occuring during demineralisation and regeneration process. *(K.U.K. Jan. 2006, Jan. 2007, June 2004)*

35. Write a detailed note on Mixed bed demineralisation. How is spent bed regenerated. *(K.U.K. Jan. 2009, Jan. 2004)*

36. Explain desalination of water by eletrodialysis method. *(K.U.K. Jan. 2009 June 2004)*

37. Name various steps involved in treatment of water for domestic use. Explain break point chlorination. *(K.U.K. Jan 2009)*

38. Discuss the factors affecting sedimentation. What are the advantages of chemically assisted sedimentation over plain sedimentation? *(K.U.K. Jan. 2004)*

39. Give names of various filters. Describe one filter process for softening of water in detail. *(K.U.K. Jan. 2004)*

40. What is osmosis? How is reverse osmosis (R.O) used for desalination of water? Write its applications, advantages and disadvantags. *(K.U.K. Jan. 2004)*

41. Name eight methods for disinfection of water. *(K.U.K. June 2004)*

PROBLEMS FOR PRACTICE

1. A water contains the following:

   H_2SO_4 = 200 mg/l, $MgSO_4$ = 20 mg/l, $CaSO_4$ = 272 mg/l and NaCl = 25mg/l. The water is to be supplied to the town of the population of one lakh only. The daily consumption of water is 50 litre per head. Calculate the cost of lime and soda required for softening the hard water for town for the month April 2002. The cost of lime is Rs. 5.00 per kg and cost of soda is Rs. 8.00 per kg.

   [**Ans.** Lime 24503.25 kg; Soda 66899.25 kg
   Cost of lime Rs. 122516.25 Cost of Soda Rs. 535194]

2. Calculate the amount of lime and soda for 10^5 L water containing HCl = 7.3 mg/L, $Al_2(SO_4)_3$ = 34.2 mg/L, $MgCl_2$ = 9.5 mg/L, NaCl = 29.25 mg/L. The purity of lime is 90% and that of soda is 98%. 10% of chemicals are to be used in excess in order to complete the reaction quickly.

   [**Ans.** Lime 3.168 kg; Soda 4.169] ***(Raj. Univ. 2001)***

3. A water sample contains the following:

   Ca^{2+} = 20 ppm, Mg^{2+} = 25 ppm, CO_2 = 30 ppm, HCO_3^- = 150 ppm, K^+ = 40 ppm. Calculate the amount of lime (80% pure) and Soda (90% pure) required to soften 1 million litres of water sample.

   [**Ans.** Lime 27.31 kg; Soda 3.675 kg]

4. Water sample on analysis gave the following results:

   $Mg(HCO_3)_2$ = 70 mg/L, $CaCl_2$ = 220 mg/L, $MgSO_4$ = 120 mg/L and $Ca(NO_3)_2$ = 164 mg/L. Calculate the quantity of lime (80% pure) and soda (90% pure) needed for softening the 10,000 litres of water.

   [**Ans.** Lime 1.81 kg; Soda 4.68 kg]

5. Calculate the amount of lime and soda required for softening 5000 litres of water sample containing Mg^{2+} 24 ppm, Ca^{2+} 30 ppm and HCO_3^- 122 ppm. [**Ans.** Lime 0.74 kg; Soda 0.398 kg] ***(P.T.U. 2003)***

6. A water sample have following results:

   Mg^{2+} 12 ppm, Ca^{2+} 50 ppm, HCO_3^- 160.8 ppm, CO_2 = 30 ppm. Calculate the amount of lime and soda required for softened of 50,000 litres of water. [**Ans.** Lime 9.25 kg; Soda 2.29 kg]

7. A completely exhausted zeolite softener requires 140 litres of brine solution having 80 gm/l of NaCl for regenerated. How many litres of water having 400 ppm can be softened by the zeolite process.

   [**Ans.** 2.393×10^4 L]

8. 1,00,000 litres of hard water has to be soft by zeolite. If the zeolite require 10 litres of NaCl solution (150 g/L of NaCl) for regeneration after softening. Calculate the amount of hardness of water.

   [**Ans.** 12.82 ppm]

9. The hardness of 1,00,000 litres of sample of water was completely removed by passing it through a zeolite softener. The softener required 400 litres of NaCl solution containing 100 g/L of NaCl for regenerate. Calculate the hardness of water sample. [**Ans.** 341.8 ppm] ***(Dibrugarh June 2000)***

10. The following data were analyzed for a water sample of Brahmputra river.

    CO_2 = 20 ppm, Ca^{2+} = 40 ppm, Mg^{2+} = 50 ppm, HCO_3^- = 366 ppm. Calculate the amount of lime and soda required for softening the 10^6 litres of water, if 139 ppm of $FeSO_4.7H_2O$ is used as coagulant.

    [**Ans.** Lime 446.79 kg; Soda 61.83 kg]

11. Water of Brahmsarovar Kurukshetra on analysis gave the following results. CO_2 = 10 ppm, HCO_3^- = 200 ppm, Mg^{2+} = 40 ppm and Ca^{2+} = 50 ppm. Total solids = 360 ppm. Calculate the amount of lime and soda required for softening the 50,000 litres of water if 400 ppm of $FeSO_4.7H_2O$ is used as coagulant. [**Ans.** Lime 18.4 kg; Soda 14.39 kg]

12. A water sample contains the following: Ca^{2+} = 120 ppm, Mg^{2+} = 120 ppm, CO_2 = 132 ppm, HCO_3^- = 122 ppm, K^+ = 40 ppm. Calculate the amount of lime 80% pure and soda 90% pure for soften 10^6 litres of water sample. [**Ans.** Lime 832.5 kg; Soda 824.4 kg]

13. A water sample containing constituents in ppm $Mg(HCO_3)_2$ = 73, $MgCl_2$ = 95, $MgSO_4$ = 12, $CaSO_4$ = 68, $Ca(HCO_3)_2$ = 81 and NaCl = 4.8. Calculate the cost of chemicals required for softening 20,000 litres of water. If purity factor for lime is 95% and for soda is 90%. The cost per 100 kg each of lime and soda are Rs.75 and Rs.2,480 respectively. [**Ans.** Lime cost Rs. 3.03; Soda cost Rs. 93.44] ***(Nagpur, 1997)***

14. Calculate the amount of lime and soda required for softening of 5000 litres of hard water which contain 72 ppm of $MgSO_4$. [**Ans.** Lime 222 g; Soda 318 g]

15. Calculate the amount of lime and soda needed for softening 10^6 litre water sample which contain Mg^{2+} 36 ppm, Ca^{2+} 20 ppm and HCO_3^{-1} 183 ppm. [**Ans.** Lime 222 kg; Soda 53 kg] ***(P.T.U. 2002)***

16. Calculate the amount of lime (88.3% pure) and soda (99% pure) required to soften, 24,000 litres of water per day for a year containing the following:

$CaCO_3$ = 1.85 ppm $\quad$ $CaSO_4$ = 0.34 ppm $\quad$ $MgCO_3$ = 0.42 ppm

$MgCl_2$ = 0.76 ppm $\quad$ $MgSO_4$ = 0.90 ppm $\quad$ NaCl = 2.34 ppm

SiO_2 = 2.32 ppm. [**Ans.** Lime = 32.3 kg; Soda = 16.88 kg] ***(R.G.T., May 2001)***

17. A water sample contains the following impurities Ca^{2+} 20 ppm, Mg^{2+} 18 ppm, HCO_3^- 183 ppm and SO_4^{2-} 24 ppm. Calculate the amount of lime and soda needed for softening. ***(K.U.K. June 2004)***

[**Ans.** Lime 185 mg/L; Soda zero mg/L]

18. 10^4 litre of hard water was softened by ion exchanger method. After treatment the cationic resin required 200 litre of $\frac{N}{10}$ HCl and anionic resin required 200 litre of $\frac{N}{10}$ NaOH solutions. Calculate the hardness of the water sample in ppm. [**Ans.** 100 ppm]

19. A Zeolite softener was used to remove the hardness of 10^4 litres of hard water. After treatment it was found that the Zeolite softener was 90% exhausted. The exhausted Zeolite bed required 200 litres of 30% NaCl solution for its complete regeneration. Calculate the hardness of water in ppm. [**Ans.** 569.8 ppm]

20. A water sample, on analysis, gave the following results in terms of $CaCO_3$ equivalent.

Total alkalinity = 290 ppm

Calcium hardness = 242 ppm

Magnesium hardness = 63 ppm

Calculate the amount of lime and soda required to soften 10^6 L water.

[**Ans.** Lime 261.22 kg; Soda 15.9 kg]

21. A water sample, on analysis, gave the following results Ca^{2+} 40 mg/L, Mg^{2+} 24 mg/L, Na^+ 8.05 mg/L HCO_3^- 183 mg/L, SO_4^{2-} 55.68 mg/L, Cl^- 6.74 mg/L

Calculate the amount of lime and soda for softening 10^6 L of water. [**Ans.** Lime 185 kg; Soda 53 kg]

22. The analytical report of a water sample gave the following result

$MgCl_2$ 47.5 mg/L. $CaCl_2$ 55.5 mg/L, $CaSO_4$ 4.06 mg/L, Turbidity 120 mg/L.

10 mg/L of alum (alum contains 7% Al) dose was found to be sufficient to remove the entire turbidity of the water sample. Calculate the total weight of dry sediment in a lime-soda softening plant for 50000 litres of water [**Ans.** 11.73740 kg]

23. A water sample, on analysis, gave the following results.

$MgCl_2$ = 95 ppm, $CaSO_4$ = 272 ppm, $MgSO_4$ = 120 ppm, SiO_2 = 4 ppm and H_2SO_4 = 49 ppm. Calculate the amount of lime 95% pure and soda 97% pure for soften 1 million litres of water.

[**Ans.** Lime 194.74 kg; Soda 491.75 kg]

24. A water sample contents the following: $MgCO_3$ = 84 ppm, $CaCO_3$ = 40 ppm, $CaCl_2$ = 5.5 ppm, $Mg(NO_3)_2$ = 37 ppm and KCl = 20 ppm. Calculate the amount of lime 86% pure and soda 83% pure for soften of 80,000 litres of water. [**Ans.** Lime 18.24 kg; Soda 3.05 kg]

25. A water sample shows the following results.

$Ca(HCO_3)_2$ = 40.5 mg/L, $Mg(HCO_3)_2$ = 36.5 mg/L, $MgSO_4$ = 30 mg/L $CaSO_4$ = 34 mg/L, $CaCl_2$ = 27.75 mg/L and NaCl = 10 mg/L.

Calculate the amount of lime 84% pure and soda 92% pure required for treatment of 20,000 litres of water. [**Ans.** Lime 1.76 kg; Soda 1.73 kg)

26. A water sample, on analysis, gave the following data $Ca(HCO_3)_2$ = 8.1 mg/L, $Mg(HCO_3)_2$ = 7.5 mg/L, $CaSO_4$ = 13.6 mg/L, $MgSO_4$ = 12 mg/L, $MgCl_2$ = 2 mg/L and NaCl = 4.7 mg/L.

Calculate the amount of lime 100% pure and soda 100% pure for soften of 1,00000 litres of water.

[**Ans.** Lime 2.026 kg; Soda 2.34 kg]

27. A water sample contains Ca^{2+} = 100 ppm, Mg^{2+} = 72 ppm, CO_2 = 88 ppm, HCO_3^- = 488 ppm and SiO_2 = 0.4 ppm. $FeSO_4 . 7H_2O$ is used as a coagulant at the rate of 139 ppm. Calculate the amount of lime and soda needed for softening 10^6 litres of water. [**Ans.** Lime 703 kg; Soda 212 kg)

28. Calculate the amount of lime and soda for softening 60,000 litres of water containing $Ca(HCO_3)_2$ = 20 mg/L $Mg(HCO_3)_2$ = 25 mg/L, CO_2 = 20 mg/L, HCl = 8.4 mg/L, $Al_2(SO_4)_3$ = 40 mg/L and $MgCl_2$ = 12 mg/L. [**Ans.** Lime 5.93 kg; Soda 2.65 kg]

29. Calculate the amount of lime 87% pure and soda 91% pure for softening 10^6 litres of water sample containing Ca^{2+} = 20 ppm, Mg^{2+} = 25 ppm, CO_2 = 30 ppm, HCO_3^- = 150 ppm and K^+ = 10 ppm.

[**Ans.** Lime 218.9 kg; Soda 33.6 kg]

30. A water sample, on analysis gave the following data Ca^{2+} = 80 ppm, Mg^{2+} = 36 ppm, K^+ = 39 ppm, HCO_3^- = 244 ppm, $FeSO_4 . 7H_2O$ (coagulant) = 69.5 ppm. [**Ans.** Lime 277.5 mg/L; Soda 185.5 mg/L]

31. A water sample contains following:

$Mg(HCO_3)_2$ = 83 mg/L; $CaCl_2$ = 222 mg/L, $MgSO_4$ = 180 mg/L and $Ca(NO_3)_2$ = 164 mg/L. Calculate the amount of lime 90% pure and soda 80% pure needed for softening of 10,000 litres of water.

[**Ans.** Lime 2.17 kg; Soda 5.96 kg] ***(M.D.U. June 2008)***

32. Water sample shows the following data:

Ca^{2+} = 250 mg/L, Mg^{2+} = 100 ppm, HCO_3^- = 300 ppm.

Calculate the amount of lime and soda for softening 10,000 litres of water.

[**Ans.** Lime 2.59 kg; Soda 0.53 kg]

33. Water having the following composition has to be softened by the lime-soda process: $Ca(HCO_3)_2$ = 220 ppm. $Mg(HCO_3)_2$ = 56 ppm, $MgCl_2$ = 130 ppm, $MgSO_4$ = 84 ppm and $CaSO_4$ = 98 ppm.

Calculate the amount of lime and soda ash required to soften 1 million litres of water.

[**Ans.** Lime 3102.82 kg; Soda 2955.81 kg]

34. A water sample contains the following:
Ca^{2+} = 100 ppm, Mg^{2+} = 95 ppm, Na^+ = 15 ppm, HCO_3^- = 160 ppm, SO_4^{2-} = 40 ppm and Cl^- = 10 ppm. Express the results in terms of salts.

35. A water sample gave the following results:
Ca^{2+} = 160 ppm, Mg^{2+} = 72 ppm, CO_2 = 88 ppm and HCO_3^- = 488 ppm. $FeSO_4 \cdot 7H_2O$ is used as a coagulant at the rate of 139 ppm. Calculate the amount of lime and soda required to soften 1,00,000 litres of water.
[**Ans.** Lime 70.3 kg; Soda 37.1 kg]

36. A water sample, using $FeSO_4 \cdot 7H_2O$ as a coagulant at the rate of 278 ppm, gave the following data:
Ca^{2+} = 240 ppm, Mg^{2+} = 96 ppm, CO_2 = 44 ppm and HCO_3^- = 732 ppm. Calculate the lime and soda required to soften 2,50,000 litres of water.
[**Ans.** Lime 272 kg; Soda = 132.5 kg]

37. A zeolite softener was completely exhuasted and was regenerated by passing 100 litres of NaCl solution, containing 120 g/L of NaCl. How many litres of a sample of water of hardness 500 ppm can be softened by this zeolite?
[**Ans.** 20,512 L]

38. The hardness of 1000 litres of water was completely removed by a zeolite softener. Later on, the zeolite softener required 30 litres of NaCl solution for regeneration. The sodium chloride solution contains 15 g/L NaCl. Calculate the hardness of water.
[**Ans.** 384.6 ppm]

39. A zeolite softener was completely exhausted to completely removed of hardness of a water sample having hardness 600 ppm. The exhausted zeolite required 150 litres of NaCl solution containing 150 g/L of NaCl. How many litres of hand water sample was treated by using the zeolite softener?
[**Ans.** 32051 litres]

40. The hardness of 10^4 litres of water sample was completely removed by a zeolite softener. The exhausted zeolite softener requried 30 litres of NaCl solution (1500 mg NaCl/L) for regeneration. Calculate the hardness of the water sample.
[**Ans.** 38.46 ppm]

Chapter 5

CORROSION AND ITS PREVENTION

5.1 INTRODUCTION

It is the process of slowly deterioration and consequent loss of a solid metallic material from the metallic surface due to unwanted attack by the atmospheric gases, soil, chemical or electrochemical environment etc. It is very slow process and starts from the surface of a metal. Its process is just "reverse of extraction of metals."

Causes

It has been found that most metals (exceptions noble metals like Au, Pt, etc.) exist in nature in their combined forms like oxides, carbonates, sulphates etc. In combined form they have lower energy. The pure or isolated metals have high energy so they are thermodynamically unstable. Due to this reason metals have tendency to acquire the thermodynamic stability *i.e.* lower energy. For this, metals easily undergo in interaction with their environment either chemically or electrochemically to form a stable compound by the process of corrosion. Due to corrosion metal has lost its malleability, ductility and electrical conductivity etc. Hence we can simply say that corrosion is an oxidation process in which metallic compound having lower energy is formed and energy liberates.

$$\underset{\text{(Higher energy)}}{\text{Metal}} \underset{\text{Metallurgy (Reduction)}}{\overset{\text{Corrosion (Oxidation)}}{\rightleftharpoons}} \underset{\text{(Lower energy)}}{\text{Metallic Compound + Energy}}$$

Examples:

1. When copper is exposed to the moist air containing carbon dioxide, a green thin film of basic carbonate $[CuCO_3 + Cu(OH)_2]$ on the surface of copper appears.

2. Most familiar and common example is the rusting of iron. When iron is exposed to the atmospheric conditions a reddish brown precipitate of iron (II, III) oxide Fe_3O_4 is formed as a result of

corrosion. Hence the iron becomes weak gradually. It has been seen that the corrosion takes place through an electrochemical reaction.

5.2 MECHANISM OF DRY AND WET CORROSION

5.2.1 Dry Corrosion (Chemical Corrosion)

It is also known as chemical corrosion. Such type of corrosion takes place due to the direct chemical action of atmospheric gases like CO_2, SO_2, O_2, H_2, etc. or anhydrous liquids on the metal surfaces. It is of three types:

(*i*) Oxidation corrosion
(*ii*) Liquid metal corrosion
(*iii*) Corrosion by other gases

(*i*) ***Oxidation corrosion.*** When corrosion takes place by direct action of oxygen on metal, is called oxidation corrosion. It occurs usually in the absence of moisture. Such type of corrosion may occur even at ordinary temperatures. For examples, alkali metals (Li, Na, K etc.) and alkaline earth metals (Be, Mg, Ca, etc.) undergo corrosion by this process.

Mechanism

When a metal is exposed to air it gets oxidized by lossing its valence electrons and reduction of oxygen takes place.

$$M \longrightarrow M^{n+} + ne^- \quad \text{Oxidation of metal}$$

$$\frac{n}{2}[\frac{1}{2} O_2 + 2e^- \longrightarrow O^{2-}] \quad \text{Reduction of oxygen}$$

$$M + \frac{n}{4} O_2 \longrightarrow M^{n+} + O^{2-}$$

At the point of contact of M^{n+} and O^{2-}, metallic oxide M_2O_n is formed. This metallic oxide scale forms a barrier to restrict further oxidation of inside metals. Since the size of cation M^{n+} is much smaller than anion O^{2-} hence M^{n+} will diffuse much faster than the O^{2-} through the scale for continuation of oxidation, it can be possible if the metallic oxide barrier is sufficiently porous. The nature of oxide film plays very important role in oxidation corrosion.

(*a*) When the oxide film is stable, impervious and tightly-adhering, it will act a protective coating and corrosion is further prevented, *e.g.* Al, Pb, Cu, etc.

(*b*) When the oxide film is unstable and has tendency to decompose back to metal and oxygen, it does not undergo in oxidation corrosion. *e.g.* Au, Ag, Pt, etc.

(*c*) When the oxide film is volatile, the inner layer of metal surface again gets exposed to air and further corrosion take place. *e.g.* Mo

$$2Mo + 3O_2 \longrightarrow 2MoO_3$$

(*d*) When the oxide film is sufficiently porous so that the diffusion of cations M^{n+} and anions O^{2-} take place smoothly then oxidation corrosion takes place continuously.

The porous nature of oxide film may be explained by Pilling-Bedworth rule.

Pilling-Bedworth rule

Pilling-Bedworth rule describes the porous and non-porous nature of metallic oxide which is formed during corrosion.

According to this rule the specific volume ratio is calculated as follows:

$$\text{Specific volume ratio} = \frac{\text{Volume of metal oxide}}{\text{Volume of metal}}$$

(*a*) If the specific volume ratio is smaller, then oxidation corrosion will take place because the oxide films will be sufficiently porous for diffusion of M^{n+} and O^{2-}.

(*b*) If volume of the metal oxide ≥ volume of the parent metal, then it will be non-porous.

(*c*) If the volume of the metal oxide < volume of the parent metal, then it will be porous.

(*ii*) ***Liquid metal corrosion.*** When a liquid metal is allowed to flow over solid metal at high temperature is called liquid metal corrosion. Due to this solid metal gets weakened.

For example, in nuclear reactor sodium metal acts as coolant it leads to corrosion of cadmium.

In fact such type of corrosion is due to the following reasons:

(*a*) The dissolution of solid in liquid metal

(*b*) Penetration of liquid metal into solid metal.

(*iii*) ***Corrosion by other gases.*** Some gases like SO_2, Cl_2, H_2S etc. react with certain metals and forms a protective or non-protective layer on metallic surface. Due to the chemically combination of metals with gases, metals undergo corrosion. The extent of corrosion depends upon the following:

(*a*) *Nature of the environment.* The environment plays very important role in corrosion because it facilates the affinity between metal and gases.

(*b*) *Chemical affinity between metal and gas.* If the affinity of metal and gas is more, corrosion will be more and more.

(*c*) *Nature of the film formed on the metal surface.* If the film formed on the metal surface is porous then more and more amount of gas will penetrate inside the metal and the corrosion will increase.

For example, (*a*) The corrosion of silver in presence of chlorine

$$2Ag + Cl_2 \longrightarrow 2AgCl$$

Here the AgCl film is protective so the further corrosion will not take place.

(*b*) The corrosion of tin in presence of chlorine

$$Sn + 2Cl_2 \longrightarrow 2SnCl_4$$

Here $SnCl_4$ film is volatile. Hence the inner layer of tin will expose to chlorine for further corrosion. As a result corrosion is increased.

5.2.2 Wet Corrosion (Electrochemical Corrosion)

It is also known as electrochemical corrosion. Such type of corrosion is due to the flow of electrons from metal surface anodic area towards cathodic area through a conducting solution. It is very common type of corrosion. Such type of corrosion take place in following conditions:

(*i*) When two dissimilar metals or alloys are in contact with eac other in the presence of a conducting medium (aqueous solution, moisture, etc.)

(*ii*) Separate anodic and cathodic areas between which the current flows the conducting medium.

(*iii*) Oxidation takes place at anode and reduction takes place at cathode.

For example, rusting of iron.

Rusting of iron

When an iron piece is exposed to moist air or mixture, a reddish brown layer is formed on the iron surface. This reddish brown layer is called rust and the phenomenon is known as rusting. Actually rust is mainly composed of hydrated ferric oxide ($Fe_2O_3 \cdot xH_2O$) or iron (II, III) oxide (Fe_3O_4). The exact composition of rust depends upon the atmospheric conditions which are responsible for rusting. The layer of rust is porous so the inside layer of iron metal further undergoes corrosion. In this way the process of corrosion of iron metal continues till it is completely oxidised.

Factors involving in rusting

Following are the factors for favouring the rusting of iron:

(*i*) Presence of air

(*ii*) Presence of moisture

(*iii*) Presence of electrolytic impurities in water

(*iv*) Presence of impurities in metals

(*v*) Nature of metal.

5.2.3 Electrochemical Theory of Corrosion (Mechanism of Rusting of Iron)

For simplicity we take an example of rusting of iron. The mechanism of rusting is based on electrochemical theory. Actually here electrochemical cell is formed at the metal surface. The water vapours on the surface of the metal dissolves CO_2 and O_2. CO_2 makes water slightly acidic in nature.

$$H_2O + CO_2 \longrightarrow H_2CO_3$$

Due to acidic nature it helps to ionize the water.

$$H_2CO_3 \rightleftharpoons 2H^+ + CO_3^{2-}$$

and

$$H_2O \rightleftharpoons H^+ + OH^-$$

The surface of iron is generally coated with a thin film of iron oxide. When this iron oxide film develops some cracks, anodic areas are created on the surfaces while the other whole parts of metal act as cathode as in Fig. 5.1. At the anodic areas metal (Fe) dissolves and Fe^{++} is formed.

Oxidation at anode

$$Fe \longrightarrow Fe^{2+} + 2e^- \qquad ...(1)$$

These electrons are consumed by H^+.

$$H^+ + e^- \longrightarrow H$$

These 'H' atoms react with the dissolved oxygen or oxygen in air to form water.

$$4H + O_2 \longrightarrow 2H_2O$$

Hence the reduction takes place at cathode.

$$[H^+ + e^- \longrightarrow H] \times 4$$

$$4H + O_2 \longrightarrow 2H_2O$$

$$\overline{O_2 + 4H^+ + 4e^- \longrightarrow 2H_2O} \qquad ...(2)$$

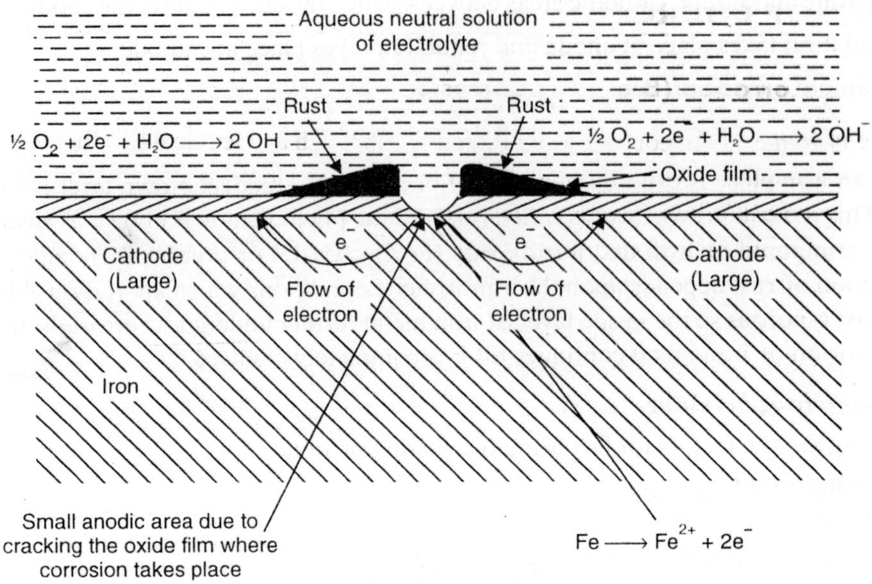

FIGURE 5.1

Therefore the overall reaction is obtained by combining equation (17) and (18) as

$$[Fe(s) \longrightarrow Fe^{2+} + 2e^-] \times 2$$

$$O_2 + 4H^+ + 4e^- \longrightarrow 2H_2O$$

$$2Fe(s) + O_2(g) + 4H^+(aq) \longrightarrow 2Fe^{2+}(aq) + 2H_2O(l)$$

This Fe^{++} react with the dissolved oxygen to form ferric oxide which undergoes hydration to form rust as

$$4Fe^{2+} + O_2 + 4H_2O \longrightarrow 2Fe_2O_3 + 8H^+(aq)$$

$$Fe_2O_3 + xH_2O \longrightarrow \underset{\text{(rust)}}{Fe_2O_3 \,.\, xH_2O}$$

The liberated electrons (e^-) from iron at anode are also consumed by dissolved oxygen as

$$\tfrac{1}{2}O_2 + 2e^- + H_2O \longrightarrow 2OH^- \qquad \text{(Reduction at cathode)}$$

The Fe^{++} from anode and OH^- from cathode diffuses and react to form $Fe(OH)_2$ as precipitate.

$$Fe^{2+} + 2OH^- \longrightarrow Fe(OH)_2\downarrow$$

In excess of oxygen it is converted into ferric hydroxide

$$4Fe(OH)_2 + O_2 + 2H_2O \longrightarrow 4Fe(OH)_3$$

This ferric hydroxide is called yellow rust and actually corresponds to $Fe_2O_3 \,.\, xH_2O$. In limited supply of oxygen the corrosion product may be even black anhydrous magnetite (Fe_3O_4).

Since the rate of diffusion of Fe^{++} is faster than that of OH^-, so corrosion occurs at the anode but rust is deposited at or near the cathode.

5.3 TYPE OF CORROSION

5.3.1 Galvanic Corrosion (Bimetallic Corrosion)

When different metals are in contact either directly or through an electrical conductor and are exposed to a corrosive atmosphere, the metal which has more negative electrode potential undergoes corrosion. Actually different metals form a galvanic cell and such type of corrosion is known as galvanic corrosion. In summary differential metal corrosion is known as galvanic corrosion. Such metal acts as anode where oxidation takes place. For example, Zn-Cu, Zn-Ag, Fe-Cu, etc. In zinc-copper galvanic cell zinc ($E° = -0.76$ V) has more negative electrode potential than copper ($E° = +0.34$ V) hence zinc undergoes corrosion.

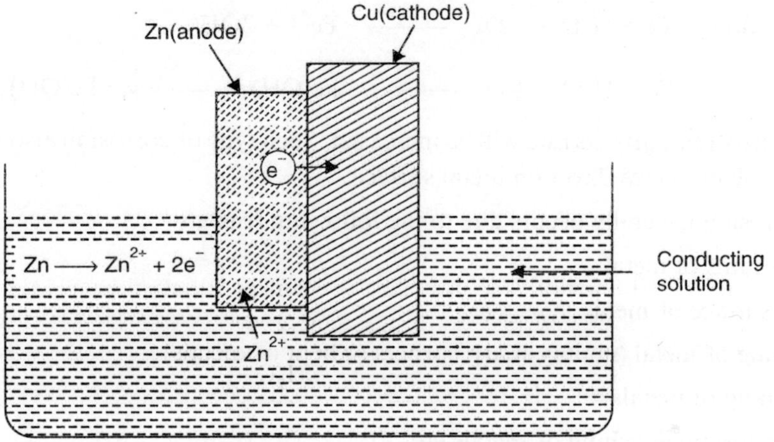

FIGURE 5.2 **Galvanic corrosion**

Mechanism

The galvanic corrosion occurs is due to the difference in the electrode potentials of metals. The metal which has lower standard reduction potential act as anode and undergoes corrosion. The electrons flow from the anode to the cathode and a galvanic cell is set up. As a result the anode undergoes oxidation and forms corresponding ions and passes into solution. The process continues and metal gets corroded gradually.

Control

Since galvanic corrosion depends upon the following factors:

(*i*) Greater the potential difference between two metals, greater is the corrosion.
(*ii*) Suitable medium for corrosion.
(*iii*) Surface area of metal.

Hence the corrosion may be controlled by the following factors:

(*i*) Avoiding the suitable medium for corrosion.
(*ii*) Minimizing the potential difference of metals *i.e.,* avoiding the galvanic couple.
(*iii*) By polishing the metals.

5.3.2 Pitting Corrosion

Pitting corrosion is a non-uniform corrosion occurs due to the attack of localized accelerators and formation of pits, cavities or pin-holes in the metal. Consider a drop of water is present on the surface of metal. The metal surface which is covered by the water droplet has low concentration of oxygen and act as anode where oxidation takes place *i.e.*, it will undergo corrosion. The rest part of metallic surface is uncovered with water droplet so it has high concentration of oxygen and act as cathode where reduction takes place. Hence there is a large cathodic area and small anodic areas develop and forms a cell. Due to the formation of small cell, corrosion starts. The reaction takes place as given follows:

$$Fe \longrightarrow Fe^{2+} + 2e^- \quad \text{(At anode)}$$

$$H_2O + \tfrac{1}{2}O_2 + 2e^- \longrightarrow 2OH^- \quad \text{(At cathode)}$$

$$\text{Overall reaction} \quad Fe + H_2O + \tfrac{1}{2}O_2 \longrightarrow Fe^{2+} + 2OH^-$$

or

$$Fe + H_2O + \tfrac{1}{2}O_2 \longrightarrow Fe(OH)_2 \xrightarrow{[O]} Fe(OH)_3$$

Once the corrosion starts, its rate will be increased. Such type of corrosion also takes place when a slightly break in the protective layer on metal surface.

Pitting corrosion generally occurs due to the following reasons:

(*i*) Local strains of metal

(*ii*) Rough surface of metal

(*iii*) Designing of metal (mainly sharp corners or bent of metals)

(*iv*) Scratching of metals

(*v*) Non-uniformly polishing of metals etc.

It may be prevented by:

(*i*) Proper designing of metal

(*ii*) Proper polishing of metal

(*iii*) Purifying the metals.

5.3.3 Differential Aeration Corrosion (Concentration Cell Corrosion)

Concentration cell corrosion is the type of corrosion which occurs due to the electrochemical attack on the metal surface when a metal is exposed to an electrolyte of varying concentrations or of varying aeration. Differential aeration corrosion is the most common type of concentration cell corrosion and occurs when a portion of metal is exposed to a different air concentration. For example, a zinc rod is partially immersed in a dilute solution of electrolyte NaCl as in Fig. 5.3, a potential difference is developed between differently aerated areas. It causes a flow of current between the two differentially aerated areas of the same metal. The part of metal in contact of air having low oxygen concentration acts as

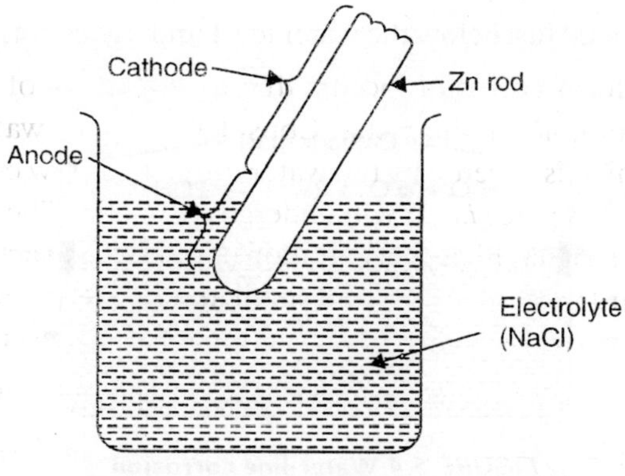

FIGURE 5.3 **Concentration cell corrosion**

anode, whereas the part of metal in contact of solution having high oxygen concentration acts as cathode. The anodic part involves in oxidation and cathodic part in reduction. The reaction proceeds as follows:

$$Zn \longrightarrow Zn^{2+} + 2e^- \quad \text{(At anode)}$$

At cathode oxygen takes up electrons to form OH^-

$$\tfrac{1}{2}O_2 + H_2O + 2e^- \longrightarrow 2OH^- \quad \text{(At cathode)}$$

The overall reaction

$$Zn + \tfrac{1}{2}O_2 + H_2O \longrightarrow Zn^{2+} + 2OH^-$$

Zinc combines with OH^- to form $Zn(OH)_2$.

$$Zn^{2+} + OH^- \longrightarrow Zn(OH)_2$$

Thus, corrosion occurs at anodic part. In a similar way, iron corrodes in water or salt solution.

Important Characteristics

(*i*) The metal having low oxygen concentration part act as anode and metal having high oxygen concentration part act as cathode.

(*ii*) It is promoted by accumulation of dirt sand, scale or other contamination, because such covered part act as anode and involves in oxidation.

(*iii*) It is a localized attack on some oxygen deficient areas, resulting in characteristics localized pitting.

5.3.4 Water-line Corrosion

It is the type of differential aeration corrosion which occurs due to the partly immersed of a metal in water. The corrosion takes place just below the water line and hence it is known as water line corrosion.

When water is stored in a steel or iron tank for a long time, the concentration of dissolved oxygen at the water surface is greater than that under surface. It forms an oxygen concentration cell. The area of low concentration of oxygen acts as anode whereas the area of high concentration of oxygen acts as cathode. Due to the potential difference a current starts flow and anodic part undergoes corrosion. Due to poor conductor water, the ions just below the water level are more readily available for corrosion. That's why the metal just below the water level undergoes corrosion. It may be represented as in Fig. 5.4.

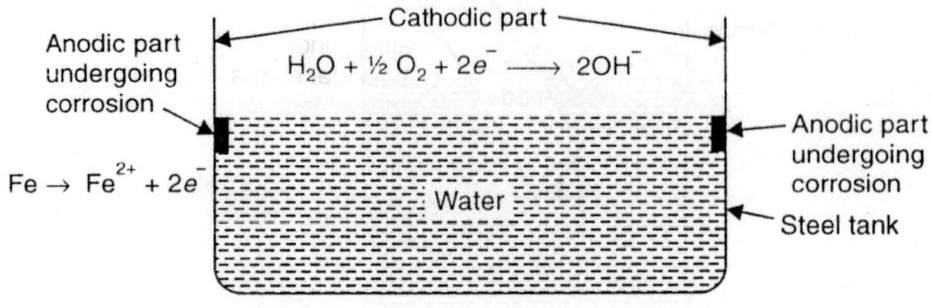

FIGURE 5.4 **Water-line corrosion**

The whole cathodic area is unaffected by corrosion. Oxidation takes place at anodic area and reduction takes place at cathodic area as given below:

At anode $\quad Fe \longrightarrow Fe^{2+} + 2e^-$

At cathode, oxygen takes up the electrons to get OH^-

$$\frac{1}{2}O_2 + H_2O + 2e^- \longrightarrow 2OH^-$$

Fe^{2+} and OH^- get combined each other to form $Fe(OH)_2$ which on further oxidation gives $Fe(OH)_3$.

$$Fe^{2+} + OH^- \longrightarrow Fe(OH)_2 \xrightarrow{[O]} Fe(OH)_3$$

The water-line corrosion is increased by the presence of salts like chloride, bromide etc. present in water but it gets retarded by the presence of anodic inhibitors like carbonates, phosphates, silicates, hydroxides, etc. If the water is relatively free from acidic impurities then the corrosion will take place very less. Water-line corrosion creates much problem to marine engineers. In the case of ships, water-line corrosion is generally accelerated by marine plants attaching themselves to the side of ships. The antifouling paints is used to minimize the such type of corrosion to some extent. It is very important for the long life and safety to ships.

5.3.5 Stress Corrosion (Stress Cracking)

Stress corrosion is the type of corrosion which occurs due to the combined effect of tensile stresses and the corrosive environment on metal when it is exposed to corrosive environment. Pure metal generally does not undergoes stress corrosion whereas fabricated metal components or articles of certain alloys like high zinc brasses and nickel brasses undergoes such type of corrosion.

Favourable Conditions for Stress Corrosion

(*i*) *Tensile stress.* It has been seen that during the manufacturing process (quenching, bending, annealing, welding, etc.) the metals get some uneven stresses which act as anode and undergo corrosion.

(*ii*) *Corrosive environment.* The specific and selective environment play very important role in stress corrosion. For examples, brass undergoes corrosion in presence of ammonia environment whereas stainless steel undergoes corrosion in acid chloride environment.

Mechanism. The mechanism of stress corrosion is also an electrochemical phenomenon. During the manufacturing process like quenching, annealing, etc. the metals get some uneven stresses which act as anode and the rest parts of the metal act as cathode. It may be represented in Fig. 5.5.

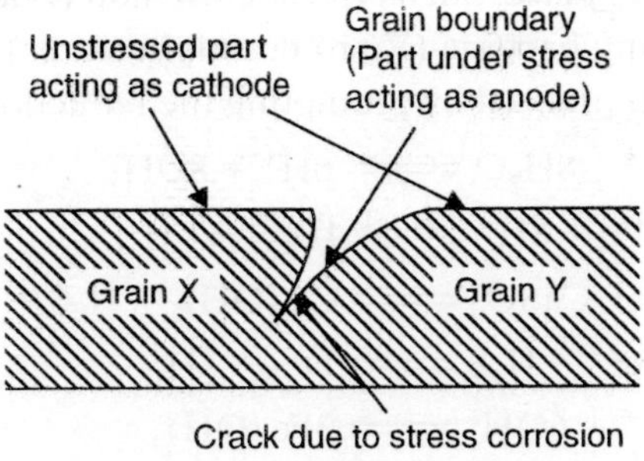

FIGURE 5.5 **Stress corrosion**

So there are number of galvanic cells performed. As a result potential difference is set up and corrosion start.

$$M \longrightarrow M^{n+} + ne^- \quad \text{(At anode)}$$

$$O_2 + 4H^+ (aq) + ne^- \longrightarrow 2H_2O \quad \text{(At cathode)}$$

It has been found that stressed areas in a metal are very reactive and undergoes corrosion even in a mild corrosive environment.

Types of Stress Corrosion

Some important types of stress corrosion are given below:

(*i*) *Season cracking.* It is generally applied to the stress corrosion of copper alloy mainly brass. It occurs mainly in the presence of ammonia or amine. Brass is a binary alloy and is made up of copper and zinc. When brass is exposed in ammonical medium, both copper and zinc form complexes $[Cu(NH_3)_4]^{2+}$ and $[Zn(NH_3)_4]^{2+}$. As a result, the dissolution of brass occurs and form cracks for stress corrosion.

(*ii*) *Caustic embrittlement.* It is covered by using highly alkaline water in high pressure boiler. It has already been discussed in water chapter (See article 3.9 of chapter 3).

5.3.6 Soil Corrosion

Soil corrosion is the type of wet corrosion but it is complex phenomenon in which number of variables involved. Chemical reactions involving almost each of the existing elements are known to take place in soils. Variations in soil properties and characteristics can have a major impact on corrosion. Soils with a high proportion of sand have very limited storage capacity for water whereas clays are excellent for retaining of water. The corrosion in soil generally depends upon its porosity, degree of aeration and electrical resistance. Soils having low electrical resistivity have high corrosivity. The low electrical resistivity depends upon the moisture contents and dissolved electrolytes which promote the rate of corrosion. At greater depths, the rate of corrosion appears to depend on the diffusion of dissolved oxygen in soil water and sometimes on sulphate reducing bacteria. Underground corrosion is also very important because oil, gas and water are transported through cast iron pipelines burried underground. The corrosion of cast iron is believed to be due to (*i*) electrochemical action in which the ferrite of the cast iron acts as anode and graphite acts as cathode (*ii*) Stray-current corrosion (The point at which the current leaves act as anode where corrosion takes place. Such type of corrosion is called stray-current corrosion.) and (*iii*) to anaerobic sulphate reducing bacteria (Vibro de sulphuricans). It is believed that the anaerobic microbiological corrosion of iron proceed by promoting the reduction of H^+ at the cathode as follow

$$8H_2O \rightleftharpoons 8H^+ + 8OH^-$$
$$4Fe + 8H^+ \rightleftharpoons 4Fe^{2+} + 8H$$
$$8H + CaSO_4 \rightleftharpoons Ca(OH)_2 + H_2S + 2H_2O$$
$$Fe^{2+} + H_2S \rightleftharpoons FeS + 2H^+$$
$$3Fe^{2+} + 6OH^- \rightleftharpoons 3Fe(OH)_2$$

Since the products FeS and Fe $(OH)_2$ do not protect the surface of iron, so localised corrosion occurs. It has been observed that microbiological corrosion is minimum in the aerated soil. In the upper layer of soil or at earth surface the corrosion takes place in presence of water and oxygen as

$$M \longrightarrow M^{n+} + ne^-$$
$$H_2O + \frac{1}{2}O_2 + ne^- \longrightarrow OH^-$$
$$\overline{M + H_2O + \frac{1}{2}O_2 \longrightarrow M(OH)_x}$$

5.3.7 Microbial Corrosion

It is the type of corrosion which takes place microbiologically. It refers that corrosion is influenced by the presence and activities of microorganisms and/or their metabolites. Metabolites are the products produced in their metabolism. Bacteria, fungi and other microorganisms can play a major part in soil corrosion. Spectacularly rapid corrosion failures have been observed in soil due to microbial action and is becoming increasingly apparent that most metallic alloys are susceptible to some form of microbial influenced corrosion. Sulphate reducing bacteria plays very important role in the corrosion. It has been noticed that anaerobic microbial corrosion of iron proceed by promoting the reduction of H^+ at the cathode as follow.

$$8H_2O \rightleftharpoons 8H^+ + 8OH^-$$
$$4Fe + 8H^+ \rightleftharpoons 4Fe^{2+} + 8H$$
$$8H + CaSO_4 \rightleftharpoons Ca(OH)_2 + H_2S + 2H_2O$$
$$Fe^{2+} + H_2S \rightleftharpoons FeS + 2H^+$$
$$3Fe^{2+} + 6OH^- \rightleftharpoons 3Fe(OH)_2$$

The corrosion products are black due to iron sulphide and ferrous hydroxide. Sulphate reducing bacterias (Sporovobiro desulphuricous) are responsible for such anaerobic corrosion of iron and steel. Their activity is maximum at 20-30°C and the range of pH 6–9.

Sulphur bacterias (thioracillus) are responsible for aerobic corrosion. They oxidise the sulphur present in the cell organelle to sulphuric acid which is sufficient for the corrosion of iron. Their activities are optimum in acid medium *i.e.,* pH < 7.

Iron and manganese microorganisms are aerobic and digest the iron and manganese ions in cells to form insoluble hydrates of iron and manganese dioxide. These are excreted by the cells. Iron bacterias grow in stagnant or running water below 40°C and pH 4-10, with a small amount of free dissolved oxygen. Film-forming microorganisms like bacterias, fungi, algae etc form a very thin layer of microbiological film on the iron surface. These films are capable of maintaining concentration gradients of dissolved salts, acids and gases at the surface of iron. Due to this local biological concentration cells, corrosion takes place

5.4 FACTORS AFFECTING CORROSION

The rate and extent of corrosion of a metal depends upon the following factors:

1. Nature of the metal and
2. Nature of the environment.

5.4.1 Nature of the Metal

(*i*) *Purity of the metal.* Impurities present in metal or alloys enhance the rate of corrosion. Impurities present in the metal or alloys form minute electrochemical cells under suitable environment conditions and undergo corrosion.

(*ii*) *Position in galvanic series.* The metal or alloy which are placed at higher up in the series are more reactive and has greater tendency to undergo corrosion.

(*iii*) *Physical state of the metal.* The physical state (orientation of crystals, grain size, stress, etc.) of a metal is very important to deciding the rate of corrosion. As the grain size is small the rate of corrosion increases. This is because solubility of the metal is increased by increasing the size of grains.

(*iv*) *Relative areas of the cathodic and anodic part.* The rate of corrosion increases with increase in the ratio of the areas of cathodic and anodic part

i.e.,
$$\text{rate of corrosion} \propto \frac{\text{Cathodic area}}{\text{Anodic area}}$$

It has been observed that corrosion takes place more rapidly in case of smaller anodic area. This is because the current density at a smaller anodic area is much greater.

(*v*) *Over voltage.* Over voltage is the difference between the voltage which is actually required for occurring the electrode reaction and expected theoretical value. Since an anodic areas having a smaller over voltage so it corrodes much faster than cathodic areas.

(*vi*) *Nature of the oxide film.* In actual practice it has been noticed that all metals undergo corrosion (oxidation) upto different extent in the aerated atmosphere. All metals get covered by a very thin film of oxide. The ratio of the volumes of the metal oxide to the metal is known as a 'specific volume ratio' or 'Pilling-Bedworth ratio' and is very important to deciding the rate of corrosion.

(*vii*) *Reactivity of metal.* Passive metals like Ni, Co, Tl, Cr etc. do not undergo corrosion frequently. They form a highly protective and very thin layer of oxide on the surface of metal or alloy. These oxide layers are 'self-healing' in nature. It means they have tendency to repair by itself during any cracking, etc.

(*viii*) *Solubility of corrosion products.* If corrosion product is soluble in the corroding medium then the inner layer of metal comes in contact of corroding medium and corrosion of metal takes place rapidly. Similarly if the corrosion product is insoluble in the corroding medium then these products act as a protective layer. As a result corrosion of metal occurs very slowly.

(*ix*) *Volatility of corrosion product.* In case of volatile corrosion product metal undergoes corrosion very rapidly. This is due to the fact that volatile product is formed, it escapes and the inner layer of metal surface exposed for further attack. As a result corrosion takes place very rapidly and continuously.

5.4.2 Nature of the Corroding Environment

(*i*) *Effect of the temperature.* At higher temperature the rate of corrosion increases.

(*ii*) *Effect of pH.* It has been seen that the corrosion takes place more in acidic media ($pH < 7$) than alkaline media ($pH > 7$) and neutral media ($pH = 7$). Amphoteric metals like Al, Zn and Pb have tendency to form complexes in alkaline media and undergo solution hence these type of metals corrodes in alkaline media.

(*iii*) *Effect of corroding medium.* It is also very important factor for deciding the rate of corrosion. The conductive corroding medium increases the rate of corrosion. Dry sandy soil has very less conductance so they are not suitable for corrosion.

(*iv*) *Effect of concentration of oxygen.* The rate of corrosion increases in presence of moisture and oxygen. In dry oxygen the rate of corrosion decreases. Differential aeration set up a number of concentration cells. Due to this electrochemical phenomenon the rate of corrosion gets increased.

(*v*) *Effect of moisture.* Moisture or humidity of air is an excellent medium for corrosion. Atmospheric moisture acts as a very good solvent for oxygen, other gases, salts, etc. and forms electrochemical cell. This increases rapidly the rate of corrosion.

(*vi*) *Effect of suspended particles in atmospheres.* Suspended active particles like NaCl, $(NH_4)_2SO_4$ and other salts present in atomosphere absorb moisture rapidly due to hygroscopic nature and act as strong electrolyte to enhance the rate of corrosion. Suspended inactive particles like charcoal etc. also gradually enhance the rate of corrosion because they can absorb moisture and sulphur gases and provide a suitable medium for corrosion.

(*vii*) *Effect of the nature of the presence of electrolyte.* The presence of electrolyte in the corroding environment play important role of corrosion. For examples, anions like Cl^- destroy the protective film on the metal surface and the inner surface is exposed for further corrosion. Hence corrosion takes place rapidly. On the other hand, anions like SiO_3^{2-} forms an insoluble layer on metal surface and prevents from further corrosion. Similarly cation like NH_4^+ increases the rate of corrosion.

(*viii*) *Effect of atmospheric gases present in air.* The gases like CO_2, SO_2, H_2S, etc. present in the atmosphere or fumes of acids HNO_3, H_2SO_4, etc. forms the medium more acidic above the metal surface because these gases are soluble in water to form acids. Acids have more conducting power hence it enhances the rate of corrosion.

5.5 PREVENTIVE MEASURES OF CORROSION (CORROSION CONTROL)

There are so many ways to control the corrosion. Some methods are explained as:

5.5.1 Using Pure Metal

Impurities always enhance the rate of corrosion because it causes heterogeneity. After purification of metals (Al, Fe, etc.) a coherent and impervious protective oxide film layer occurs on the metal surface and corrosion is controlled as a result.

5.5.2 Using Metal Alloys

Corrosion resistance of most metals may be increased by alloying them with suitable elements like Cr, Ni, etc. in a homogeneous medium.

5.5.3 Proper Designing

The metallic apparatus must be proper designed to minimize the corrosion.

(*i*) Avoid the contact of dissimilar metals in the presence of a corroding solution.

(*ii*) When two dissimilar metals are to be in a contact, the anodic material should be in a large area as much as possible for minimum corrosion.

(*iii*) If two dissimilar metals in contact to be used, they should be as close as possible to each other in the electrochemical series.

(*iv*) An insulating fitting may be applied for joining two dissimilar metals. Never join direct metal-metal electrical contact.

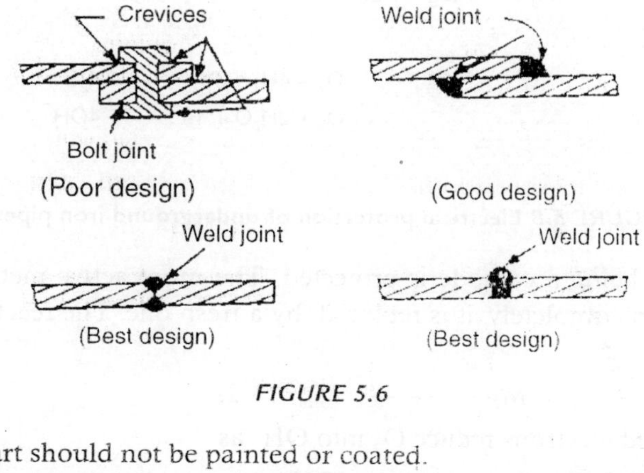

FIGURE 5.6

(*v*) The anodic part should not be painted or coated.

(*vi*) Always prevent the occurrence of inhomogeneities in metal and in the corrosive environment. Since crevices permit concentration differences. Hence bolts and rivets should preferably be replaced by a butt-weld as in Fig. 5.6.

(*vii*) Sharp corners and recesses in apparatus should be avoided by proper designing as in Fig. 5.7.

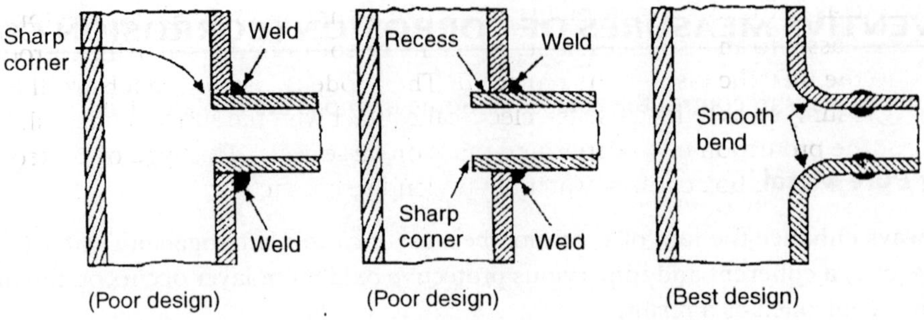

FIGURE 5.7

5.5.4 Cathodic Protection (Electrical Protection)

In this method we use the metal as cathode so the corrosion does not occur. These are of two types:

(*i*) *Sacrificial anodic protection method (Galvanic protection).* This method is useful for the protection of underground pipes, tanks, etc. In this method the more active metal like Mg, etc is used as anode and the connection is made as in Fig. 5.8.

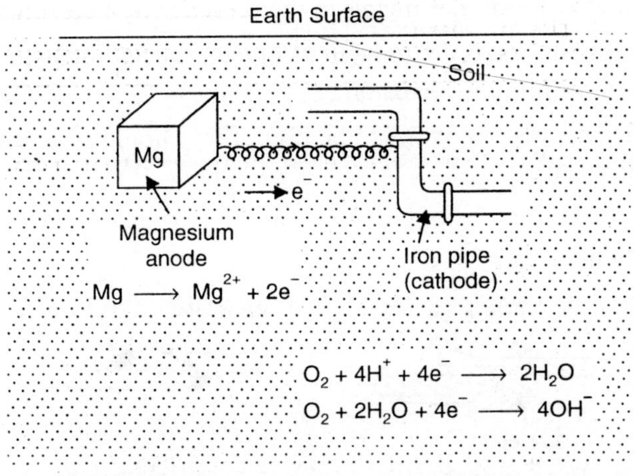

FIGURE 5.8 **Electrical protection of underground iron pipes**

The parent metal being a cathode is protected. The more active metal gets corroded slowly. When anode is consumed completely, it is replaced by a fresh one. The reaction takes place as given below:

At anode, $Mg \longrightarrow Mg^{2+} + 2e^-$

At cathode released electrons reduce O_2 into OH^- as:

$$O_2 + 2H_2O + 4e^- \longrightarrow 4OH^-$$

Thus cathode (iron etc.) gets protected. Since the reactive metals (Mg, etc.) sacrify itself during the protection of other metal (cathode) hence it is termed as sacrificial anode protection.

(*ii*) *Impressed current cathodic protection.* An impressed current from DC source is applied in opposite direction to nullify the corrosion current and convert the corroding metal from anode to cathode. It may be represented as in Fig. 5.9. An insoluble anode like platinum, silica, etc. is used for this purpose. A direct current is passes to an insoluble anode, buried in the soil or immersed in the corroding medium and connected to the metallic lustre to be protected. The anode is, usually, in a backfill (composed of coke, breeze or gypsum) so as to increase the electrical contact with the surrounding soil. This method is very useful for the protection of underground pipes on large scale. This type of protection has also been applied to open water box coolers, water tanks, condensers, etc.

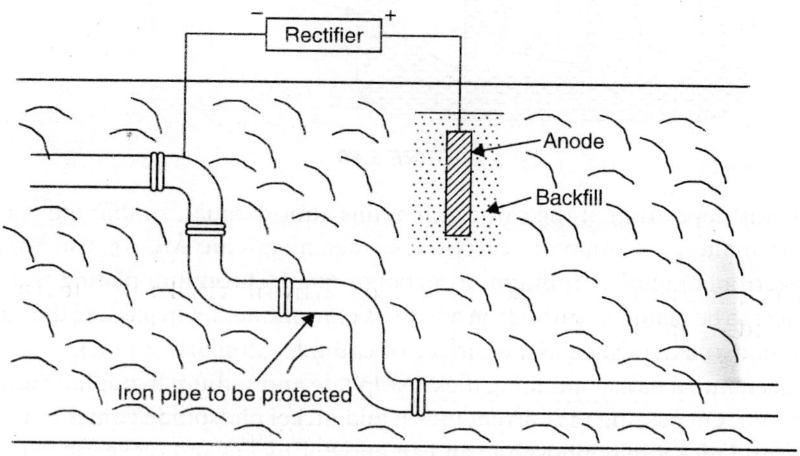

FIGURE 5.9 **Cathodic protection by impressed current method**

5.5.5 Protective Coatings

It is very common and the simplest method for prevention of metals from corrosion. It is also known as barrier protection.

(*a*) *By surface coatings.* The metal surface may be coated with paints enamels, layers, oil, grease or non corroding metals like Ni, Cr, etc. by electroplating for preventing the metal surface in contact with moisture, oxygen and CO_2. In this way corrosion may be controlled.

(*b*) *Using antirust solutions.* Alkaline phosphates and alkaline chromates prevent the metal from rusting because they deposit an insoluble protective layer on metal.

(*c*) *Anodic coatings.* The metal which is used for coating is more anodic than the metal which is to be protected, for example, the coating of Zn or Al on steel surface prevents the steel from corrosion.

5.5.6 Electroplating

It is the most widely used method of coating metals. It is the method for prevention of metals from corrosion. The process is carried out in an electrolytic cell represented in Fig. 5.10.

The metal to be plated is made the cathode in a plating both containing the ions of metal to be deposited. Anodes are usually, either the metals to be deposited or an inert material of good electrical conductivity like graphite. Generally, conditions are so maintained that the rate of dissolution at the

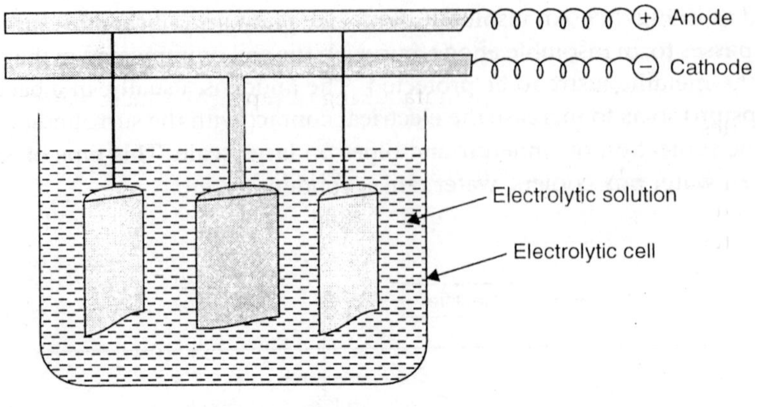

FIGURE 5.10

anode and the rate of deposition at the cathode remains almost equal, so that the composition of the electrolyte remains unaltered. Commonly electrodeposited metals are Au, Ag, Cu, Sn, Zn, Ni, etc. Iron is protected by electroplating of chromium or nickel or zinc. Chromium plating is done using a lead anode and the object to be plated as cathode in a bath of chromic acid. On passing the current chromium goes from the solution and deposited at the surface of cathode. Similarly for nickel plating, the metal to be plated is immersed into a bath containing nickel sulphate and sodium hypophosphite at about 100°C and pH range 4.5–5.0. On passing the current nickel and nickel phosphide comes out from the solution of nickel sulphate and nickel hypophosphite and an alloy of nickel and nickel phosphide is plated as a strong adherent thin film at the surface of metal. This method has a great importance in industries for the prevention of metals part in machines from corrosion.

5.5.7 Galvanization and Tinning

It is the method of coating by metals and alloys having low melting point like Zn, Sn, Pb etc. on the surface of metals to be coated. The metal to be coated is placed in a bath containing the molten metal. The thickness of the coating is adjusted by squeezing out the excess of coating metal by using hot rollers. When iron sheet is dipped into a molten zinc or tin solution, a thin layer of zinc or tin is coated as a thin adhere on the surface of iron sheet. Since zinc or tin is more anodic than iron hence iron sheets get prevented from corrosion. Coating of zinc on iron by this process is called galvanising and tin on iron surface is called tinning,

Process of Galvanization

Metal to be galvanised is first pickled for about 15 minutes in a pickling solution consisting of 5–10% H_2SO_4 at 60°C. Now it is treated with 5% HF to remove the impurities on the surface of metal like steel. It is washed with water and then passed through 5–20% zinc-ammonium chloride solution, act as cleaning solution. The above method is called preliminary treatment. After that steel is passed through melt zinc metal tank at about 450°C. Finally, it is annealed at 650°C and then slowly cooled. Actually zinc combines with iron (steel) to form an alloy.

Applications

(*i*) It is used for protecting the metals especially iron from corrosion.

(*ii*) Galvanised iron in used in the manufacturing of pipe, wire, sheet ware, wire cloth, nails etc. in industries.

(*iii*) It is also used in aircrafts, machines, roofing sheet etc.

Note: It should be noted that zinc coated utensils cannot be used for preparing and storage of food, especially, acidic in nature. It may be poisoned.

Process of Tinning

Tinning is very similar to galvanising. Now-a-days tinning process is very popular. First of all iron sheets are descaled by pickling in 4–8% H_2SO_4 at 60°C for about 5 minutes as a preliminary treatment. Now it is passed through a tinning bath (tank).

Description of Tinning Bath

A line diagram of tinning bath may be represented as in Fig. 5.11.

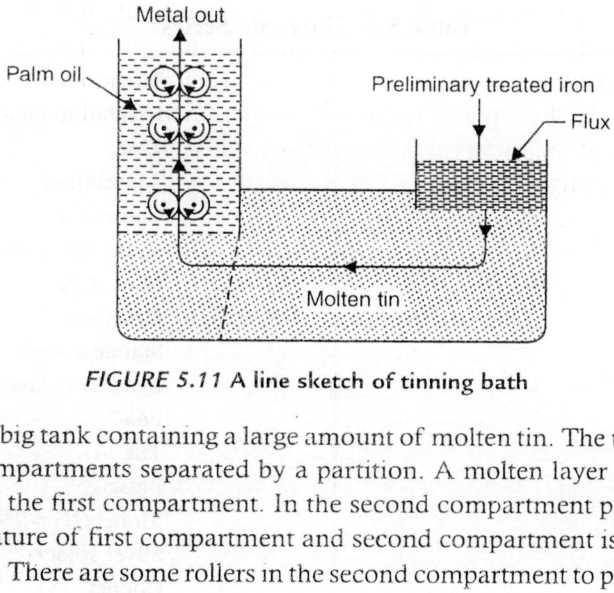

FIGURE 5.11 **A line sketch of tinning bath**

Tinning bath is a big tank containing a large amount of molten tin. The upper part of the tinning bath consists of two compartments separated by a partition. A molten layer of flux of zinc chloride floats over molten tin in the first compartment. In the second compartment palm oil is floated on the molten tin. The temperature of first compartment and second compartment is kept at 300–350°C and 230°–250°C respectively. There are some rollers in the second compartment to pass out the metal outside from the chamber.

Working

Preliminary treated iron is allowed to pass through the flux layer of $ZnCl_2$ and then through the molten tin. After that the metal goes to the palm oil compartment. The oil keeps tin as molten state and protects the tin layer from oxidation during the solidification of tin in air. The tinned metal comes out from the palm oil chamber with the help of rollers. Palm oil is removed from the surface of tinned metal by adsorbing it in bran, saw dust or similar materials.

Applications

(*i*) It is used in the manufacture of cans used for good packing.
(*ii*) Tinned utensils can be used for preparing food.
(*iii*) Tinned copper wire facilitate soldering.
(*iv*) Tinned copper does not react with sulphur, so it is used in the rubber insulation in industries.

5.6 GALVANIC SERIES

Electrochemical series is very helpful to understand the standard reduction potential. According to this series a metal placed at high in the series is more anodic and undergoes corrosion rapidly than the metal below in the series. For example, zinc is placed at higher level than iron, so zinc corrodes faster than iron and so on. It has been found that some metals in the electrochemical series do not follow this trend. For example, the position of titanium (Ti) is higher than silver (Ag) but titanium is less reactive towards corrosion. Similarly aluminium (Al) is above zinc (Zn) but zinc corrodes faster. Hence, a new series came into exist which is based on relative oxidation potentials in sea water. This series is known as Galvanic series and may be represented as in Table 5.1.

Table 5.1. Galvanic Series

More anodic (Active) ↑	Magnesium
	Magnesium alloys
	Zinc
	Aluminium
	Cadmium
	Aluminium alloys
	Mild steel
	Cast iron
	Stainless steel
	Lead-Tin alloy (Solder)
	Lead
	Tin
	Brass
	Monel (Ni = 7%, Cu = 30%, Rest Fe)
	Silver solder
	Copper
	Nickel
	Bronze
	Copper-Nickel alloys
	Silver
	Chromium stainless steel
	Graphite
	Titanium
	Gold
More cathodic (Noble)	Platinum

According to this series the metal or alloy higher up the position in the series is more anodic and undergoes corrosion very rapidly. For example, the position of zinc is higher than aluminium hence zinc undergoes corrosion rapidly not aluminium.

EXERCISE

1. Define corrosion. Explain the mechanism of electrochemical corrosion. *(K.U.K. June 2009, Jan 2009)*
2. Explain the mechanism of dry corrosion.
3. Write a short note on following:
   (*i*) Pilling-Bedworth rule
   (*ii*) Electroplating of metal.
4. Define corrosion of metals. Explain the electrochemical theory of wet corrosion. *(N.I.T.K. 2003)*
5. Explain the terms Galvanisation and Tinning? How galvanisation is differ from tinning? *(K.U.K June 2009)*
6. Define following:
   (*i*) Wet corrosion (*ii*) Tinning
   (*iii*) Galvanic corrosion (*iv*) Microbial corrosion.
7. Write short notes on the following:
   (*i*) Water line corrosion (*ii*) Galvanic series
   (*iii*) Rusting of iron (*iv*) Stress corrosion. *(K.U.K June 2009, June 2008)*
8. What is corrosion of metals? What are the factors which affect corrosion? How much rust ($Fe_2O_3 . 3H_2O$) will be formed, when 100 kg of iron have completely rusted away? *(U.P. Tech. Jan. 2001)*
   [**Hint.** Mol. wt. of $Fe_2O_3 . 3H_2O = 214$ gm
   one molecule $Fe_2O_3 . 3H_2O$ contains 112 g iron *i.e.*, 2 atoms of Fe
   Since 0.112 kg iron produces 214 g of rust
   $\therefore$ 100 kg iron produces $\frac{214}{0.112} \times 100$ g of rust.]
9. How dry corrosion is differ from wet corrosion?
10. What do you mean by chemical corrosion (Dry corrosion). How is dry corrosion differ from wet corrosion?
11. Explain dry corrosion in details. Why is it known as chemical corrosion?
12. What do you mean by Galvanic corrosion? What are the factors which influences the galvanic corrosion? How is it controlled?
13. What is Pitting corrosion? Write the reactions involve in Pitting corrosion. How is it prevented?
14. What is concentration cell corrosion? Write the reactions involve in differential aeration corrosion. What are its important characteristics?
15. Explain water line corrosion in details. How water-line corrosion creates problem to marine engineers? How is it minimized?
16. What are the factors affecting corrosion? How is it prevented? *(K.U.K. June 2006, C.D.L.U. Dec. 2003, K.U.K. Jan 2007)*
17. What are the effects of temperature, pH, over voltage and reactivity of metals influences the corrosion?

18. Write short notes on the following:
    (*a*) Sacrificial anodic protection (Galvanic protection)
    (*b*) Impressed current cathodic protection
    (*c*) Galvanization
    (*d*) Galvanic series.
19. How nature of the corroding envioronment affects the corrosion? Explain Pilling-Bedworth rule for corrosion.
20. Write short notes on the following:
    (*a*) Soil corrosion
    (*b*) Microbial corrosion
    (*c*) Stress cracking **(*K.U.K. Jan. 2006, Jan. 2007*)**
    (*d*) Electrochemical corrosion
    (*e*) Electrochemical theory of corrosion
    (*f*) Caustic embrittlement.
21. Discuss the role of nature of oxide formed in oxidation corrosion. State and explain Pilling-Bedworth rule. **(*K.U.K. June 2004*)**
22. (*a*) Describe the mechanism of electrochemical corrosion. **(*K.U.K. June 2004*)**
    (*b*) Write short notes on the following:
    (*i*) Pitting corrosion **(*K.U.K. Jan. 2009, Jan. 2004*)**
    (*ii*) Role of sacrificial anode in corrosion control. **(*K.U.K. June 2004*)**
    (*iii*) Water line corrosion **(*K.U.K. Jan. 2007, Jan. 2009*)**
23. (*a*) Giving two examples of each, explain the difference between Galvanic corrosion and concentration cell corrosion. **(*K.U.K. Jan. 2004*)**
    (*b*) Discuss the factors influencing corrosion. **(*K.U.K. Jan. 2004, June 2008, Jan. 2009, June 2006*)**
    (*c*) Explain how can corrosion be controlled by proper designing? **(*K.U.K. Jan. 2004*)**
    (*d*) How are the metals protected against corrosion by modifying the environment? **(*K.U.K. Jan. 2006*)**
24. Write short notes on the following: **(*C.D.L.U. Dec. 2003*)**
    (*i*) Wet corrosion
    (*ii*) Dry corrosion
    (*iii*) Pitting corrosion. **(*K.U.K. Jan. 2007, Jan. 2009*)**
25. What is meant by corrosion? How does it differ from errosion? **(*K.U.K. Jan. 2005, Jan. 2007, June 2008*)**
26. Explain the mechanism of hydrogen evolution and oxygen absorption in electrochemical corrosion. Illustrate your answer with figures. **(*K.U.K. Jan. 2005, Jan. 2006, June 2006*)**
27. What happens and why?
    (*i*) Iron sheets riveted with copper rivets.
    (*ii*) An iron pole is partly burried under earth.
    (*iii*) Zinc plate fixed below the ship. **(*K.U.K. Jan. 2005*)**
28. How rate of corrosion is affected by: Position of metal in galvanic series, nature of oxide film and ratio of cathodic to anodic area? **(*K.U.K. June 2009*)**
29. What is electrochemical corrosion? With the help of rusting of iron, discuss the electrochemical theory of corrosion. **(*K.U.K. Jan. 2009*)**

30. How material selection and design can prevent corrosion? Which protective coatings can be used for corrosion prevention? *(K.U.K. Jan. 2008)*
31. What is meant by corrosion inhibitors? Give two examples. *(K.U.K. Jan. 2008, Jan. 2006)*
32. Metal under water drop undergoes accelerated corrosion, why? *(K.U.K. Jan. 2008)*
33. What do you mean by galvanic series? How is it differ from electrochemical series? Discuss the role of galvanic series in corrosion. *(K.U.K. Jan. 2006)*
34. Define galvanisation. How is cathodic protection different from galvanisation? *(K.U.K. June 2006)*
35. What do you understand by microbial corrosion? How is it different from other type of corrosion?
36. What is soil corrosion? How sulphate reducing bacteria affect the rate of corrosion in soil?
37. Define microbial corrosion. Explain the mechanism of microbial corrosion. Discuss the role of sulphur bacteria and other microorganisms in the microbial corrosion.

Chapter 6

LUBRICATION AND LUBRICANTS

6.1 INTRODUCTION

Lubricants are those type of substances which are used to reduce the frictional forces between two surfaces contact each other. It does not allow the direct contact between two rubbing surfaces because it reduces the co-efficient of friction between two surfaces. The process of decreasing the frictional forces between the surfaces is called 'lubrication'. Lubricants play very important role in machines, tools and many apparatus. It is very useful to non-living as well as living things. We (human beings) also use oil, creams etc. for maintenance the smoothness of our skins. For examples, grease, oil, vaseline, cream, etc. are widely used as lubricants.

Functions

Following are the important functions of lubricants:

(*i*) It reduces the frictional forces between two rubbing surfaces.

(*ii*) It reduces the cost of maintenance of the machines and tools.

(*iii*) It reduces the loss of heat energy produced by frictional forces between two surfaces. Hence it acts as a coolant.

(*iv*) It enhances the efficiency of machine by reducing the frictional forces.

(*v*) It also acts as a seal. In internal combustion engine (ICE) it is used as a seal between piston and the cylinder wall and prevents the leakage of gases under high pressure from the cylinder. In chemical laboratory it is used in burettes, vacuum pumps, condensers, rotovapours, etc.

6.2 FRICTION

When two surfaces are come close to each other during motion, a resisting force comes into existence, which tend to retard their motion. The resisting force is known as frictional force and the phenomenon is termed as friction. It is a common phenomenon and occurs in all type of machines as well as living things. When we rub our palms then frictional force appears. It has been observed that frictional force always appear in all type of matter how smooth they are. In actual practice smooth surfaces are not smooth in the real sense. Microscopic studies prevails that a number of peaks and valleys are formed on the metal surfaces. Peaks and valleys are termed as asperities. Due to frictional force wearing and tearing always occur. When two metallic surfaces are close to each other the real area of contact is always smaller than the apparent area. This is due to the formation of asperities. As a result the material loss of their some moving parts which is termed as wear. Hence wear is the progressive loss of substance from the surface of a body during motion due to frictional force.

Classical Laws of Friction

It states that:

(*i*) The frictional force is directly proportional to the applied load on the surface.

Exceptions. Very hard and very soft materials may not obey this law.

(*ii*) The frictional force is independent of apparent area of contact. It is totally depend upon the real area of contact. Real area of contact is always smaller than the apparent area of contact between two surfaces.

(*iii*) The coefficient of friction (frictional force) depends upon the nature of materials.

(*iv*) Static coefficient is always greater than the kinetic coefficient.

(*v*) The frictional force is independent of the sliding speed of surface up to an appreciable extent.

(*vi*) The coefficient of friction is directly proportional to mean shearing strength of the contact and inversely proportional to yield pressure of softer metal.

There are a number of exceptions have been noticed by experience against classical law.

6.3 PRINCIPLE (THEORY) AND MECHANISM

It has been noticed that when two metallic surfaces are contact each other in motion then a frictional force is developed. Due to frictional force wearing and tearing of soft metal takes place hence a number of peaks and valleys are formed on the surface of metal. Peaks and valleys are known as 'asperities'. Due to these asperities the efficiency of machine is decreased. When lubricants are used between the metallic surfaces, this hinders the formation of asperities and reduces the frictional force and minimizes the wear and tear of metal. When two sliding surfaces are in contact, the real area of contact (A) is smaller than apparent area and is represented as:

$$A = \frac{W}{P_m} \quad \Rightarrow \quad W = AP_m \qquad \ldots(1)$$

where W is the load applied and P_m is the yield pressure of softer metal.

For metal-metal contact, it may be represented as:

$$F = AS_m \quad \text{...(2)}$$

where F = force required to cause motion and S_m = mean shearing strength of the contacts *i.e.*, metal junctions.

The coefficient of friction (f) is given by

$$f = \frac{F}{W} = \frac{S_m}{P_m} \quad \text{...(3)}$$

Equation (3) indicates that coefficient of friction (f) is independent of the real area of contact and is determined by shearing strength of a solid lubricant and the yield pressure. Coefficient of friction (f) should be as small as possible for effective lubrication and it may be reduced by lowering the shear strength or increasing the yield pressure. It has been found that solid lubricants have very low shear strength and cause low coefficient of friction. Generally three type of principles and mechanism of lubricants have been proposed.

1. Fluid film lubrication (Hydrodynamic lubrication)
2. Boundary lubrication (Thin film lubrication)
3. Extreme-pressure lubrication

1. Fluid Film Lubrication

It is known as thick film lubrication or hydrodynamic lubrication. It is done with liquid lubricants. It may be represented as in Fig. 6.1. The moving surfaces are separated from each other by a thick film of fluid (~ 1000 Å thickness) so that the direct contact of two surfaces of metals or junctions may be minimized. Since thick film of lubricants (~ 1000 Å thickness) are used, hence it is also known as thick-film lubrication. As a result wearing and tearing of metals get minimized due to less friction. In such type of cases the coefficient of friction is very low (0.001 to 0.03). Hydrodynamic friction occurs in the case of a shaft running at a fair speed as well as in well-lubricated bearing with not very high load. In a journal bearing, the film of lubricating oil covers the irregularities of shaft and the bearing surfaces and do not allow to come in contact the metallic surfaces to each other as in Fig. 6.2. Such type of lubrication is useful in delicate and light machines like watches, clocks, guns, sewing machines, scientific equipments, etc.

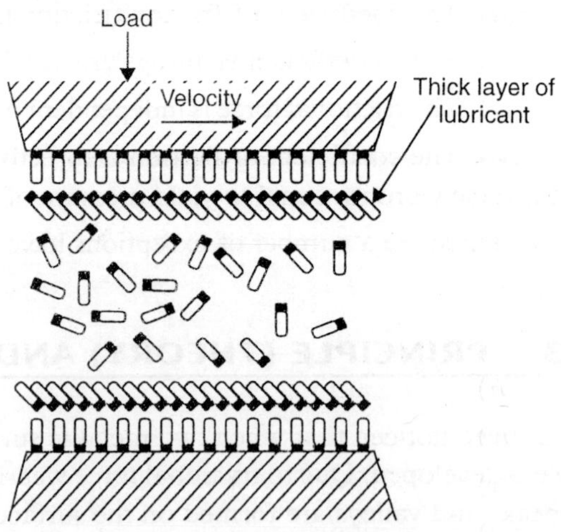

FIGURE 6.1 **Fluid film lubrication**

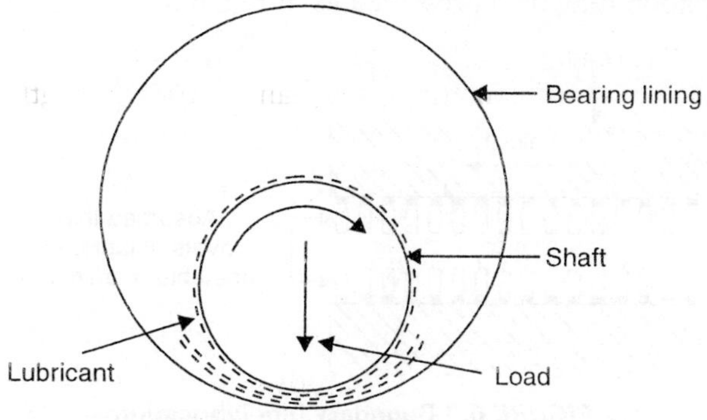

FIGURE 6.2

2. Boundary Lubrication

It is also known as thin film lubrication because in this lubrication the thickness of lubricating film may not exceed one or two molecular layers. Boundary lubrication is necessary when fluid film lubrication fails to maintain the lubrication. This happens due to the following reasons:

(*i*) Shaft comes into motion (action) from rest.

(*ii*) The load is very high.

(*iii*) The viscosity of oil is very low.

(*iv*) The speed is very slow.

At this stage the lubrication is maintained by boundary film lubrication (thickness less than 1000 Å). It may be possible that some of the asperities (peaks and valleys) in metallic surface may be higher than the film thickness so wearing and tearing take place. Hence it is necessary that a thin layer of oil be adsorbed by physical or chemical forces on some metal surfaces, which would avoid the direct contact of metals. *The property by virtue of oil sticks to the surface of machine parts even at high temperature and heavy loads is called oiliness.* Due to oiliness the coefficient of friction becomes very low—0.05 to 0.15. For boundary lubrication, the lubricant molecules should have:

(*i*) long chain hydrocarbon.

(*ii*) polar groups to promote spreading and orientation over the metallic surfaces at high pressure.

(*iii*) lateral attraction between the chains.

(*iv*) active groups or atoms for forming chemical linkages with the metals or other surfaces.

(*v*) high viscosity index.

(*vi*) good oiliness.

(*vii*) low pour point.

(*viii*) resistance to heat and oxidation.

Such type of lubrication may be represented as in Fig. 6.3.

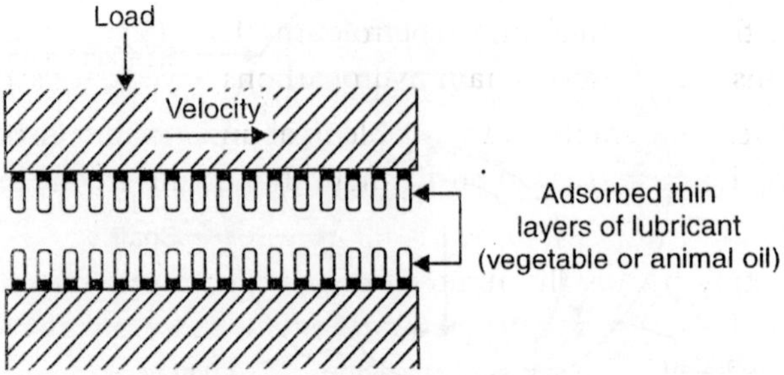

FIGURE 6.3 **Boundary film lubrication**

On the basis of above characteristics the following types of lubricants are useful:

(*i*) Mineral oils, which are thermally stable, mixed with fatty acids or fatty oils.

(*ii*) Solid lubricants graphite or molybdenum disulphide as emulsion in oil.

(*iii*) Vegetable and animal oils having greater oiliness than mineral oils.

3. Extreme Pressure Lubrication

When the moving surfaces are working under very high pressure and high temperature the ordinary liquid lubricants either vapourize or decompose. In such cases extreme pressure lubrication is done. For this special additives (extreme pressure additives) are used along with the liquid lubricants. Chlorinated esters, sulphurized oils and tricresyl phosphates are examples of extreme pressure additives. These additive (compounds) combine with the metallic surfaces at high temperature to form metallic chlorides, sulphides or phosphides in the form of durable film. These film can withstand very high loads and high temperature due to their high melting points. Extreme pressure lubricants have great advantages. They are used as lubricants in wire drawing, machining of tough metals, etc.

Comparison between fluid film lubrication and boundary lubrication. Following are the differences between fluid film lubrication and boundary lubrication:

Sr. No.	*Fluid Film Lubrication*	*Boundary Lubrication*
1.	The thickness of lubricating oil film is more than 1000 Å.	The thickness of lubricating oil film is less than 1000 Å.
2.	These are used as such no metallic surfaces required for adsorption.	The thin film of lubricating oil is adsorbed by physical or chemical forces at the metallic surfaces.
3.	Lubricants have less viscosity.	Lubricants have high oiliness.
4.	The load applied is sufficient to keep apart the moving surfaces. Hence it is known as hydrodynamic lubrication.	The load applied is carried by the layers of adsorbed lubricants.
5.	These are used in the machines working under light load and high speed.	These are used in the machines working under heavy load and low speed.
6.	For example, watches, clocks, gums, sewing machines and scientific instruments etc. require such type of lubrication.	For example, rollers, gears, tractors, railway track joints, etc. require such type of lubrication.

Purification of Petroleum or Mineral Oils

These are obtained by fractional distillation of petroleum. The length of hydrocarbon chain varies between 12–50 carbon atoms. Less carbon chain hydrocarbons have less viscosity.

These are inexpensive, quite stable and available in abundance but also contain lot of impurities like wax, asphalt etc. Hence, it is necessary to purify them. Following are the processes for purifying the mineral and petroleum oil.

(*i*) *Dewaxing.* Dewaxing means the removal of wax from oil. Wax rises the pour point and makes it unfit for lubrication. For dewaxing the oil is mixed with a solvent like propane, trichloroethylene, mixture of benzene and ethylenedichloride etc. and then refrigerated. The wax is precipitated which is removed from the oil by passing the oil-wax suspension through a continuous filter or centrifuge (1700 rpm). The solvent present in the oil is then recovered by distillation.

(*ii*) *Acid refining.* Acid refining means removal of impurities by acids from oil. The dewaxed oil contains a number of undesirable constituents like naphthenic and asphaltic impurities. The dewaxed oil is treated with conc. H_2SO_4 and then agitated. Some of the unwanted impurities dissolve in acid while others are converted into tarry sludge. The sludge is removed by filtration. The filtrate is neutralized with a calculated amount of NaOH to remove excess of acid. Finally the oil is decolourized by passing through fuller's earth maintained at 100–140°C.

(*iii*) *Solvent refining.* The oil is mixed with a suitable solvent like furfural, dichloroethylether, nitrobenzene, mixture of propane and cresol etc. Oil is immiscible with these solvents but the undesirable impurities like naphthenic, asphaltic and resinous components are highly soluble. Now it is allowed to stand for sometimes. The liquid separates into two layers:

(*a*) *Oil layers.* It is free from impurities but containing some solvent.

(*b*) *Solvent layer.* It contains all dissolved impurities.

The oil layer is separated out and distilled. The refined oil is left behind. Similarly solvent layer is distilled separately for removing the solvent.

Merits of solvent refining

(*i*) It is cheaper.

(*ii*) It gives higher yield of refined oil.

(*iii*) It removes asphaltic materials satisfactorily.

(*iv*) It gives a refined product which shows less change in viscosity with temperature *i.e.*, high viscosity index.

Demerits

(*i*) The refined product has less resistant to oxidation due to the removal of natural oxidation inhibitors.

(*ii*) The product has less oiliness so it is necessary to mix some additives.

Note. No single oil is used as lubricant for many of the modern machines. Hence some oils having typical properties are mixed to improve the properties of lubricating oil. The resultant oil is called blended oil.

6.4 CLASSIFICATION OF LUBRICANTS

According to the state of lubricants these are classified into the following classes:

1. Solid lubricants,
2. Semi-solid lubricants, and
3. Liquid lubricants.

1. Solid Lubricants

Lubricants which exist in solid form are called solid lubricants, for example, graphite, molybdenum disulphide, soap stone, wax, mica, chalk, talk etc. These are used in heavy machinery which operate under very heavy load and low speed. Recently some stearates and palmitates of aluminium and mineral acid have also been employed as solid lubricants. Graphite is more expensive solid lubricants but it is widely used because it is non-inflammable and resists oxidation even at higher temperature. The solid lubricants are used either as such or suspended form in oil, grease or water.

Conditions for using solid lubricants

(*a*) In such type of machines where semisolid or liquid lubricants are not suitable.

(*b*) In the heavy machines which are operated under very heavy loads and at low speeds.

(*c*) When the contamination of lubricating oil or grease due to the presence of dust or grit particles etc. is noticed.

(*d*) When semisolid or liquid lubricants are highly combustible.

2. Semi-solid Lubricants

These are neither solid nor liquid. They exist in the gel form like paste or grease. They consist of a soap dispersed throughout a liquid lubricating oil may contain specific additives for specific purposes.

These type of lubricants have higher frictional resistance than oils and therefore can support much heavier loads at lower speeds. For examples, greases, vaseline, waxes, creams, etc.

Conditions for using semi-solid lubricants

(*a*) When it is necessary to seal the bearing or joint against the dirty and dust particles or moisture.

(*b*) When the machine is worked at low speed under high load.

(*c*) When the contamination of lubricating oil is unacceptable and harmful for products.

(*d*) When the lubricating oil is not suitable for machines.

The main function of soap in semi-solid lubricants acts as thickening agent so that the lubricant (grease) sticks firmly to the metal surfaces.

3. Liquid Lubricants

They exists in liquid form and reduce friction or wear between two moving metallic surfaces by providing a continuous film in between them.

These are also very important and are widely used in several machines, tools etc. because they act as: (*a*) sealing agent (*b*) corrosion preventer and (*c*) cooling medium. A good lubricating oil must possess the following properties:

(*i*) Low freezing point,
(*ii*) Thermal stability,
(*iii*) Sufficient viscosity,
(*iv*) High boiling point,
(*v*) Low pressure,
(*vi*) High oxidation resistance, and
(*vii*) Safe storage and handling, etc.

For example, Animal oil, vegetable oil, mineral oil, emulsion, etc.

Conditions for using liquid lubricants

(*i*) Where solid or semisolid lubricants are not suitable.

(*ii*) In light machines like watches, clocks, sewing machines, etc.

(*iii*) Where the machines are operated at low speed under ordinary load.

(*iv*) In such type of machines whereas less amount of heat is produced during motion so that the produced heat is insufficient to ignite or burn the lubricant.

Liquid lubricants are of the following types:

(*i*) Animal oils and vegetable oils,
(*ii*) Mineral oils (Petroleum oil),
(*iii*) Blended oils (compounded oil), and
(*iv*) Emulsions.

(*i*) ***Animal oil and vegetable oils.*** Animal and vegetable oils are very important having good oiliness and obtained from the animal and vegetable kingdom contain glycerides of higher fatty acids. Before developing the petroleum industries these were widely used. For example,

Animal oils—lard oil, whale oil, seal oil, etc.

Vegetable oils—palm oil, mustard oil, olive oil, cotton-seed oil, etc.

Although animal and vegetable oils have good oiliness property yet they have following disadvantages.

(*a*) They are costly and undergo oxidation easily.

(*b*) They hydrolyse in aqueous medium or moist air.

Hence their properties may be improved by adding some chemicals. They are used as blending agents to improve the oiliness.

(*ii*) ***Mineral oils (Petroleum oil).*** These are obtained by fractional distillation of petroleum. Although they have poor oiliness but they are widely used as lubricants because they are cheap, stable and available in abundance. The length of hydrocarbon chain varies from 12 to 50 carbon atoms. The oil containing higher carbon is more viscous than the oil containing lower carbon. The oiliness is improved by the addition of high molecular weight compounds like oleic acid, stearic acid, etc.

(*iii*) ***Blended oils (Compounded oil).*** Since a number of oils have poor oiliness and viscosity hence no single oil is used as lubricant for many of the modern machines. So some specific additives are incorporated into the oils to improve their typical properties. The resultant oil is called blended oil. Some additives are listed below:

(*a*) *Oiliness carrier*—coconut oil, caster oil and fatty acids.

(*b*) *Antioxidants (Inhibitors)*—Aromatic, phenolic or amino compounds.

(*c*) *Antifoaming agents*—Glycols and glycerol.

(*d*) *Emulsifiers*—sodium salts of sulphuric acid.

(*e*) *Thickeners*—polystyrene, polyesters, etc.

(*f*) *Viscosity index improvers*—hexanol and other high molecular weight compounds.

(*g*) *Extreme pressure additives*—fatty esters or fatty acids, organic compounds containing sulphur or phosphorus or chlorine, etc.

(*h*) *Abrasion indicators*—tricresylphosphate.

(*i*) *Corrosion preventers*—organic compound of phosphorous or antimony.

(*iv*) **Emulsions.** An emulsion is a two phase system of two immiscible liquids in which one liquid act as the dispersion medium and the other as dispersed phase. For example, a mixture of oil and water. Emulsions are prepared by vigorously mixing the two immiscible liquids in the presence of a stabilizing substance known as emulsifier or emulsifying agent in ultrasonic vibrators. Emulsions are important lubricant which are used in several machines like boring, milling etc. Emulsions are of two types:

(*a*) *Oil in water type (cutting emulsion or cutting oils).* Oil in water type emulsions are prepared by mixing the oil with sufficient amount of water in the presence of water soluble emulsifying agent like alkyl sulphate or alkyl/aryl sulphonate etc. Sometimes glycol, glycerol, etc. are also added. Since oil present in the emulsion act as lubricant whereas water acts as the coolant hence it is known as coolant cum lubricants. They are used generally in diesel motor piston, boring machines, milling machines, internal combustion engines, etc.

(*b*) *Water in oil type (cooling liquids).* Water in oil type emulsions are prepared by mixing the water in sufficient amount of oil in the presence of water-insoluble emulsifying agents like calcium metal soaps. In such type of emulsion water acts as dispersal phase and oil acts as dispersion medium. Sometimes certain additives are mixed in emulsions for specific applications. For example, sulphonated additives are mixed with emulsions for cutting oils. An emulsion of water and oil (1: 1) is used for lubrication of steam cylinders to keep the wall cool with less oil consumption.

Q. 1. Why graphite is preferred as solid lubricant?

Ans. Graphite is widely used as solid lubricants due to the following reasons:

(*i*) It is soapy in touch.

(*ii*) It is non-inflammable.

(*iii*) It is not oxidized in air up to 375°C.

(*iv*) It is used either in powdered form or in suspension form.

(*v*) When graphite is dispersed in oil it is called 'oildag' and when it is dispered in water it is called 'aquadag'. Oildag is useful in internal combustion engine and aquadag is useful in foodstuffs industry.

(*vi*) It has unique layer structure like in Fig. 6.4. Each carbon atom is sp^2 hybridysed and covalently attached to three neighbouring carbon atoms lying in the same plane. As a result planar hexagonal rings are formed. The bond length of C-C in ring is 1.42 Å. These rings constitute a number of layers (sheets) of atoms. The layers or sheets are held together by weak Van der Waals forces and are separated by a large distance of 3.4 Å. Due to weak Van der Waals force it is soft and smooth. Hence it is useful as lubricant.

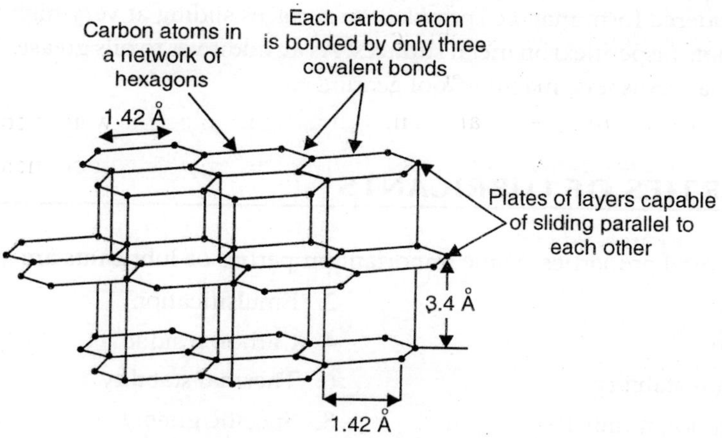

FIGURE 6.4 **Layered structure of graphite**

Q. 2. Why Molybdenum disulphide is used as solid lubricant?

Ans. Molybdenum disulphide (MoS_2) is used as solid lubricants due to the following reasons:

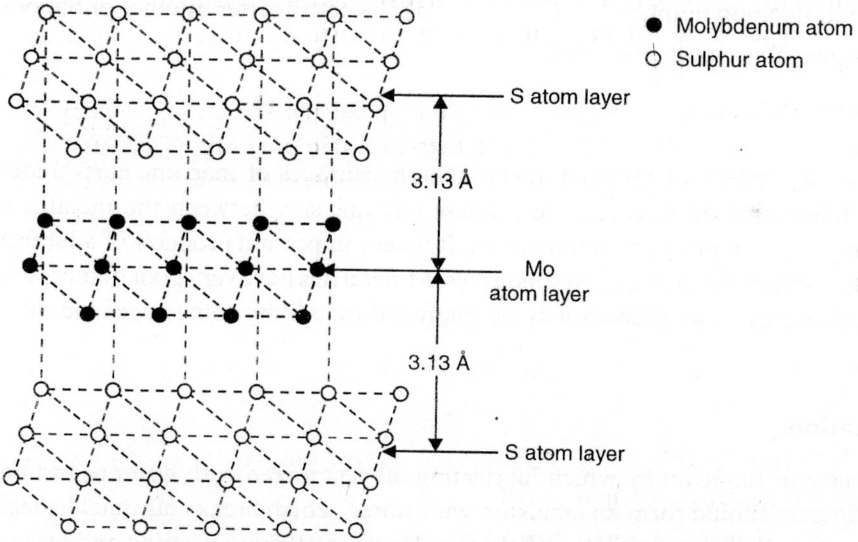

FIGURE 6.5 **Sandwich like structure of molybdenum disulphide**

(*i*) It has sandwich like structure as in Fig. 6.5. The layer of molybdenum atoms lie between two layer of sulphur atoms. The two layers of sulphur atoms are separated by a distance of 6.26 Å and the distance between molybdenum layer and sulphur layer is 3.13 Å. The layers are held together by very weak Van der Waals forces. Due to Van der Waals forces, it is soft and smooth in nature.

(*ii*) It is stable in air upto 400°C.

(*iii*) It possesses very low coefficient of friction.

(*iv*) It has a higher specific gravity than graphite.

(*v*) It is used as either in powdered form or in additives.

(*vi*) The powdered form may be sprinkled on surfaces sliding at very high velocities. It forms a film and fills low spots (asperities) on metal surfaces. As an additives to oils greases it is used as lubricant in shaft bearing of jaw crushers, machine tool gearing etc.

6.5 PROPERTIES OF LUBRICANTS

Lubricants have several properties. Some important properties of lubricants are given below:

1. Oiliness
2. Emulsification
3. Volatility
4. Carbon residue
5. Corrosion stability
6. Thermal stability
7. Precipitation number
8. Specific gravity
9. Mechanical stability
10. Colour and fluorescence
11. Flash and fire point
12. Viscosity
13. Viscosity index
14. Ash content
15. Cloud point and pour point
16. Aniline point
17. Neutralization number
18. Koettsdoerfer number or Saponification value
19. Iodine value (Iodine number).

1. Oiliness

It is a measure of capacity of lubricant to stick on the surfaces of machine parts under high pressure and load. A lubricating oil having good oiliness remains stay between the metallic surfaces of the machine even at high pressure and temperature. It is very important property of a lubricant particularly under boundary or thin film lubrication conditions. Mineral oils have very poor oiliness whereas vegetable oils have good oiliness. The oiliness may be improved by adding some vegetable oils or higher fatty acids.

2. Emulsification

It is the property of lubricant by which lubricating oil gets mixed with water to form an emulsion. A good lubricating oil should form an emulsion with water, which breaks off quickly because emulsions have tendency to collect dirt, grit etc. and protect the machine from wearing and tearing.

3. Volatility

It is the important property of lubricating oil by which oil evaporates at high temperature leaving behind a residual oil. A good lubricant should have low volatility. The volatility of a lubricant is determined by vapourimeter, as in Fig. 6.6.

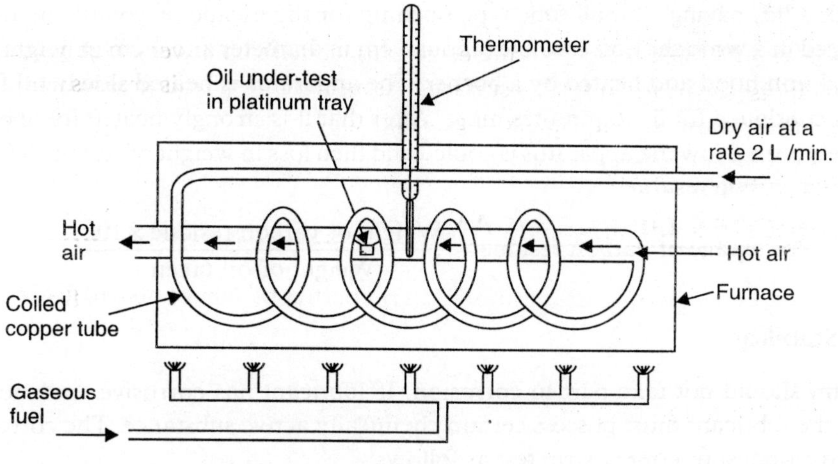

FIGURE 6.6 **Vapourimeter**

It consists of a furnace heated by some fuel gas. There is coiled form of copper tube at the centre of the furnace for passing dry air. A known amount of testing oil is taken in a weighted platinum tray, which is introduced into the copper tube. Dry air is passed through the copper tube for an hour at the rate of 2 litres/minute. After heating 1 hour the tray is cooled and weighted. The percentage loss in weight gives the result of volatility.

$$\text{Volatility of lubricating oil} = \frac{\text{Loss in weight of oil} \times 100}{\text{Weight of experimental oil taken}}$$

4. Carbon Residue

A good lubricant should deposit least amount of the carbon in use. The carbon residue is estimated by 'Conradson Method'. The apparatus is as shown in Fig. 6.7. A known weight of experimental oil is taken in a silica crucible which is placed in another Skidmore iron crucible. The Skidmore iron crucible

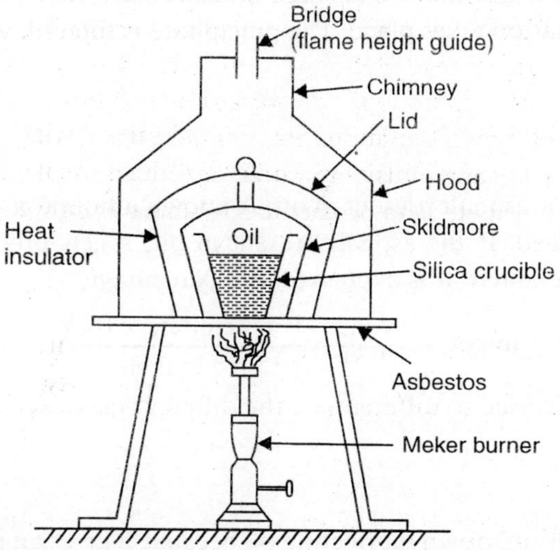

FIGURE 6.7 **Conradson's apparatus for carbon residue estimation**

is provided with a lid, having a small tube type opening for the escape of volatile matter. The whole assemble is placed in a wrought iron crucible (about 8 cm in diameter and 6 cm in height) covered with chimney-shaped iron hood and heated by a burner. The apparatus is heated slowly till flame appears. Slow heating is continued for 5–10 minutes more. After that it is strongly heated for about 15 minutes till the completely burn. Now the apparatus is cooled and then loss in weight of residue left is determined as a percentage of carbon residue.

$$\text{Percentage of carbon residue} = \frac{\text{Weight of carbon residue} \times 100}{\text{Weight of oil taken}}.$$

5. Corrosion Stability

A good lubricant should not take part in corrosion. If lubricant has corrosive tendency towards the metal it means the lubricant must possess certain chemically active substance. The corrosive tendency is tested by corrosion test or copper strip test as follows:

A polished copper strip is placed in the experimental lubricating oil at a specified temperature for a specified time. After a specified time the copper strip is taken out and examined. If the strip is tarnished, it shows that oil contains any chemically active substance. The corrosive tendency of oil may be reduced by adding some corrosive inhibitors like organic compound containing phosphorous, arsenic, antimony, etc.

6. Thermal Stability

A good lubricant must be stable towards heat. It should not be decompose during operation at high temperature. Lubricating oils are mainly decomposed due to oxidation, hydrolysis by moisture and pyrolysis due to heat. Oxidation stability is tested by slight-oxidation test method. A known amount of oil is taken in a specific flask and air is removed inside the flask by oxygen. The flask is then heated in an oil-bath maintained at 200°C. After $2\frac{1}{2}$ hours of heating the flask is cooled and the contents are diluted with petroleum naphtha and allowed to stand undisturbed. After 1–2 hours if any precipitate is performed then it means oxidation takes place. The precipitate is filtered, washed and weighted and is expressed as a percentage.

7. Precipitation Number

It determines the percentage of asphalt present in oil. A known amount of the lubricant is dissolved in petroleum ether and centrifuged. If the asphalt present in oil, it gets precipitated. The precipitate is filtered, washed, dried and weighted. It is expressed as a percentage.

$$\text{Precipitation number} = \frac{\text{Weight of precipitate} \times 100}{\text{Weight of oil taken}}$$

Precipitation number is used to differentiate the different classes of the lubricants.

8. Specific Gravity

It is helpful in identification of unknown lubricating oil because it gives an idea of the type of crude oil from which it has been obtained. It is expressed in terms of API (American Petroleum Institute) degree.

$$\text{API gravity} = \frac{141.5}{\text{Specific gravity at } 60° \text{ F}} - 131.5$$

(141.5 is the modules of the API scale)

The apparatus is known as API hydrometer. The API gravity of water at 60°F is 10. API hydrometer starts at 10 which is equivalent to 1. Lubricants lighter than water have the values lesser than 10. Heavier naphthalenic-base mineral oils have lower API gravity than those of lighter paraffin base oils.

9. Mechanical Stability

A good lubricant must be great stable under very high pressure during different mechanical tests. It may be tested by 'four ball extreme pressure lubricant test' as in Fig. 6.8. The experimental oil or lubricant is poured in a machine containing four balls. The lower three balls are stationary while the upper ball is rotated. The load is gradually increased and the balls are examined at specific intervals. If the balls come out clean, the lubricant is satisfactory under the load and if the balls are welded together due to the formation of heat by increasing the load then the lubricant is not satisfactory. It is also a good and important property of a lubricant.

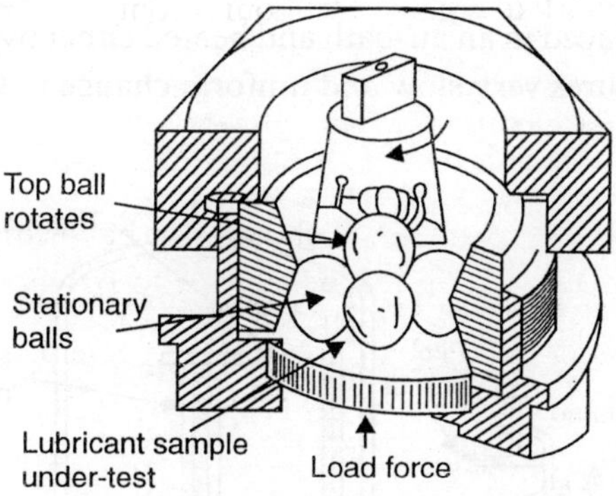

FIGURE 6.8 **Four-balls extreme-pressure lubricant tester**

10. Colour and Fluorescence

Each type of lubricant has some specific colour or even colourless. Colour is very helpful to identify the lubricant. If the lubricant has not its own characteristics colour then it must contain impurities or contamination *i.e.*, the lubricant is not fit for lubrication. The colour of lubrication is checked against a reference (standard) lubricant. Although it is not a sure test but is applicable widely. Similarly fluorescence of a lubricant is also helpful to find out adulteration or contamination in lubricant. Fluorescence of a lubricant is measured by a standard fluorometer.

11. Flash and Fire Point

Each type of lubricant has its own flash and fire point. If the lubricant contains contamination then flash and fire point is affected.

Flash point. It is the lowest temperature of lubricant at which the lubricant gives off sufficient vapour that ignite for a moment, when a flame is brought near it.

Fire point. It is the lowest temperature of lubricant at which the lubricant gives off sufficient vapour that ignite (burn) continuously for atleast five seconds, when a flame is brought near it.

Spontaneous Ignition Temperature (SIT) of a liquid oil is the temperature at which ignition occurs without introducing of a flame, when an inflammable liquid is allowed to fall in drops into a hot metal crucible.

Generally the fire point is 5–10°C higher than the flash point for any lubricant. These properties are very important for using the lubricants in machines or industries. *A good lubricant should have flash point atleast above the temperature at which it is to be used otherwise fire will catch in industries.* The flash and fire point is determined by Pensky-Marten's apparatus as shown in Fig. 6.9.

It consists of following parts:

(*a*) *Oil cup.* It is a metallic cup having diameter 5 cm and depth 5.5 cm fitted with a lid. The lid is provided with opening holes of standard sizes. A thermometer is introduced into one hole. The test flame is introduced by another hole. Similarly other holes are used to passing air and introducing stirrer in it. There is a mark inside the cup at the top to fill the lubricating oil.

(*b*) *Shutter.* It is provided at the top of the cup. It acts just like liver mechanism. Shutter may be opened or closed with the help of a handle known as shutter controller. The holes on lid are opened for bringing the flame over the oil surface with the help of a shutter.

(*c*) *Air bath.* An oil cup is placed in an air-bath and heated either by a gas burner or electricity. Air is a bad conductor of heat. It ensures very slow and uniform change in temperature whether the oil is being heated or cooled.

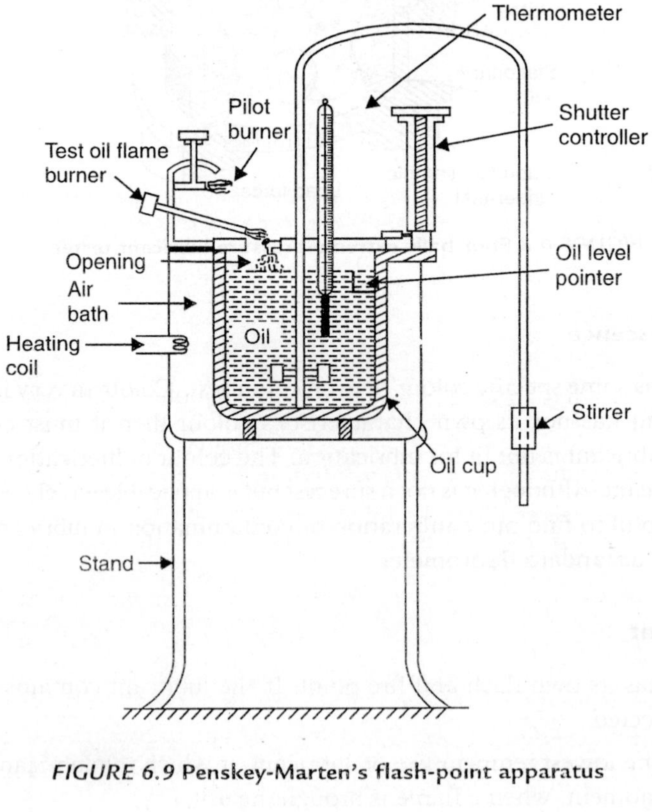

FIGURE 6.9 Penskey-Marten's flash-point apparatus

(*d*) *Pilot burner.* Pilot burner is always lighted to support the burning of test flame burner. As the test flame is introduced into the oil surface through the hole, it gets extinguished but when the test flame is returned to its original position, it is automatically lighted by the pilot burner.

(*e*) *Flame test burner.* It is a burner by which the test flame is introduced into cup containing lubricating oil. It is supported by a pilot burner.

Working. First of all experimental lubricating oil is filled up to the mark in the oil cup and is heated with constant stirring at the rate of about 1–2 revolutions per second. The heating of oil cup is adjusted in such a way that the temperature of the oil rises at the rate of about 5°C per minute. At every 1°C rise in temperature, test flame is introduced for a moment with the help of a shutter. The temperature at which a distinct flash (a combination of weak sound and light) appears inside the cup is recorded as the flash point of lubricating oil. The heating is continued further and the test flame introduced as before. The temperature at which the experimental lubricating oil catches fire at least 5 seconds is recorded as its fire-point.

Factors Affecting the Flash and Fire Point of an Oil

Following are the factors that affecting the flash and fire point of lubricating oil.

(*i*) *Presence of water or moisture.* Presence of water may raise or lower the flash point. On heating, the water forms steam. It may prevent the vapour from igniting and hence raise the flash point. However, steam-distillation of low molecular weight constituents of the oil will tend to decrease the flash point. Oils containing water may split and determination of flash point will very difficult. So free water present in oil should be removed by following methods.

(*a*) Settlement and decantation.

(*b*) Centrifugal action.

(*c*) Absorption by dehydrating agents like anhydrous $CaCl_2$, anhydrous Na_2SO_4, etc.

(*d*) Filtration through a filter paper containing Plaster of Paris.

(*ii*) *Vapour pressure of the oil.* High vapour pressure of the lubricants have lower flash and fire points.

(*iii*) Whether the flash and fire point test is made by open cup or closed cup method. Open cup methods give higher flash and fire point.

(*iv*) Frequency of application of test flame.

(*v*) Rate of heating affects the time available for the vapour to diffuse into air and hence affects the flash and fire points.

(*vi*) Variations in the time of opening the shutter.

(*vii*) Variations in the size of the test flame.

(*viii*) Variations in the rate of stirring.

Contamination of an oil with small amounts of volatile organic substances produces irregular flashes below the true flash point of an oil. It is called 'Freaky' flash point. Flash points of fatty oils vary between 300–500°F. When fatty oils are heated, they may undergo thermal decomposition. Evolution of vapours of decomposition products and traces of free fatty acids present in the oils gives rise to freaky flash point. Flash point 'Closed' is determined by heating the oil in a closed cup. The test flame is injected into the cup through an opening, produced temporarily and ignition of vapour takes place inside the cup. The vapour is not free to diffuse to the atmosphere. Hence, accurate and reproducible results are obtained. Flash point 'Open' is determined by heating the oil in a cup without cover. The surface of the oil is thus exposed to atmosphere. There is loss of vapour which is enhanced by any drafts over the testing device. Hence the results are not very accurate and reproducible. In practice open cup is the common accepted device for measuring flash points above 175°F.

Generally Abel's apparatus and Pensky-Marten's apparatus both are used to determine the flash point and fire point. The results obtained by Abel's apparatus are very precise. The method is however slow and is recommended for oils flashing upto 120°F. In Abel's apparatus, the oil cup is surrounded by an air jacket which is heated by a water bath. The temperature rises slowly at a rate of about 5°F per minute. In the Piensky-Marten's method, the air jacket surrounding the oil cup is heated directly and the rate of heating is higher about 10 ± 1°F per minute. It is the most suitable and commonly used for all oils flashing above 120°F. Since air is a bad conductor of heat, so it helps in controlling the cooling rate.

12. Viscosity

It is the property of a liquid or a fluid by virtue of which it offers resistance to its own flow. If two layers of a liquid separated by a distance '*d*' and moving with a relative velocity difference 'n' then force per unit area (F) required to maintain this velocity difference is given by

$$F = \frac{\eta v}{d}$$

where η is the coefficient of viscosity.

If $d = 1$ cm and $v = 1$ cm/s

Then $F = \eta$

Hence coefficient of viscosity may be defined as the force per unit area required to maintain a unit velocity gradient between two parallel layers. For every degree rise in temperature, there is a decrease of roughly 2% in the coefficient of viscosity of most of liquids.

The unit of viscosity is poise and poise may be defined as:

"A force of 1 dyne is required to maintain a relative velocity difference of 1 cm/sec between two parallel layers 1 cm apart".

Viscosity is the most important single property of any lubricating oil. If the viscosity of the oil is too low, a liquid oil film cannot be maintained between two moving surfaces and excessive wearing takes place.

"Absolute viscosity is the tangential force in dynes per square centimeter required to maintain a velocity difference of 1 cm/sec in two parallel layers of fluid 1 cm apart."

In case of heavy pressure and low speed machines less viscous oils will be easily squeezed out hence thick viscous oils are used. Similarly in light machines less viscous oils are used. The apparatus which is used to determine the viscosity is known as viscometer. Viscosity is determined by following methods:

(*a*) Ostwalds Viscometer Method

(*b*) Redwood Viscometer Method

(*c*) Saybolt Viscometer Method

(*d*) Engler Viscometer Method.

Determination of viscosity. In industry viscosity of lubricating oil is determined by Redwood viscometer or Saybolt viscometer or Engler viscometer. The Redwood viscometer measures the time in seconds for 50 ml of lubricating oil to flow through standard orifice under a given set conditions. Saybolt viscometer measures the time in seconds for 60 ml of lubricating oil to flow through standard orifice under a given set conditions. In the case of Engler viscometer it is expressed as the ratio of the time for oil to the time taken by water at 20°C.

By Redwood viscometer. It is of two types: (*i*) Redwood Viscometer No. 1 and (*ii*) Redwood Viscometer No. 2. The difference between these two are viscometer No. 1 is used for determining the viscosities of thin lubricating oils. It has a jet of bore diameter 1.62 mm and length 10 mm. Viscometer No. 2 is used for determining the viscosities of thick lubricating oils. It has a jet of diameter 3.8 mm and length 15 mm. It may be represented as in Fig. 6.10.

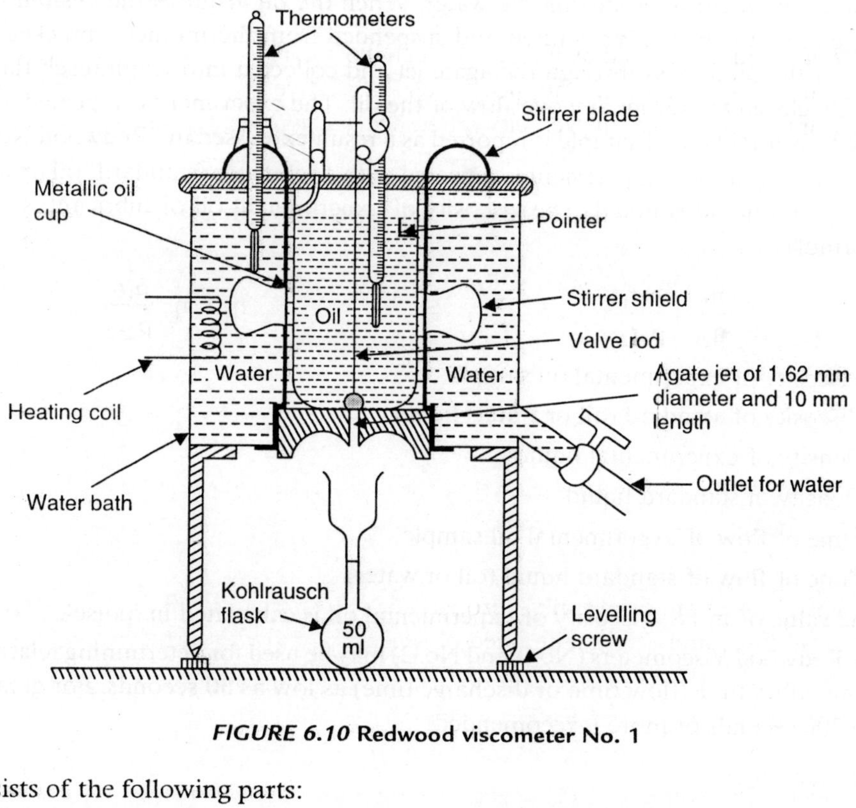

FIGURE 6.10 **Redwood viscometer No. 1**

It consists of the following parts:

(*a*) *Oil cup.* It is a silver plated brass cylinder (having 90 mm in height and 46.5 mm in diameter). The cup is open at the upper end. The bottom of the cylinder is fitted with a agate jet with bore of diameter 1.62 mm and length 10 mm. The jet is opened or closed by a valve rod. The level of the oil to be filled is indicated by a pointer. The lid of the cup is filled with a thermometer for measuring the temperature of the oil.

(*b*) *Heating bath.* Oil cup is surrounded by a cylindrical copper bath containing water. A thermometer is fitted in this bath to maintain the temperature.

(*c*) *Stirrer.* A stirrer having four blades is provided in water bath to maintain uniform temperature of water bath as well as oil cup.

(*d*) *Spirit level.* The lid of the cup is provided with a spirit level for vertical levelling of the jet.

(*e*) *Levelling screw.* The entire apparatus stands on three legs provided at their bottom with levelling screw.

(*f*) *Kohlrausch flask.* It is specially shaped 50 ml flask for receiving the oil from the jet outlet.

Working

The apparatus is leveled and water bath is filled with water. A thermometer is placed in water bath. The oil cup is cleaned and ball of valve rod is placed on the agate jet to close it. An empty cleaned Kohlrausch flask is kept just below the jet. The experimental oil is filled in oil cup up to a pointer. A thermometer is also placed in oil cup to read the temperature of oil. Now water bath is heated up to a certain temperature with constant stirring the water. When the oil acquires the desired temperature, heating is stopped and the ball valve is lifted and suspended from thermometer bracket. The time is noted for 50 ml of the oil to pass through the agate jet and collected into Kohlrausch flask. Now the valve immediately closed to prevent any overflow of the oil. The experiment is repeated and the mean value of time of flow for 50 ml oil sample is reported as a result expressed in "Redwood No. 1 seconds" at a particular temperature. The experiment is repeated with a reference (standard) oil or water and the time of flow of 50 ml liquid is noted. The viscosity of experimental oil or lubricant is calculated by applying the formula

$$\frac{\eta_1}{\eta_2} = \frac{\rho_1 t_1}{\rho_2 t_2} \quad \Rightarrow \quad \eta_1 = \eta_2 \frac{\rho_1 t_1}{\rho_2 t_2}$$

where η_1 = Viscosity of experimental oil sample
η_2 = Viscosity of standard (oil or water) liquid
ρ_1 = Density of experimental liquid
ρ_2 = Density of standard liquid
t_1 = Time of flow of experimental oil sample
t_2 = Time of flow of standard liquid (oil or water)

Thus the value of η_1 *i.e.*, viscosity of experimental oil is calculated in 'poise'.

Though Redwood Viscometers (No. 1 and No. 2) may be used for determining relative viscosities of oils having an efflux time (flow time or discharge time) as low as 30 seconds. For greater accuracy the efflux time 200 seconds or more is recomended.

Unit of viscosity

C.G.S. System: 1 poise = 1 dyne/sec/cm^2

and 1 centipoise = $\frac{1}{100}$ poise

S.I. System: Newton Second/m^2 (N S/m^2)

1 N S/m^2 = 10 poise

High viscosity oils have less penetrating power in rubber, so the deteriorating action of mineral oil on rubber diminishes with increase in viscosity. However, if sufficient or long time is allowed for penetration then swelling in rubber occurs.

Relationship between Absolute/Relative Viscosity and Kinematic Viscosity

Absolute viscosity (η) is the tangential force per unit area (1 cm^2) required to maintain the velocity difference of 1 cm/sec in two parallel layers of fluid which are 1 cm apart.

Kinematic viscosity (ν) is the ratio of absolute viscosity to density of any fluid. Its unit is stokes or centistokes.

Relationship $$\nu = \frac{\eta}{\rho}$$

where. ν = Absolute kinematic viscosity

η = Absolute dynamic viscosity

ρ = Density of fluid

[**Note.** For academic purpose viscosity is usually expressed in 'centipoise' or 'centistoke'.]

13. Viscosity Index

The arbitrary scale which measures the variation of viscosity with temperature is called viscosity index. Generally the viscosity of an oil decreases with rise in temperature. The viscosity of a good lubricant should not change very much with the rise in temperature. If the viscosity of oil is very much affected with the rise in temperature it is called low viscosity index. Similarly if the viscosity of oil is slightly affected with the rise in temperature, it is called high viscosity index lubricant. A good lubricant should have high viscosity index. Such lubricants can therefore be used over widely varying temperatures and are known as 'all weather lubricants'.

Some lubricants having high viscosity Index are following :

Silicones, Polyglycol ethers, Diesters, Or triesters, etc.

Determination of viscosity index. For the determination of viscosity index of experimental oil the viscosities of testing oil at 100°F and 210°F are first found out. If the difference between the two values is low then the oil is good and if the difference is high then the oil is poor. The viscosity at 100°F of the oil under test is represented by U. Now we compare the viscosity of oil under test with two standard oil, one with the highest viscosity index (VI = 100) and another with the lowest viscosity index (VI = Zero). Paraffinic-base Pennysylvanian oils have highest viscosity index (VI = 100) and naphthanic base Gulf oils have lowest viscosity index (VI = Zero). Against each of these is marked their viscosities at 100°F and 210°F. The former are known as H-oils and the latter as L-oils. We get a series of these two types of oils.

Now from the list of H-oils (VI = 100) we choose the oil which has the same viscosity at 210°F as the oil under test and its corresponding viscosity at 100°F is noted. It is represented by H. Similarly from the list of L-oils (VI = zero) we choose the oil which has same viscosity at 210°F as the oil under test and its corresponding viscosity at 100°F is noted. It is represented by L.

Now, Viscosity Index (VI) may be calculated as

$$\text{Viscosity Index (VI)} = \frac{L - U}{L - H} \times 100$$

where U = Viscosity of experimental oil at 100°F

L = Viscosity of low viscosity index standard oil (Gulf oil having VI = Zero) at 100°F and also having the same viscosity of experimental oil at 210°F

H = Viscosity of high viscosity index standard oil (Pennysylvanian oil having VI = 100) at 100°F and also having the same viscosity of experimental oil at 210°F.

Viscosity index of lubricating oil may be increased by adding of certain organic polymers, which are partially soluble in the oil. The produce oil-polymer blends, which have a very slight temperature coefficient of viscosity.

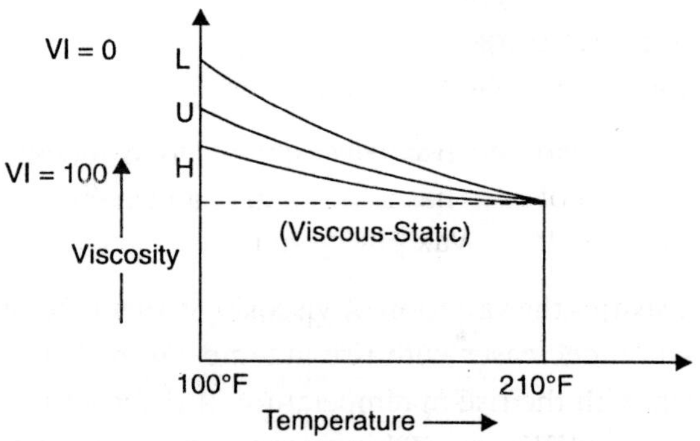

FIGURE 6.11

The effect of temperature on viscosity may be represented as in Fig. 6.11. The viscosity-temperature curves of the experimental oil along with two standard oils having high and low VI are represented in Fig. 6.11. Lubricating oils with small variation in viscosity along with temperature exhibit flatter curve as in case of high viscosity index oil. Linear molecules of the lubricant having flexibility through free rotation about the chemical bonds shows generally a high viscosity index. Hence we prefer a high viscosity index lubricant in practice due to the same viscosity over a range of temperature. The state of a lubricant when its viscosity does not change with rise in temperature is called Viscous-Static. Such type of a lubricant can be prepared by adding appropriate amount of a suitable linear polymer or viscosity-index improver.

Numerical. *An oil of unknown viscosity index has a Saybolt universal viscosity of 58 seconds at 210°F and of 580 seconds at 100°F. The high VI standard (Pennysylvanian oil) has Saybolt viscosity of 58 seconds at 210°F and 430 seconds at 100°F. The low VI standard (Gulf oil) has a Saybolt universal viscosity of 58 seconds at 210°F and 780 seconds at 100°F. Calculate the VI of unknown oil.*

Solution.

Here L = 780 sec, U = 580 sec and H = 430 sec

$$\text{VI of experimental oil} = \frac{L - U}{L - H} \times 100 = \frac{780 - 580}{780 - 430} \times 100 = 57.14$$

14. Ash Content

It is determined by heating strongly the known amount of lubricant in a weighted crucible in the presence of air. After cooling the weight of ash is calculated. The process is repeated so many times till the constant weight. The percentage of ash in a lubricant is very useful to determine the material that may cause abrasion and wear. Ash content of a good lubricant should be less as much as possible.

$$\text{Percentage of ash} = \frac{\text{Weight of ash}}{\text{Weight of lubricant}} \times 100$$

15. Cloud Point, Pour Point and Solid Point

During the slow cooling of an oil, the temperature at which an oil becomes cloudy in appearance is called its 'cloud point', while the temperature at which the oil ceases to flow or pour is called 'pour point'. Cloud point and pour point indicates the suitability of lubricants in cold conditions. Pour point is considered 5°F higher than the solid point. Solid point is the highest temperature at which an oil does not move when the standard jar containing the oil is kept in a horizontal position for 5 seconds. The pour point of oil free from wax is called 'Viscosity pour point'. For oil containing wax, pour point is the temperature at which crystallization of wax has gone to such an extent that the oil will stop flowing if cooled further. This temperature is called 'Wax pour point'.

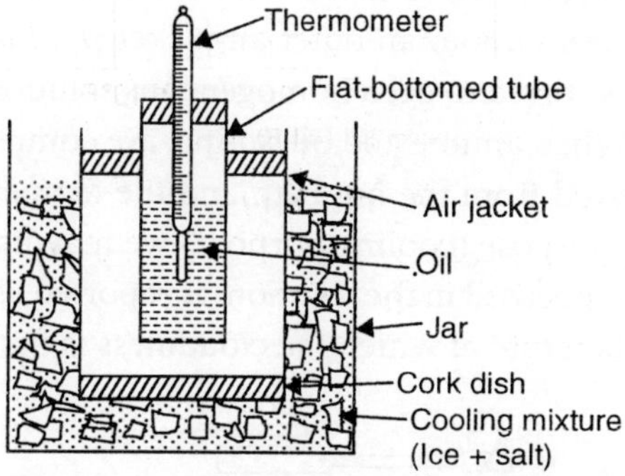

FIGURE 6.12 **Cloud point and pour point**

Determination. It is determined with the help of pour point apparatus as shown in Fig. 6.12. It consists of a flat-bottomed tube for taking lubricating oil. This tube is enclosed by a air jacket. The air-jacket is surrounded by freezing mixture (ice + $CaCl_2$) contained in a jar. The flat tube is half filled with the experimental oil. A thermometer is placed in the oil. The oil gets start cooling and the temperature decrease slowly. At an interval of fall in temperature every 1°C, the tube is withdrawn from the air jacket for a moment (2–3 seconds) and examined and then replaced in ice-bath immediately. The temperature at which cloudness is noted is recorded as the cloud-point. After this cooling is continued and the test tube is removed from the cooling bath after every 3°C fall of temperature and tilted to observe the flow or pour of oil. The temperature at which oil does not flow in the test tube, even when kept horizontal for 5 seconds, is recorded as the pour point. If oil shows any movement, it is at once replaced in the jacket till acquiring the pour point.

The 'Viscosity pour point' of an oil can be lowered by lowering the viscosity of the oil. This can be done either by removing some of the more viscous constituents of the oil or by adding a component of lower viscosity. 'Wax pour point' can be lowered either by dewaxing or by adding a suitable pour point depressant. 'Paraflow' is a poly-alkyl naphthalene, used as an important pour point depressant. It is prepared by condensing chlorinated wax with naphthalene in the presence of catalyst (anhydrous aluminium chloride). The presence of paraflow (1–2% concentration) reduces the wax pour point of an oil by 50°F or more. Pour point depressant get adsorbed on the surface of wax crystals during the initial stages of crystal formation. Hence, they reduce the size of the wax crystals. They also alter the crystal structure in such a way that the amount of oil held by the crystals by adsorption is reduced. These effects enhances the oil flow at lower temperature and thus reduce the 'Wax pour point' of the oil.

16. Aniline Point

Aniline point is the minimum temperature at which the equilibrium exit between equal volume of aniline and oil sample. Aniline is readily soluble in those type of lubricants which are rich in aromatic and naphthenic compounds.

Determination. The whole apparatus may be represented as in Fig. 6.13. The apparatus is cleaned and dried at 100–110°C. An equal volume of pure and dried aniline and dried oil sample have been taken in the test tube (B). This test tube is fitted with electrically operated glass rod stirrer and a thermometer. The test tube is inserted into an outer air-jacket (A) made of heat resistance glass. The aniline and oil sample mixture is stirred to get a homogeneous solution. For this sometimes we use hot bath and stirring is continued. When aniline and oil sample are completely mixed to get homogeneous mixture then the jacket is removed from the hot bath and the temperature is allowed to fall at a rate below 1°C per minute. For this purpose (cooling purpose) we may use cold bath. The temperature, at which cloudiness or haziness is appeared in the solution, is reported as aniline point of the sample. It is generally 1–2°C below the temperature at which the cloudiness is first observed.

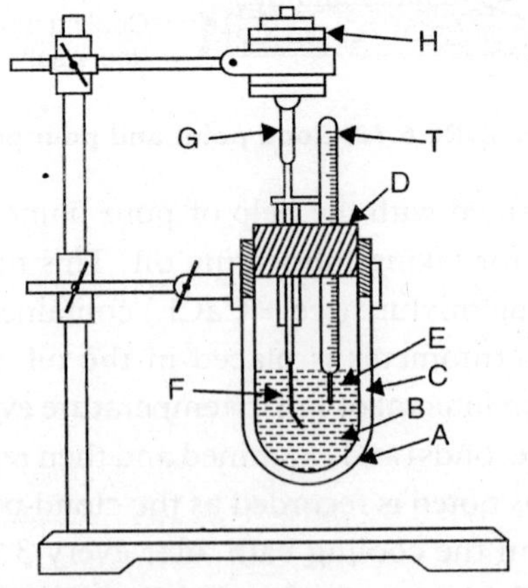

FIGURE 6.13 **Aniline point apparatus**
A: Outer air jacket, B: Test tube, C: Air gap, D: Corks,
E: Equal volume of oil and aniline,
F: Glass stirrer with auger tip,
G: Rubber tubbing, H: Variable speed motor,
T: Aniline point thermometer.

Precautions. (*a*) All apparatus should be dry and cleaned.

(*b*) Aniline is hygroscopic, so water should not be used in hot or cold baths. Non aqueous, non volatile solvents should be used.

(*c*) Aniline is highly toxic, so we should handled carefully.

(*d*) Always avoid air and water throughout the experiment.

Mixed aniline point. There are certain lubricants which have high aromatic contents. When these are mixed with equal volume of aniline, they formed a homogeneous mixture but during cooling or solidification, separation of different phases may not be observed. For determination of such type of aniline points one volume of oil sample is mixed with two volumes of aniline and one volume of suitable solvent (diluent) like *n*-hexane or *n*-heptane. Addition of diluent lowers the miscibility of aniline with the sample. So with decrease in temperature, separation of phases can be easily observed. The equilibrium solution temperature observed under these conditions is known as mixed aniline point.

Paraffin-base type lubricating oils have the highest aniline points. The range of their aniline point depends on the degree of refinement. The aniline point of an oil decreases with the increase in the percentage of its aromatic content. Hence, the aniline point is inversely related to the aromatic content of the oil. A diesel fuel with a high aniline point have low aromatic content. This gives an easy start to the engine and reduces knocking. The ignition quality of a diesel fuel is also reported in terms of Diesel Index which is related to the aniline point by the following:

$$\text{Diesel Index} = \text{Aniline Point in °F} \times \frac{\text{API gravity}}{100}$$

For high-speed diesel fuel, the aniline point should be above 160°F. Aniline point thermometers are available in three different ranges.

(*i*) + 38 to + 42°C

(*ii*) 25 to 105°C

(*iii*) 90 to 170°C.

Significance. Aniline point gives an indication of the possible deterioration of oil in contact with rubber sealings, packings etc. A high aniline point means a higher percentage of paraffinic hydrocarbons and hence a lower percentage of aromatic hydrocarbons. Actually aniline point of lubricant is a measure of its aromatic content. A lubricant having low aniline point has tendency to attack the rubber seals, used in the system to prevent leakage. That's why a lubricant of high aniline point is recommended for systems in which rubber seals are being used.

17. Neutralization Number

Neutralization number determines the acidity or alkalinity of lubricating oil. It represents either the total acid value or total base value. During refining the oil some free organic acids are always associated with the lubricating oil, so acidity is more common and is expressed in acid value or acid number.

Acid value. It is the number of milligrams of KOH required to neutralize the free acid in 1 gram of the oil. Acid value should less than 0.1 for lubricating oil. Acid value greater than 0.1 indicates that the oil is oxidized and as a result sludge formation and corrosion may occurs.

Determination. Take a known weight of few ml (say 3–4 ml) of oil sample in a titration flask. The oil sample is dissolved in minimum quantity of suitable solvent like ethyl methyl ketone. Now few drops of phenolphthalein indicator is added. After shaking the contents, it is titrated against standard $\left(\text{say } \frac{N}{100}\right)$ alcoholic KOH solution taken in a burette. The appearance of light pink colour represents the end point. The experiment is repeated 3–4 times with oil sample for getting concordent reading. The whole process is also repeated 3–4 times with the solvent alone as a blank titration.

Calculation. Let weight of oil sample = W g

$$\text{Normality of alc. KOH} = \frac{N}{100}$$

Volume of $\frac{N}{100}$ alcohlic KOH used in titration = A ml

Volume of $\frac{N}{100}$ alcoholic KOH used in blank titration = B ml

$\therefore$ Exact volume of $\frac{N}{100}$ alcoholic KOH used in titration = (A – B) ml

$$\text{Acid Value} = \frac{\text{Volume of KOH used (ml)} \times \text{Normality} \times \text{Eq. Wt. of KOH}}{\text{Weight of sample}}$$

$$= \frac{(A-B) \times \frac{1}{100} \times 56}{W} = \frac{(A-B) \times 56}{100 \,.\, W}$$

[**Note:** If the reaction mixture is heated before titration then the esters in the oil may be hydrolysed by the moisture present in it and thus enormous results will be obtained. Sometimes pink colour fade away within 2–3 minutes. This is due to the presence of some impurities (materials), other than the free acids, in oil sample. These materials may slowly and easily attacked by the alkali.].

Precautions

(*a*) Alcoholic solution of KOH should be freshly prepared and standardized before use.

(*b*) For pale oils, phenolphthalein indicator is used but for dark coloured oils *p*-naphtholbenzoin should be used as an indicator. It is better to determine end point potentiometrically.

18. Koettsdoerfer Number or Saponification Value

"It is the number of milligram of KOH required to saponify fatty materials present in one gram of oil". It is determined by refluxing a known weight of oil sample with a known amount of standard alcohlic KOH solution. The unreacted KOH is estimated by titrating against standard HCl solution.

Determination. In a titration flask, a known weight of oil sample (say W gram) is taken and dissolved in minimum quantity of suitable solvent like ethyl-methyl ketone. Now exact known amount of standard $\left(\text{say } \frac{N}{10}\right)$ alcoholic KOH is also added. The same amount of solvent ethyl-methyl ketone and alcoholic KOH is taken in another flask for blank titration. Both the flasks are fitted with air condenser and refluxed for 45 minutes. After cooling both the flasks few drops of phenolphthalein indicator is added till the appearance of pink colour. Now it is titrated against standard $\left(\text{say } \frac{N}{10}\right)$ HCl until the disappearance of pink colour. The experiment is repeated 3–4 times to get concordent reading.

Calculation. Let weight of oil sample taken = W g

Volume of $\frac{N}{10}$ alc. KOH added to both flask = 50 ml

Volume of solvent added to both flask = 50 ml

Volume of $\frac{N}{10}$ HCl used for oil sample titration = A ml

Volume of $\frac{N}{10}$ HCl used for blank titration = B ml

Exact volume of $\frac{N}{10}$ HCl used for oil sample titration = (B – A) ml

$$\text{Saponification Value} = \frac{\text{Volume of KOH used (ml)} \times \text{Normality} \times \text{Eq. wt. of KOH}}{\text{Weight of sample}}$$

$$= \frac{(B-A) \times \frac{1}{10} \times 56}{W} = \frac{(B-A) \times 56}{10\,W}$$

Precautions

(*a*) Alcoholic solution of KOH should be standard before use.

(*b*) The titration flask should be occasionally shaken during refluxing.

(*c*) For coloured oil *p*-naphtholbenzoin indicator should be used.

(*d*) Ethyl methyl ketone should be preferred as a solvent due to high boiling point and increases the rate of saponification.

Significance

(*a*) It is the alkaline hydrolysis of fatty oil. After hydrolysis glycerol and soaps are obtained.

(*b*) It distinguish between vegetable and animal oils, fatty oils and mineral oils. Mineral oil is a mixture of hydrocarbon so they do not react with NaOH or KOH and hence cannot saponify.

(*c*) It is used to determine the extent of compounding in a blended oil to improve the oiliness.

(*d*) It gives the value of extent of adulteration in oil.

(*e*) It is used to identify the given fatty oil, because each fatty oil has its own characteristic value.

19. Iodine Value

It is also known as iodine number and may be defined as the number of grams of iodine equivalent to the amount of iodine monochloride (ICl) consumed by 100 g of the oil. It is determined by dissolving known weight of oil sample in CCl_4 and treating with a known excess of Wij's solution. (Wij's solution is the solution of ICl in glacial acetic acid). One molecule of ICl adds on each double bond present in oil.

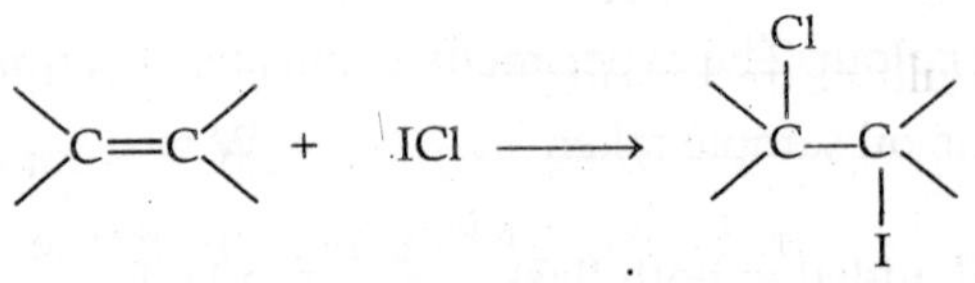

The unreacted ICl oxidizes the KI solution to I_2

$$ICl + I^- \longrightarrow I_2 + Cl^-$$

The liberated iodine is titrated iodometically with standard hypo solution using freshly prepared starch solution indicator.

$$\underset{\text{Thiosulphate ion}}{2S_2O_3^{2-} + I_2} \longrightarrow \underset{\text{Tetrathionate ion}}{S_4O_6^{2-} + 2I^-}$$

The consumed amount of hypo solution gives the value of unreacted iodine.

Determination. In iodine titration flask, the known amount (say W g) of lubricating oil is taken and dissolved in minimum quantity of suitable solvent like CCl_4. A known volume of Wij's solution is also added carefully. Similarly same amount of solvent CCl_4 and Wij's solution is taken in another iodine titration flask for blank titration. Now exact known volume (few ml) of 10% KJ solution is added to both the flasks and fitted with glass stoppers and then is shaked well. Now both the flasks are kept in dark place for about 1 hour. Now it is titrated against standard $\left(\text{say } \frac{N}{20}\right)$ hypo solution till the colour of solution changes to pale yellow. After that few drops of freshly prepared starch solution is addded as an indicator. It gives blue colour. Again it is titrated until the disappearance of blue colour. The end point is recorded and experiment is repeated 3–4 times to get the concordent reading. Blank experiment is also done on the same pattern.

Calculation. Let weight of oil sample taken = W g

Volume of $\frac{N}{20}$ hypo used in oil sample titration = A ml

Volume of $\frac{N}{20}$ hypo used in blank titration = B ml

Exact volume of $\frac{N}{20}$ hypo used for oil sample titration = (B – A) ml

It is equivalent to iodine or ICl consumed by W g of oil sample.

$$\text{Iodine present in (B – A) ml of } \frac{N}{20} \text{ iodine solution} = \frac{(B - A) \times \text{Eq. wt. of } I_2 \times \text{Normality}}{1000}$$

$$\text{Iodine Value} = \frac{\text{Vol. of } I_2(\text{ICl}) \text{ used} \times \text{Normality} \times \text{Eq. wt. of } I_2 \times 100}{\text{Weight of oil} \times 1000}$$

$$= \frac{(B - A) \times 127 \times 1 \times 100}{W \times 20 \times 1000} = \frac{(B - A) \times 0.635}{W}.$$

Precautions

(*a*) All apparatus and chemicals should be dry and cleaned otherwise the presence of moisture spoils the Wij's solution by decomposing ICl as follows:

$$ICl + H_2O \longrightarrow HCl + HIO.$$

(*b*) Starch solution should be freshly prepared.

(*c*) Wij's solution should be added excess with the help of burette.

(*d*) Glass stoppers of the iodine titration flask should be moistened with KI solution.

It is necessary to prevent the leakage of Cl_2 or I_2 vapours that may be formed by decomposition of ICl.

(*e*) The reaction mixture should be kept in dark place, otherwise sunlight catalyses the decomposition of ICl into I_2 and Cl_2 as follow:

$$2ICl \xrightarrow{h\nu} I_2 + Cl_2$$

Significance

(*a*) It determines the degree of unsaturation of oils and fats. Hence fatty oil has tendency to absorb oxygen.

(*b*) If an oil is exposed to air, it combines with oxygen to form chemical compounds which are unstable for use as lubricants. As a result acids and gummy sludge formation takes place.

(*c*) Iodine value of a good lubricating oil should be as low as possible.

(*d*) It indicates the oldness of oil.

6.6 ADDITIVES FOR LUBRICANTS

Generally it has been found that no any lubricating oil is very much suitable as above. In industries lubricants are very important in machines and tools. There are lack of properties of single lubricating oil so the properties of lubricants may be improved by adding some specific suitable compounds, such type of compounds which are used to improve the properties of lubricants are known as additives for lubricants. Some important additives are listed below.

(*a*) *Rust inhibitors.* Alkyl succinic acids, fatty acids, etc. are useful for anti-rusting.

(*b*) *Antifoam additives.* Silicones, glycerols etc. are used as antifoam additives.

(*c*) *Anti-oxidants.* Organic sulphides, amines, phenols, etc. are used as anti-oxidants.

(*d*) *Oiliness improves.* Fatty amines, fatty acids, vegetables oils are used as oiliness improver because they increase the strength of the oil film and prevent its rupture.

(*e*) *Corrosion inhibitor.* Phosphorized or sulphurized terpenes, organo-metallic compounds are used in lubricants as corrosion inhibitor.

(*f*) *Viscosity index improvers.* Long chain polymers like polyiso-butylene, polyesters, polystyrene etc. are used as viscosity index improver. They prevent the oil from thickening at low temperature and thining at higher temperature.

(*g*) *Extreme pressure additives.* Organic phosphorous compounds, sulphurized fats, metallic soaps etc. are used as extreme pressure additives in lubricants. They (additives) either get absorbed on the metal surface or react with the metal to form a layer having low shear strength on the metal surface and prevent wearing and tearing. They also protect them from welding and seizure under extreme pressure conditions.

(*h*) *Emulsifiers.* Monoesters of polyhydric alcohols, sodium salts of carboxylic acid or sulphonic acids etc. are very good emulsifier. They promote the mixing of oil and water and helps in the formation of emulsion.

(*i*) *Pour point additives.* Polymethylmethacrylate, wax alkylated naphthalene, etc. are important additives for depressed of pour-point of lubricants. They maintain the characteristics of lubricating oil in liquid form even at lower temperature.

6.7 SYNTHETIC LUBRICANTS

The lubricants which have been synthesized in laboratory are known as synthetic lubricants. Synthetic lubricants are much better than ordinary lubricants because they are able to do work under extreme conditions. The properties of synthetic lubricants are much better than the ordinary lubricants.

Characteristics of Synthetic Lubricants

(*a*) They have high thermal stability *i.e.,* they do not decompose under high temperature.

(*b*) They prevents from rusting, oxidation, hydrolysis, etc.

(*c*) Some are used as excellent solvents for additives.

(*d*) They possess high flash point or fire point.

(*e*) They possess high viscosity index value *i.e.,* the variation of temperature does not affect its viscosity.

(*f*) Such type of lubricants may be used over a wide range of temperature (– 50°C to 270°C).

(*g*) They prevents from corrosion, etc.

Some important synthetic lubricants are given below:

Dibasic acid esters, organophosphate esters, chlorinated or fluorinated hydrocarbons, silicate esters, silicones, polyalkylene glycols and their derivatives, etc.

6.8 SEMI-SOLID LUBRICANTS (GREASES)

Greases are semi-solid lubricants like thixotropic gels consisting of metallic soaps dispersed in lubricating oil. The liquid lubricant may be a petroleum oil or synthetic oil having some specific additives.

Preparation

These are usually prepared by saponification of fat with alkali followed by adding hot lubricating oil. Lubricating oil is generally a petroleum oil but synthetic oil is also used for the preparation of special type of greases. The metallic soap acts as thickening agent and facilitate the gel formation. Metallic soap also enhances the oiliness property *i.e.,* it support the grease to stick on the metal surface firmly. Generally soap content in grease is about 5–20%. Soap present in grease determines its resisting power towards water, oxidation, corrosion etc.

Conditions for use. (*a*) When it is necessary to seal the bearing against the entry of dust, dirt or moisture etc.

(*b*) When the machine is required to work at low speed under high load.

(*c*) When the solid lubricants or liquid oil is not efficient.

(*d*) When the spilling of oil is highly undesirable in industries.

Important Classification

(*a*) *Lithium based greases.* These are petroleum oils containing lithium soaps. They are water resistant and suitable for use at low temperature only (upto 15–20°C). It is very costly.

(*b*) *Calcium based greases.* These are emulsions of petroleum oils with calcium soaps. These are the cheapest and most commonly used. They are known as cup-greases and insoluble in water. They are used at low temperature (below than 80°C). At higher temperature oil and soap begins to separate out.

(*c*) *Aluminium based greases.* These are emulsions of petroleum oils with aluminium soaps. They are good adhesive and water resistant. They are used upto 90°C.

(*d*) *Sodium soap greases.* These are emulsions of petroleum oil with sodium soaps. They are not water resistant but are used up to 175°C because they possess good high temperature.

(*e*) *Axle greases (Resin soap greases).* These are very cheap and known as resin soap greases. These are prepared by adding lime or any heavy metal hydroxide to resin and fatty oils. The mixture is treated with fillers like talc and mica etc. The mixture is allowed to stand when the grease floats as stiff mass. They are water resistant and useful for less delicate equipments working under high loads and at low speeds.

(*f*) *Smooth and Fibrous greases.* During the formation of grease, the soap crystallizes in the form of fibres (threads) having lengths 20 times or more than their thickness. Most soap fibres are microscopic in size and the grease appears very smooth. It is called smooth grease. However, when the fibrous bundles are large enough like macroscopic, to be seen with naked eye, the grease acquires a fibrous appearance. The fibrous structure is also noticeable when the grease is pulled apart. For example, sodium-base grease, petrolatum etc.

Petrolatum is a mixture of oils and waxes stabilized by a third component and is obtained from still residues of paraffin base crudes after fractionation. It does not leave an oily stain on paper as wax is the external phase and oil is the internal phase. It can be used as a grease.

Properties

(*a*) ***Consistency (Yield value).*** It is expressed in term of penetration and is measured by 'penetrometer'. Penetration is defined as "the distance in tenth of millimeter that a standard cone penetrates vertically in the sample under the standard conditions of load (150 g), temperature (25°C) and time (5 sec). The apparatus of 'penetrometer' is as shown in Fig. 6.14.

It consists of following parts:

(*i*) *Heavy base.* It is the base made of cast iron alloy and provided with spirit level, levelling screw and a plane surface table on which a box containing experimental grease is placed.

(*ii*) *Vertical support.* It is an iron rod fitted to the base. These are slotted marks on it, around which a holder can be moved up and down. The holder has a screw which can be lightened in any of the slots.

(*iii*) *Moving dial rod.* It is behind the dial by a mechanical mechanism and is provided with a clutch arrangement for dis-connecting or connecting it to the circular dial.

(*iv*) *Circular dial.* The holder carries a circular dial gauge which is graduated in millimeters.

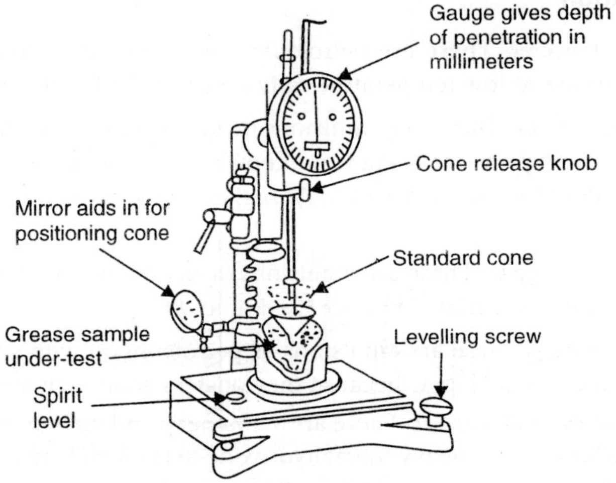

FIGURE 6.14 **Penetrometer**

(*v*) *A Mirror.* Vertical rod is provided with an adjustable mirror for removing the parallax while positioning the cone in contact with grease sample surface.

Working

At first the apparatus is levelled and the sample of experimental grease in a box is placed below the cleaned cone. The height of the cone is so adjusted with the help of mirror, that tip of the cone just touches the sample. The initial reading is noted. The cone is then released for exact five seconds by pressing a button provided for this purpose. The cone penetrates into the grease. After 5 seconds, press button is releases and the final reading is noted. The difference of the two dial readings gives the penetration.

Factors Affecting the Penetration Number

Following are the factors having great much affected the penetration number.

(*i*) The presence of air voids in the sample affect penetration number. Due to this the cone will penetrate to a greater depth and thus the observed penetration number of the sample will be higher.

(*ii*) If the movement of holder shaft is not frictionless the cone will not fall freely and so the test will show lower penetration number.

(*iii*) The penetration number of a grease depends mainly on the structure and interaction of the gelling element, the amount of soap and non-soap thickener present in it.

(*iv*) It also depends upto some extent on the viscosity of the lubricating oil present.

(*v*) It is also affected by the handling procedure during manufacture of grease, the temperature of filling the containers and the rate of cooling.

(*b*) ***Drop Point.*** It is the temperature at which grease is converted from the semi-solid to the liquid state. This temperature actually determines the upper temperature limit of the applicability of the grease. The apparatus which is used to determine the drop-point of lubricant is as shown in Fig. 6.15.

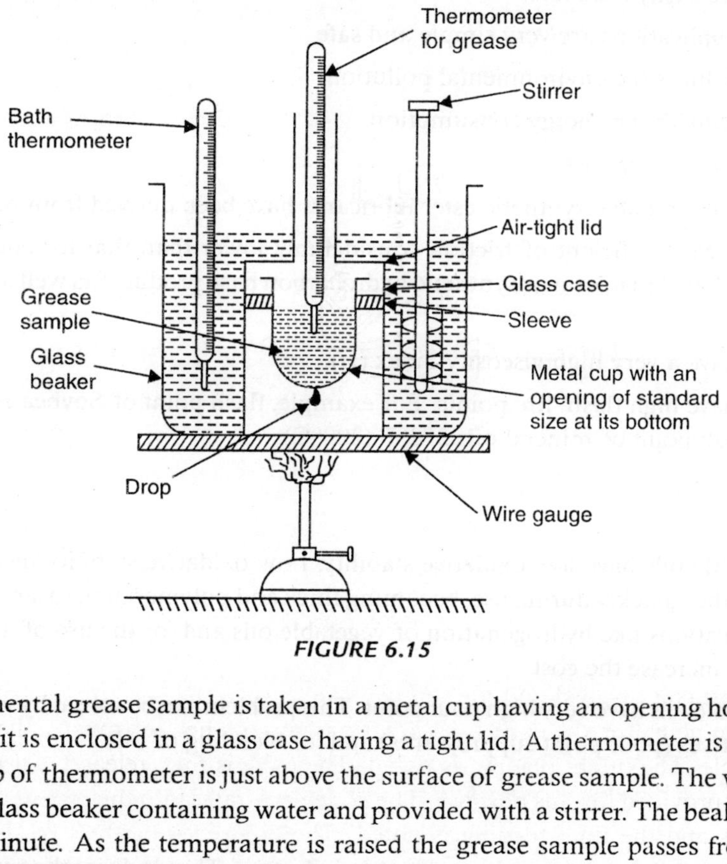

FIGURE 6.15

The experimental grease sample is taken in a metal cup having an opening hole of standard size in its bottom. Now it is enclosed in a glass case having a tight lid. A thermometer is also inserted in the cup, so that the bulb of thermometer is just above the surface of grease sample. The whole arrangement is then placed in a glass beaker containing water and provided with a stirrer. The beaker is heated slowly at a rate of 1°C/minute. As the temperature is raised the grease sample passes from a semi-solid to liquid state. At a particular temperature first drop of grease falls from the opening hole in the bottom of the cup. The temperature is recorded as drop-point of grease.

Dropping point of a grease is increased by adding a complexing agent. Complexing agents are metallic salts of short chain organic acids like acetic acid or lactic acids or the inorganic salts like carbonates or chlorides etc. When soap crystal or fibre is formed by co-crystallization of a normal soap with a complexing agent, is called complex soap. The separation of a liquid lubricant from a lubricating grease by any means is called bleeding of a grease.

6.9 BIODEGRADABLE LUBRICANTS

Biodegradable Lubricants are such type of lubricants which are easily decomposed or destroyed when spilled on to open land or into water, without leaving behind harmful substances. For examples, plant based oils or lubricants like sunflower oil, mustard oil, Soybean oil etc. All vegetable oils have been used as lubricants in their natural forms. They have several advantages and disadvantages when considered for industrial and machinery lubrication.

Advantages

(*i*) Plant based biodegradable lubricants are less toxic.

(*ii*) They are easily available.

(*iii*) Their applications are very simple and safe.

(*iv*) They reduces the environmental pollution.

(*v*) They reduces the energy consumption.

(*vi*) They have lower cost.

(*vii*) The biodegradable synthetic ester lubricants have been derived from renewable resources.

(*viii*) They have coefficient of friction considerably lower than that for equiviscous mineral oil based lubricants. Polyalphaolefin or synthetic hydrocarbon base products as well as polyglycols are well known lubricants.

(*ix*) They have a very high viscosity index (*vi*).

(*x*) They have high flash/fire points. For example, flash point of Soybean oil is 610°F (326°C), higher than of flash point of mineral oils 392°F (200°C).

Disadvantages

(*i*) Vegetable oils have low oxidative stability. Low oxidative stability means, if untreated, the oil will oxidise rather quickly during use, becoming thick and polymerizing to a plastic like consistency. Chemical modifications like hydrogenation of vegetable oils and/or the use of antioxidants can solve this problem, but increase the cost.

(*ii*) Vegetable oils have very high pour point. This problem can be solved by winterization, addition of chemical additives (pour point suppressants) and/or blending of oils.

Importance

Biodegradable lubricants have a great importance in industries. They reduces the environmental pollution and energy consumption. They are easily available with lower cost. They have been derived from renewable sources. So Environmental agencies, American standard for testing and materials (ASTM) and organization for economic cooperation and development (OECD) recomended such type of lubricants. High oleic varieties of canola oil, rapeseed, sunflower and Soybean oil are now becoming standard base oils for biodegradable lubricants and greases. Now-a-days a Soybean seed is developed through Dupont technology. The product obtained by this technology contains more than 83 percent oleic acid as compared to only 20 percent oleic acid content in conventional Soybean oil. According to ASTM and OECD, the oil is inoculated with bacteria and is kept under controlled condition for 28 days. The percentage of oxygen consumption or carbon dioxide evolution is monitored to determine the degree of biodegradability. Most vegetable oils biodegrade over 70 percent within that period as compared to petroleum oils biodegrading at about 15–35 percent. Due to this factor the importance of biodegradable oils are increasing day by day in industry and machines.

EXERCISE

1. What are lubricants? Discuss their important functions. ***(K.U.K. Jan. 2004, June 2006, June 2008)***
2. Define friction. Give an account of the laws govern friction.
3. Define lubrication. Discuss its mechanism. What are the differences between fluid film lubrication and bounding lubrication mechanism? ***(K.U.K. Jan. 2005)***
4. Why is hydrodynamic lubrication not suitable or possible under boundary lubrication condition? Discuss about boundary lubrication.
5. What do you mean by purification of lubricants? Discuss solvent refining.
6. How are lubricants classified on their physical status? What are the conditions for use of semi-solid lubricants?
7. Define and explain the term 'Emulsion'. Why graphite and MoS_2 preferred as solid lubricants?
8. Define viscosity and viscosity index of lubricants. How viscosity of lubricating oil is determined? Explain it. ***(K.U.K. June 2006)***
9. Define flash and fire point. Determine the fire point of lubricants by Pensky-Marten's apparatus. ***(K.U.K. Jan. 2006)***
10. What are lubricating oils and how are they classified? Give suitable example.
11. Write short notes on the following:
    (*i*) Aniline point
    (*ii*) Iodine number
    (*iii*) Cutting oils
    (*iv*) Cooling liquids
    (*v*) Blended oils
12. Define viscosity index. How is it determined for a lubricant?
13. What are synthetic lubricants? Write its characteristics.
14. Why additives are used in lubricants? Give some examples of additives, which are commonly used in lubricants. ***(K.U.K. June 2004)***
15. What is grease? How is it prepared? Discuss its classification.
16. How lubricants are selected for used in industries? What type of lubricants are used for very high pressure and low speed machines, internal combustion engines, transformers?
17. How consistency and drop point values of lubricants are determined? Explain clearly. Give their significance. ***(K.U.K. Jan. 2007, June 2008)***
18. What do you understand by 'asperities'? Name four organic and four inorganic lubricants. Under what conditions grease are used?

19. Write short notes on the following:
    (*i*) Oiliness
    (*ii*) Neutralization number
    (*iii*) Saponification value
    (*iv*) Cloud and pour point of lubricant.
    (*v*) Synthetic lubricant. ***(K.U.K. Jan. 2004)***

20. Discuss the mechanism of hydrodynamic lubrication. Under what operating conditions are greases preferred to lubricating oils? ***(K.U.K. Jan. 2004)***

21. What is Frictional resistance? Why is it desirable to reduce it? In what ways can it be achieved? ***(K.U.K. June 2004)***

22. Describe the penetration test for greases. What is meant by consistency? Describe the mechanism of Boundary lubrication. ***(K.U.K. June 2004)***

23. What are different types of greases? How are they made? ***(C.D.L.U. Dec. 2003)***

24. Describe the various factors involved in the selection of a lubricant. Write a short note on solid lubricant. ***(C.D.L.U. Dec. 2003)***

25. Explain clearly the importance of the following in selecting a lubricating oil for a particular use?
    (*i*) Viscosity
    (*ii*) Flash point
    (*iii*) Acidity.

26. What is Wij's solution? Where is it used? Why does graphite act as an excellent lubricant on the surface of moon? ***(K.U.K. Jan. 2005)***

27. What do you mean by 'all weather lubricants'? Give some examples of lubricants that have high Viscosity Index.

28. How does the viscosity of a liquid vary with rise in temperature?

29. How is the Viscosity Index of a lubricating oil improved?

30. What is meant by 'Viscous-Static'? How can a viscous static lubricant be prepared?

31. What is 'Solid point' of an oil? What is the difference between 'Wax pour point' and 'Viscosity pour point'?

32. What are pour point depressants?

33. Define 'paraflow'. How can the pour point of an oil be lowered?

34. Define Spontaneous Ignition Temperature (SIT) of a liquid. What is the purpose of the air jacket Surronding the oil cup?

35. Explain the factors that affect the flash and fire points of oils? How is free water removed from an oil?

36. What is meant by 'Freaky' flash, Flash point 'Closed' and Flash point 'Open'? Why fatty oils donot have sharp flash point?

37. Describe the main difference between the Abel's closed-cup and Pensky-Marten's closed-cup methods for flash point determination.

38. Discuss the effect of the viscosity of a mineral oil on its action on rubber?

39. Name the oils having highest aniline points. How aromatic content of a lubricating oil is related with its aniline point?

40. Write the ranges of aniline point thermometers. How is aniline point related to the ignition quality of a diesel fuel?

41. During the determination of acid number by titration method, sometimes the end point pink colour fades away in a few seconds. Explain its cause. What will happen if the reaction mixture is heated before titration?

42. What is the effect of the present of moisture in the sample or glassware during the determination of iodine value of a lubricating oil?

43. Why is the glass stopper moistened with KI solution?

44. Why the reaction mixture kept in dark?

45. Enumerate the important factors on which the penetration number of a grease depends.

46. What do you understand by following:

(i) Bleeding of a grease

(ii) Smooth and Fibrous grease

(iii) Petrolatum.

47. What is meant by a complex soap? What is the effect of complexing agent on the dropping point of a grease?

48. Discuss extreme pressure lubrication and additives to improve the lubricant properties. ***(K.U.K. Jan. 2006)***

49. Give the classification of lubricants with examples of each type. Specify the conditions under which each type is used. ***(K.U.K. Jan. 2006)***

50. What are the factors that should be considered in selecting a liquid lubricant? Define saponification value of a lubricant. ***(K.U.K. June 2006)***

51. Define the terms Lubrication and Lubricants. What are the different types of lubricants? Discuss the basic principle of lubrication. ***(K.U.K. Jan. 2008)***

52. Write a short note on Extreme pressure lubrication. ***(K.U.K. Jan. 2008, June 2009)***

53. What are greases? Under which conditions are they used? Discuss drop point test for greases. ***(K.U.K. June 2009)***

54. Which additives are used in lubrication as

(i) viscosity index improvers

(ii) Pour point depressants

(iii) Oxidation resistance. ***(K.U.K. June 2009)***

55. Define and give significance of

(i) Saponification value

(ii) Flash and fire points. ***(K.U.K. June 2009)***

56. With respect to lubricants explain the following. Flash and fire point, viscosity and viscosity index, cloud and pour point, Acid value. ***(K.U.K. Jan. 2009)***

57. Define lubrication and explain fluid film lubrication. ***(K.U.K. June 2009)***

58. Define visocity and Saponification value. Write short notes on following.

(*i*) Hydrodynamic lubrication

(*ii*) Acid value and Saponification value

(*iii*) Flash point and fire point

(*iv*) Extreme pressure lubrication. ***(K.U.K. June 2009)***

59. What are biodegradable lubricants? What are advantages of biodegradable lubricants over general lubricants? Discuss its importance.

60. Write a short note on

(*i*) Biodegradable lubricants

(*ii*) Iodine value.

Chapter 7

POLYMERS AND POLYMERIZATION

7.1 INTRODUCTION

Polymer is a Greek word (Poly = many and mers = units) meaning thereby that polymer is made up of a large number of repeating units of monomers. For example, polyethene is a polymer of monomer ethene.

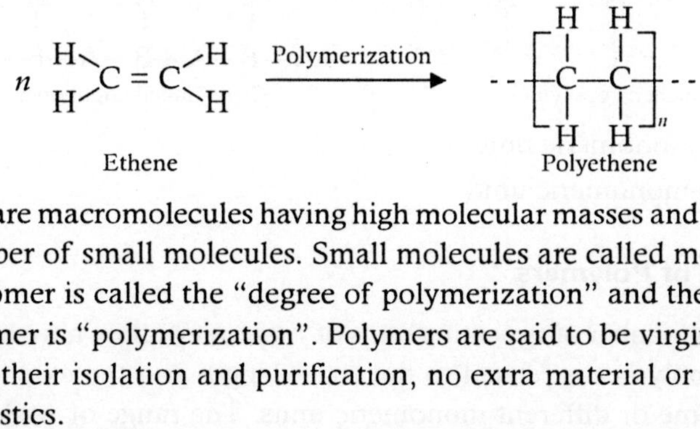

Ethene Polyethene

Hence polymers are macromolecules having high molecular masses and formed by the combining together of a large number of small molecules. Small molecules are called monomers. The number of repeating units of monomer is called the "degree of polymerization" and the process of formation of polymer from its monomer is "polymerization". Polymers are said to be virgin if they are in their pure form. Besides that after their isolation and purification, no extra material or compound is required to improve their characteristics.

All polymers are macromolecules but all macromolecules are not polymers because they do not contain the repeating units of monomers. For example, cholesterol, chlorophyll etc. are macromolecules not polymers. Polymerization always takes place in monomers which have at least two reacting sites or multiple bond. The reacting sites may be active functional groups. For example, HOOC—$(CH_2)_2$— COOH, H_2N —$(CH_2)_5$ —COOH etc. are bifunctional monomers. If the monomer is polyfunctional it forms a three-dimensional network polymer *i.e.,* cross-linked polymer. The number of reactive sites in a monomer is called its functionality. If the functionality is two *i.e.,* the monomer is bifunctional then it forms a long chain unbranched polymer like

- - —A—A—A—A— - -   - - —A—B—A—B— - -

where 'A' and 'B' are monomeric unit.

If the functionality is more than two then it is called polyfunctional and in this case monomeric units form a cross-linked chain.

For examples,

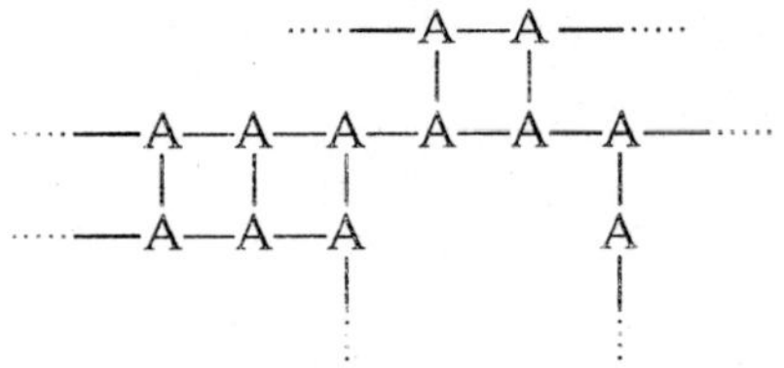

Cross-linked homopolymer

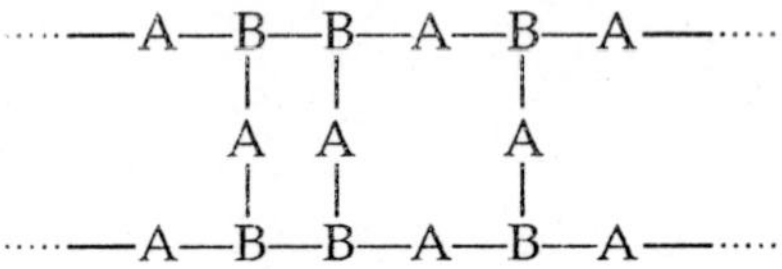

Cross-linked copolymer

(where A = Trifunctional (polyfunctional) monomeric unit)

(where A = Bifunctional monomeric unit
and B = Trifunctional (polyfunctional) monomeric unit)

If we take a mixture of bifunctional and trifunctional monomers, their ratio decides whether the polymer shall be branched or cross-linked.

For examples,

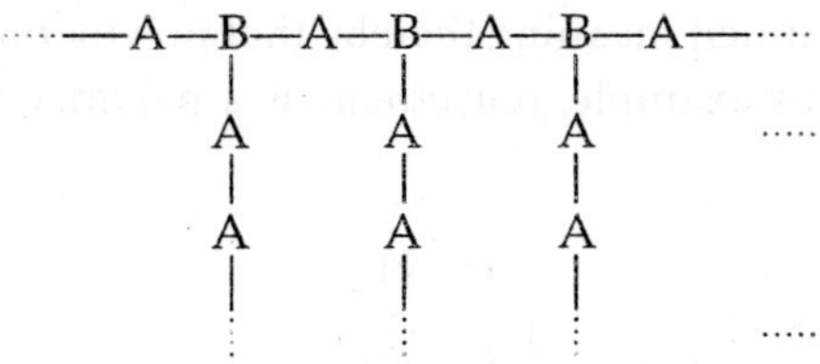

Branched copolymer

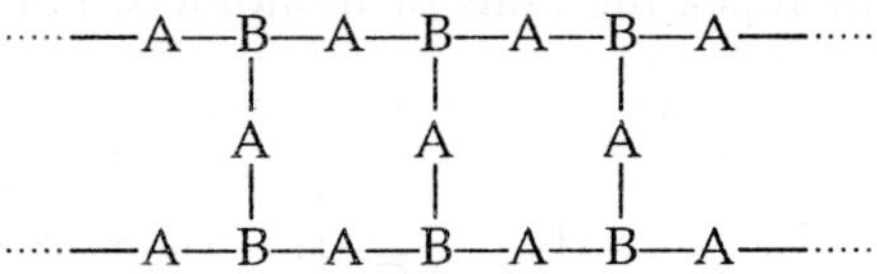

Cross-linked copolymer

(where A = Bifunctional monomeric unit
and B = Trifunctional monomeric unit)

7.1.1 Molecular Weight of Polymers

Since polymer is made of a number of repeating units of monomers, they have high molecular weights. The molecular weight of polymer is related to the chain length and the extent of cross-linking. The polymer is made of the same or different monomeric units. The range of molecular weights is fairly wide. Therefore, the molecular weight of polymer is considered as an average of the molecular weight of the molecules present. The molecular weight of a polymer is determined in terms of the following:

(*i*) Number-average molecular weight, ($\overline{M_n}$)

(*ii*) Weight-average molecular weight, and ($\overline{M_w}$).

(*i*) ***Number-average molecular weight*** ($\overline{M_n}$). The number-average molecular weight ($\overline{M_n}$) of a polymer is calculated by

$$\overline{M_n} = \frac{n_1M_1 + n_2M_2 + n_3M_3 + \ldots\ldots}{n_1 + n_2 + n_3 + \ldots\ldots}$$

$$= \frac{\sum n_i M_i}{\sum n_i}$$

$$= \frac{\text{Total mass of the polymer sample}}{\text{Number of molecules present in the sample}}$$

where $n_1, n_2, n_3, \ldots\ldots, n_i$ are the number of molecular species having molecular masses $M_1, M_2, M_3, \ldots\ldots, M_i$ respectively.

An exact weight of a polymer sample is dissolved into a suitable solvent to get a polymer solution of known concentration. Each molecule of the sample in spite of different size or weight, makes an equal contribution to the depression in freezing point (Cryoscopy), elevation in boiling point (Ebullioscopy), osmotic pressure (Osmometry) and lowering in vapour pressure. The measurement of an appropriate colligative property thus affords the number average molecular weight ($\overline{M_n}$). The number average molecular weight ($\overline{M_n}$) is also determined by end-group analysis method.

(*ii*) ***Weight-average molecular weight*** ($\overline{M_w}$). The weight-average molecular weight ($\overline{M_w}$) of a polymer is given by

$$\overline{M_w} = \frac{w_1 M_1 + w_2 M_2 + w_3 M_3 + \ldots\ldots}{w_1 + w_2 + w_3 + \ldots\ldots}$$

where $w_1, w_2, w_3, \ldots\ldots$ represents the total mass of the species having molecular weights $M_1, M_2, M_3, \ldots\ldots$ respectively.

$$\therefore \qquad \overline{M_w} = \frac{n_1 M_1^2 + n_2 M_2^2 + n_3 M_3^2 + \ldots\ldots}{n_1 M_1 + n_2 M_2 + n_3 M_3 + \ldots\ldots}$$

$$(\because \quad w_1 = n_1 M_1, \quad w_2 = n_2 M_2, \quad w_3 = n_3 M_3, \quad \ldots\ldots\ldots\ldots)$$

$$\overline{M_w} = \frac{\sum n_i M_i^2}{\sum n_i M_i} = \frac{\sum n_i M_i^2}{w}$$

where w = total weight of the polymer sample. (*i.e.*, $w = \sum n_i M_i$).

Weight-average molecular weight of a polymer is determined by light scattering and ultracentrifuge/sedimentation-velocity techniques which depend mainly on the size/weight of the polymer molecules and only to a small extent on their number. These methods are absolute and better than number-average molecular weight, and serve as the ultimate basis for a third type of average termed as viscosity-average molecular weight ($\overline{M_v}$).

Poly Dispersity Index (PDI)

It is the ratio of weight average and number average of a polymer.

i.e.,

$$\text{PDI} = \frac{\overline{M_w}}{\overline{M_n}}$$

Natural polymers are also known as monodisperse polymers since their molecules have same or a narrow range of molecular masses.

For natural polymer PDI = 1

Synthetic polymers are also known as polydisperse polymers since their molecules have a wide range of molecular masses. For synthetic polymer PDI > 1.

7.1.2 Glass Transition Temperature (T_g)

The temperature at above which the polymer is soft and below which it is hard and brittle is known as glass transition temperature. It has been found that there is a temperature boundary for almost all amorphous polymers and many crystalline polymers only above which the substance remains soft, flexible, brittle and glassy. For example, on cooling the ordinary ruober ball below – 70°C, it becomes so hard and brittle that it will break into pieces like a glass ball falling on a hard surface but at above – 70°C it remains soft and flexible. This temperature is known as glass transition temperature of rubber ball. The hard, brittle state is known as the glassy state and the soft flexible state is known as viscoelastic state. On further heating the uncross linked polymer becomes a highly viscous liquid and starts flowing. This state is known as viscofluid state and the required temperature is termed as flow temperature (T_f).

The softening temperature of polystyrene is lower than that of polyethene although the former contains polar group. It is due to the fact that in polystyrene, the bulky phenyl groups attached to the chain prevent a close approach of the polymer chain and hence the attractive forces between them decrease.

7.2 CLASSIFICATION OF POLYMER AND POLYMERIZATION

These are of the following types:

1. On the basis of their occurrence in nature polymers have been classified into two types:
   (*i*) Natural polymer and (*ii*) Synthetic polymer.
   (*i*) *Natural polymers.* The polymers which occur in nature are called natural polymers, for example, proteins, cellulose, starch, natural rubber etc.
   (*ii*) *Synthetic polymers.* The polymers which have been prepared in the laboratory by synthetic routes are called synthetic polymers. These are also known as man made polymers, for example, bakelite, rubber, PVC, polyethene etc.
2. On the basis of monomer unit, it is of two types:
   (*i*) Homopolymer and (*ii*) Copolymer.
   (*i*) *Homopolymer.* If the polymer consists of the same monomer, it is called homopolymer like —A—A—A— where A is monomer unit. For example, Polyethene, polypropene etc.
   (*ii*) *Copolymer.* If the polymer consists of different type of monomers, it is called copolymer like —A—B—A—B— where 'A' and 'B' are monomeric units. For example, nylon 6: 6, bakelite, styrene-butadiene rubber etc.
3. On the basis of polymeric structure, it is of three types:
   (*i*) Linear, (*ii*) Branched chain, and (*iii*) Cross-linked.

(*i*) *Linear polymer.* When monomeric units are combined in a linear fashion it is called 'linear polymer'.

—A—A—A—	—A—B—A—B—
Linear Homopolymer	Linear Copolymer
or	or
(Homochain Polymer)	(Heterochain Polymer)

(*ii*) *Branched polymer.* When monomeric units are joined together in a branched way it is called 'branched polymer'.

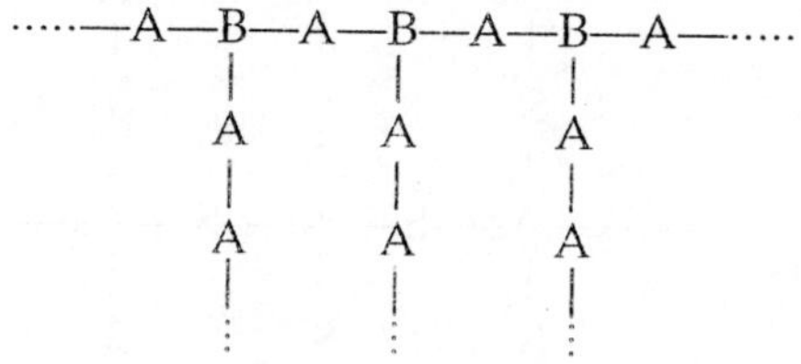

Branched Copolymer

(where A = Bifunctional monomer and B = Trifunctional monomer)

(*iii*) *Cross-linked polymers.* When monomeric units are joined together in a cross-chain fashion, it is called 'cross-linked polymer'.

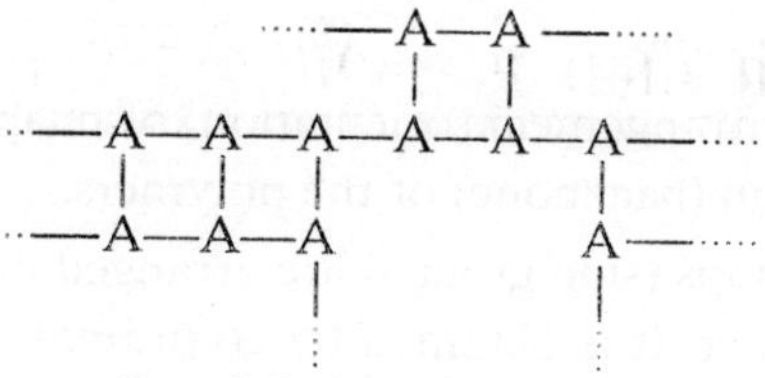

Cross-linked homopolymer

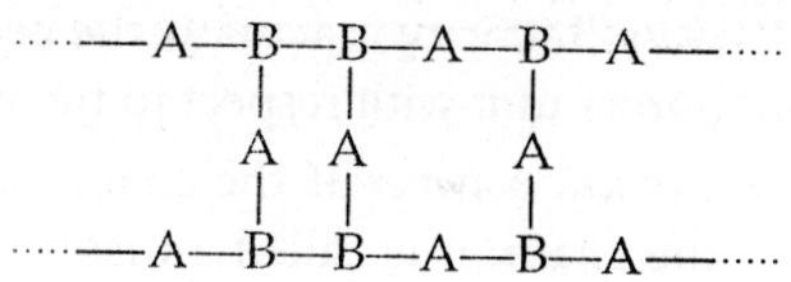

Cross-linked copolymer

(Where A = Trifunctional monomeric unit)

(Where A = Bifunctional monomeric unit and B = Trifunctional monomeric unit)

4. On the basis of monomeric unit in the backbone of polymer, it is classified into two types:

(*i*) Organic polymer, and (*ii*) Inorganic polymer.

(*i*) *Organic polymer.* If the backbone (main chain) of polymer is made of only carbon atom, it is called organic polymer. For example,

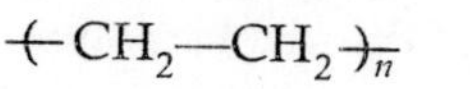

Polyethene

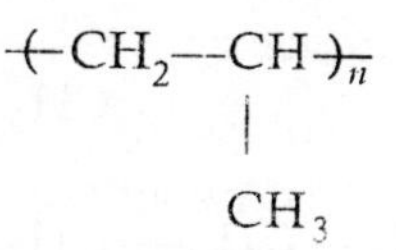

Polypropene

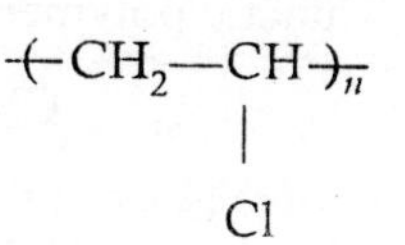

Polyvinylchloride (PVC)

(*ii*) *Inorganic polymer.* If the backbone of polymer is made of inorganic (other than carbon) atom like oxygen, nitrogen, sulphur, etc., it is called inorganic polymers. For examples, Polyphosphazine, silicones, chalcogenised glasses etc.

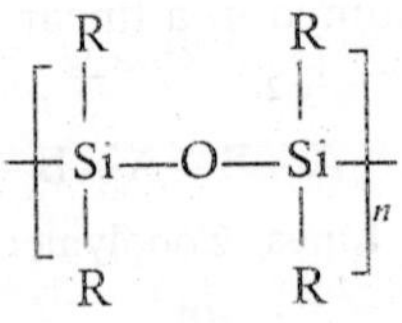

Silicones

where R = Alkyl or halide groups

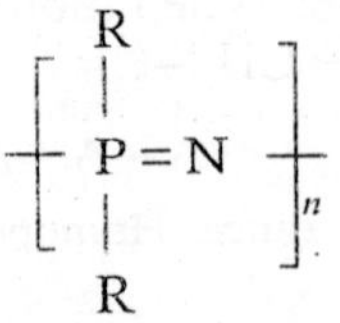

Polyphosphazine

where R = —Cl, —OCH_3, —OC_2H_5

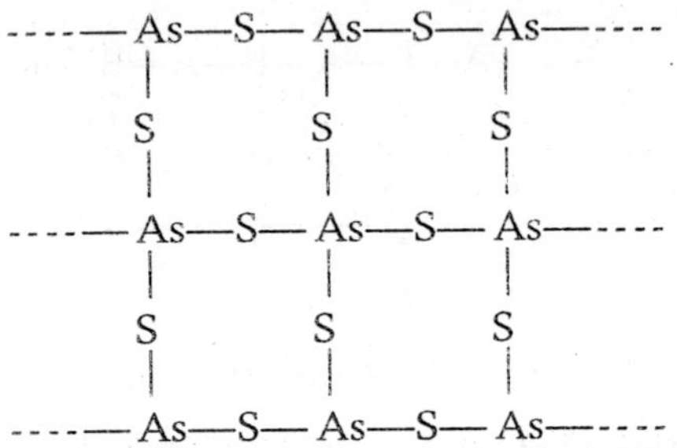

(Chalcogenised glasses)

5. On the basis of tacticity the polymers are classified into three classes:

(*i*) Isotactic, (*ii*) Syndiotactic and (*iii*) Atactic.

Tacticity. Tacticity is actually the geometric arrangement (orientation) of characteristic group of monomer unit with respect to the main chain (backbone) of the polymers.

(*i*) *Isotactic polymer.* If the characteristic groups (side groups) are arranged on the same side of the chain, it is called an isotactic polymer. It is obtained by co-ordination polymerization (using Ziegler-Natta catalyst). It is more crystalline and strong due to the presence of regularity of the chain structure.

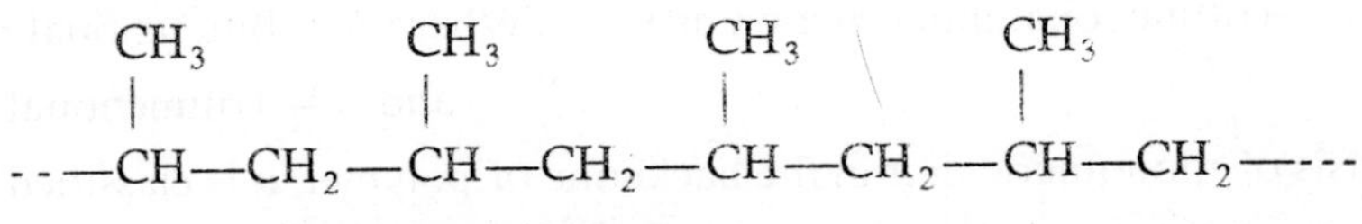

Isotactic polypropene

(*ii*) *Syndiotactic polymer.* If the side groups (characteristic groups) are arranged in an alternate fashion, it is called the syndiotactic polymer. It is also stronger and more crystalline than atactic polymer.

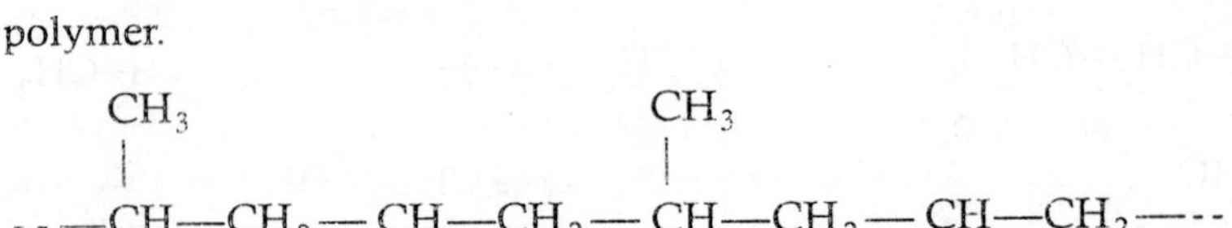

Syndiotactic polypropene

(*iii*) *Atactic polymer.* If the characteristics groups (side groups) are arranged in irregular fashion (randomness) around the main chain, it is called atactic polymer. It has poor strength and more elasticity and is obtained by free radical polymerization process.

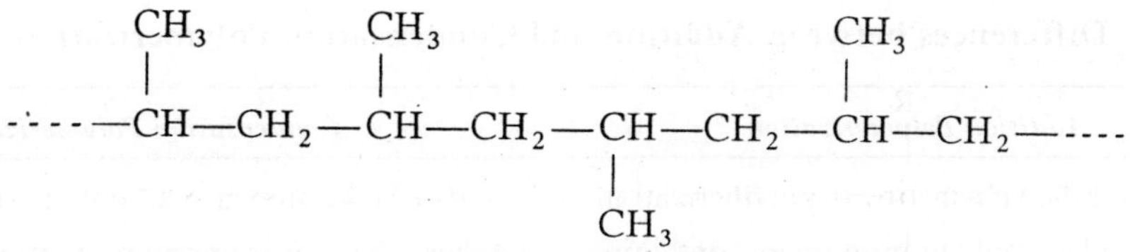

Atactic polypropene

Syndiotactic and isotactic polymers are more crystalline than the atactic polymer because in syndiotactic and isotactic polymers the characteristics groups are in a regular orientation which allow the adjacent chains to approach each other so that the attractive force may be effective.

6. On the basis of synthesis (Chemical combination) polymers are of two types:

(*i*) Addition polymer and (*ii*) Condensation polymer.

(*i*) *Addition polymer.* If the polymer is formed by the repeating of an exact multiple of the original monomeric molecule it is called 'addition polymer' and the process of polymerization is known as 'addition polymerization'. Such polymerization always occurs through free radicals or carbocations or carbanions intermediates and favoured by heat, light, pressure or a catalyst.

For examples,

1. $$\underset{\text{Ethene}}{nCH_2=CH_2} \longrightarrow \underset{\text{Polyethene}}{\left(CH_2-CH_2\right)_n}$$

2. $$\underset{\text{Vinyl chloride}}{nCH_2=CH-Cl} \longrightarrow \underset{\text{Polyvinylchloride}}{\left(CH_2-\underset{Cl}{\underset{|}{CH}}\right)_n}$$

(*ii*) *Condensation polymer.* If the polymer is formed by the combination of two or more similar or different molecules of monomeric units usually with the loss of simple molecules like NH_3, H_2O etc., it is called 'condensation polymer' and the process of polymerization is known as 'condensation polymerization'. The basic requirements for these types of polymerization are the presence of some reactive functional groups like —OH, —COOH, etc. in the monomeric units.

For examples,

1. $$\underset{\text{Hexamethylenediamine}}{n[H_2N-(CH_2)_6-NH_2]} + \underset{\text{Adipic acid}}{n[HOOC-(CH_2)_4-COOH]} \xrightarrow[-\,2nH_2O]{\Delta} \underset{\text{Nylon 6:6}}{\left[NH-(CH_2)_6-NH-CO-(CH_2)_4-CO\right]_n}$$

2. 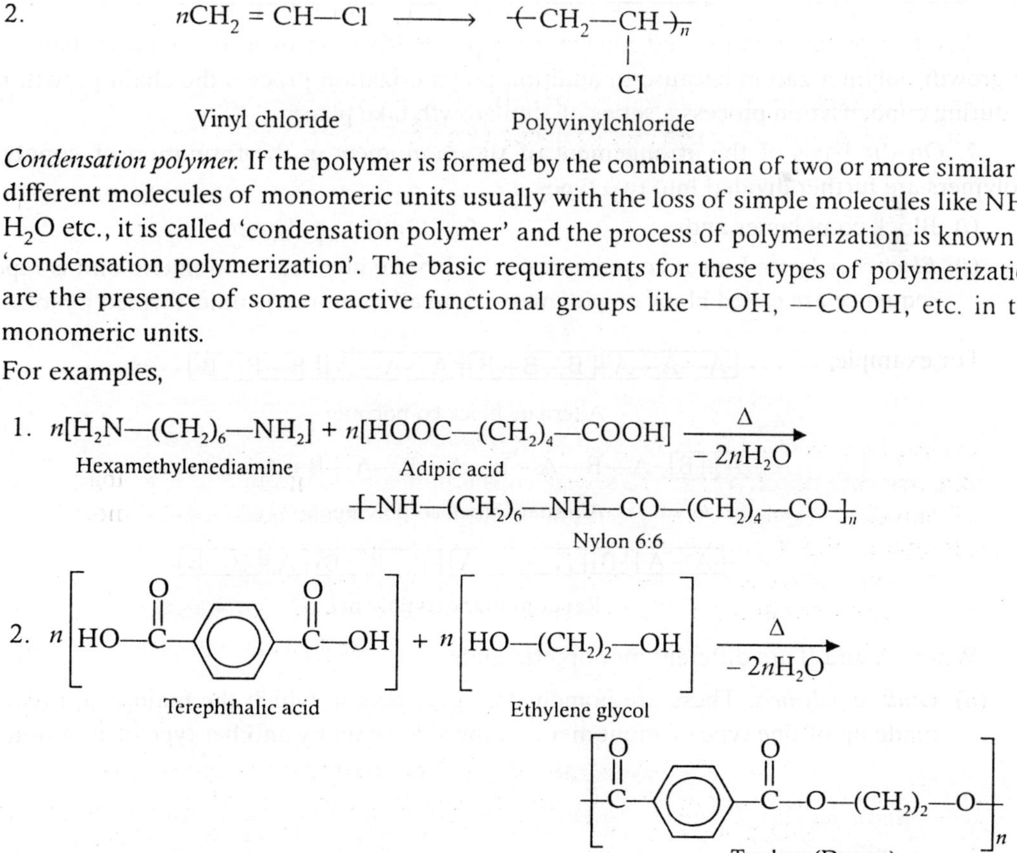

Differences between Addition and Condensation Polymerization

Sr. No.	*Addition Polymerization*	*Condensation Polymerization*
1.	It is also called chain growth polymerization.	It is also called step growth polymerization.
2.	It takes place only in monomers containing multiple bonds.	It takes place in monomers containing functional groups.
3.	The concentration of monomer decreases steadily throughout reaction.	The concentration of monomer disappears early in the reaction.
4.	Elimination of simple molecule does not take place.	Elimination of simple molecule like H_2O, NH_3, HCl etc., are essential.
5.	High molecular weight polymer is formed at once.	The molecular weight of polymer rises steadily throughout the reaction.
6.	Long reaction time give high yields but molecular weight of polymer is affected slightly.	Long reaction time are essential to obtain high molecular weights.
7.	The product is generally a thermoplastic.	The product may be either thermoplastic or thermosetting plastic.
8.	Polyethylene, polystyrene, polypropene, poly (vinyl chloride) etc., are its examples.	Bakelite, urea formaldehyde resin, epoxy resins, etc., are its example.

Addition polymerization and condensation polymerization are also termed as chain growth and step growth polymerization because in addition polymerization process the chain growth take place and during condensation process a series of step growth take place.

7. On the basis of the arrangement of the monomers in the formation of copolymers, the copolymers are further divided into two types:

(*i*) Block co-polymer and (*ii*) Graft co-polymer.

(*i*) *Block co-polymer.* Linear co-polymers in which the identical monomeric unit is repeated in a sequence are called block co-polymers. Here all monomeric units formed a block.

For example, - - -[A—A—A]-[B—B—B]-[A—A—A]-[B—B—B]- - -

Alternate block co-polymer

- - -[A]-[B]-A—B—A—B—A—B—A—B—A—B— - - - -

Alternate block co-polymer

- - -[A—A]-[B]-[A—A—A]-[B—B—B]-[A]-[B—B]-

Random block co-polymer

Where A and B are different monomeric units.

(*ii*) *Graft co-polymers.* These are branched co-polymers in which the main chain (backbone) is made up of one type of monomer and the side chain by another type of monomer.

For example,

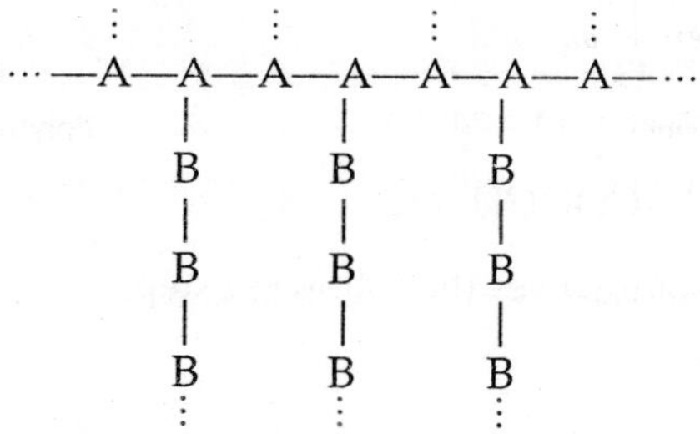

Graft co-polymers are very useful and play very important role in industries.

It has been observed that in several polymerization it is very difficult to identify whether the polymer has occurred through an addition or a condensation process. For example, Nylon-6 is prepared by both processes, *i.e.*, addition polymerization as well as condensation polymerization process. During addition polymerization nylon-6 is prepared by caprolactum as follows:

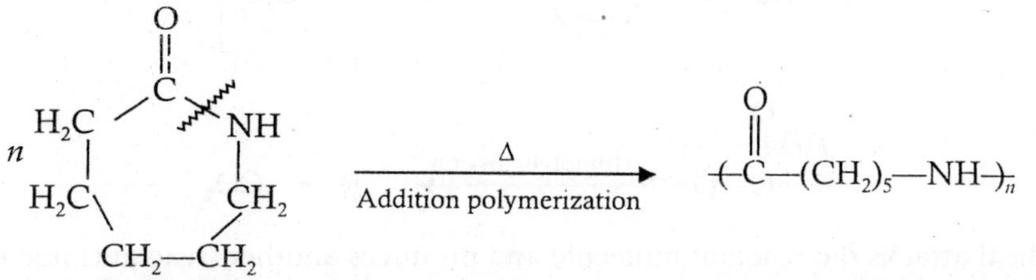

Caprolactum Nylon 6

During condensation process nylon-6 is also prepared by monomer 6-aminohexanoic acid as follow:

$$n\mathrm{H_2N{-}(CH_2)_5{-}COOH} \xrightarrow[-n\mathrm{H_2O}]{\Delta} \mathrm{+\!\!\!-\!NH{-}(CH_2)_5{-}\overset{\overset{\displaystyle O}{\|}}{C}\!-\!\!\!+_n}$$

6-aminohexanoic acid Condensation polymerization Nylon 6

Here the problem is to identify such type of polymerization. Due to this reason on the basis of the way (mode) of addition of monomeric units in a polymers, these (polymers) have been classified into two types:

(A) Chain growth polymers and (B) Step growth polymers.

(A) Chain Growth Polymers

It is the type of polymers which are formed by the successive addition of monomer units to the growing chain through the generated reactive intermediate like free radical, carbocation or carbanion and the process is called chain growth polymerization. According to the intermediate formation these are of the following types:

(*a*) ***Free radical polymerization.*** Polymerizations initiated by catalysts, called initiators producing free radicals are known as free radical polymerization. Generally organic peroxides (benzoylperoxide or acetylperoxide, etc.) are used as free radical initiators. For example, polymerization of substituted alkene takes place as

$$n\,CH_2{=}\underset{\underset{\displaystyle G}{|}}{CH} \xrightarrow{\text{Organic peroxide}} \left(CH_2{-}\underset{\underset{\displaystyle G}{|}}{CH} \right)_n$$

Monomer Polymer

where G = —Cl, —CN, —C_6H_5, —$COOCH_3$ etc.

Mechanism. The mechanism involves the following steps:

(*i*) ***Chain initiation***

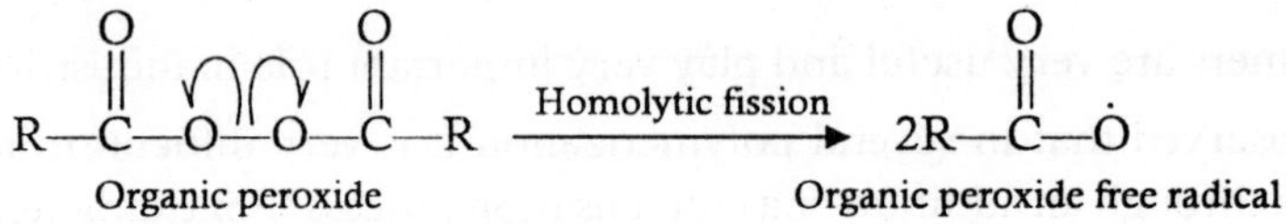

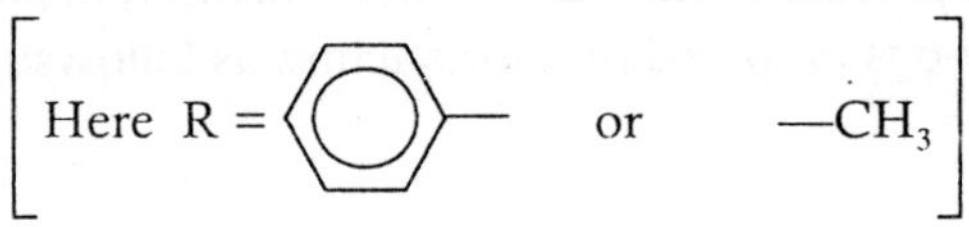

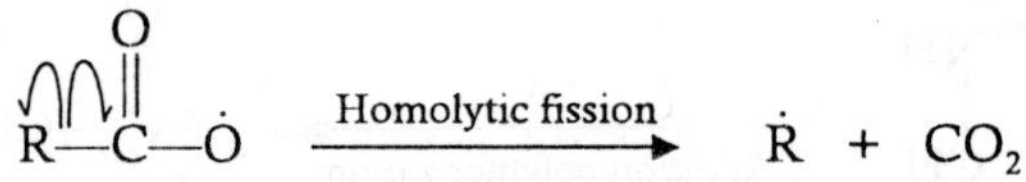

Free radical attacks the reactant molecule and produces another (reactant) free radical.

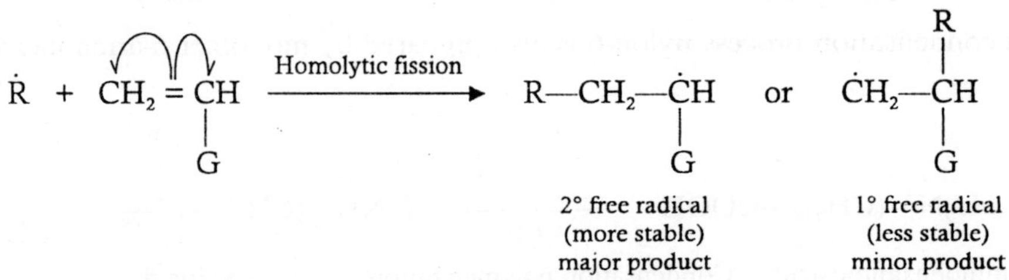

(*ii*) ***Chain propagation.*** The generated free radicals attack another molecule of reactants (alkene) and propagates the process.

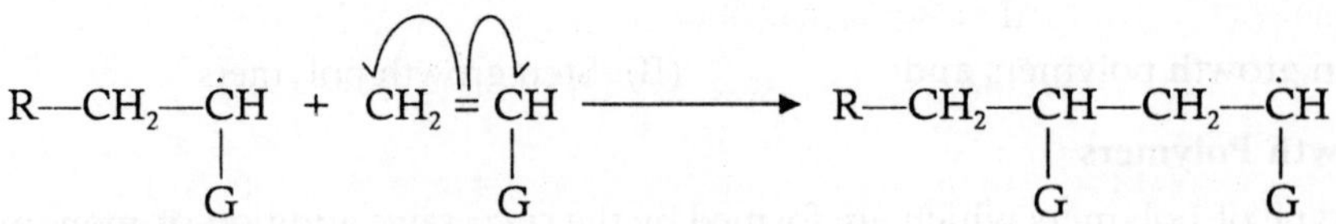

This step is repeated many times to form a desired long chain.

(*iii*) ***Chain termination.*** This step may take place due to coupling or disproportionation or chain transfer reactions.

(*a*) *Coupling.* It involves collision between free radical chains or between growing free radical chain and a catalyst radical.

(*i*)

$$\mathrm{R{+}CH_2{-}\underset{G}{\underset{|}{CH}}{+}_n CH_2{-}\underset{G}{\underset{|}{\dot{C}H}} + \underset{G}{\underset{|}{\dot{C}H}}{-}CH_2{+}\underset{G}{\underset{|}{CH}}{-}CH_2{+}_n R \longrightarrow}$$

(free radical growing chain) (another free radical growing chain)

$$\mathrm{R{+}CH_2{-}\underset{G}{\underset{|}{CH}}{+}_n CH_2{-}\underset{G}{\underset{|}{CH}}{-}\underset{G}{\underset{|}{CH}}{-}CH_2{+}\underset{G}{\underset{|}{CH}}{-}CH_2{+}_n R}$$

(*ii*)

$$\mathrm{R{+}CH_2{-}\underset{G}{\underset{|}{CH}}{+}_n CH_2{-}\underset{G}{\underset{|}{\dot{C}H}} + \dot{R}' \longrightarrow R{+}CH_2{-}\underset{G}{\underset{|}{CH}}{+}_n CH_2{-}\underset{G}{\underset{|}{CH}}{-}R'}$$

(free radical growing chain) (initiator free radical)

(*b*) *Disproportionation.* It involves the accepting of one hydrogen by one free radical from the other resulting in mutual deactivation.

$$\mathrm{R{+}CH_2{-}\underset{G}{\underset{|}{CH}}{+}_n CH_2{-}\underset{G}{\underset{|}{\dot{C}H}} + \underset{G}{\underset{|}{\dot{C}H}}{-}CH_2{+}\underset{G}{\underset{|}{CH}}{-}CH_2{+}_n R \longrightarrow}$$

(free radical growing chain) (another free radical growing chain)

$$\mathrm{R{+}CH_2{-}\underset{G}{\underset{|}{CH}}{+}_n CH_2{-}\underset{G}{\underset{|}{CH_2}} + R{+}CH_2{-}\underset{G}{\underset{|}{CH}})_n{-}CH = \underset{G}{\underset{|}{CH}}}$$

(*c*) *Chain transfer.* It involves the abstraction of one atom of some impurity present in the monomer.

$$\mathrm{R{+}CH_2{-}\underset{G}{\underset{|}{CH}}{+}_n CH_2{-}\underset{G}{\underset{|}{\dot{C}H}} + AsH \longrightarrow R{+}CH_2{-}\underset{G}{\underset{|}{CH}}{+}_n CH_2{-}\underset{G}{\underset{|}{CH_2}} + \dot{A}s}$$

(impurity in monomer)

Examples of free radical polymerization.

(*i*)

$$n\mathrm{CH_2{=}\underset{Cl}{\underset{|}{CH}}} \longrightarrow \mathrm{{+}CH_2{-}\underset{Cl}{\underset{|}{CH}}{+}_n}$$

Vinyl chloride Poly (vinylchloride) (PVC)

PVC has a good flame resistance and electrical resistance. It is used in the preparation of electrical wire coatings, rain wears, soles of shoes, etc.

(*ii*)

$$n\mathrm{CH_2{=}\underset{Cl}{\underset{|}{CH}}} \longrightarrow \mathrm{{+}CH_2{-}\underset{CN}{\underset{|}{CH}}{+}_n}$$

Acrylonitrile Poly (acrylonitrile) (PAN) or Orlon

Orlon is used for making blankets, synthetic carpets etc.

(*iii*)

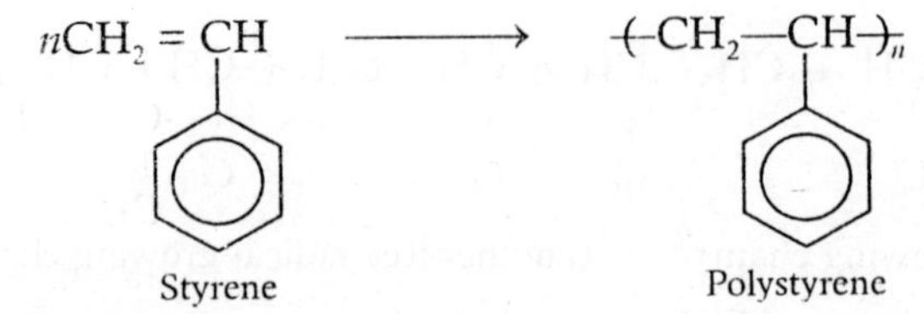

Polystyrene is used for making cups for hot drinks, toys, radio and television boxes etc.

(*b*) ***Cationic addition polymerization.*** This type of polymerization is initiated by acids like H_2SO_4, HF, etc., or Lewis acids like $AlCl_3$, BF_3, $SnCl_4$ etc., along with some traces of water. Alkenes carrying electron releasing substituents generally undergo cationic addition polymerization.

Role of water. In case of Lewis acids it produces proton (H^+)

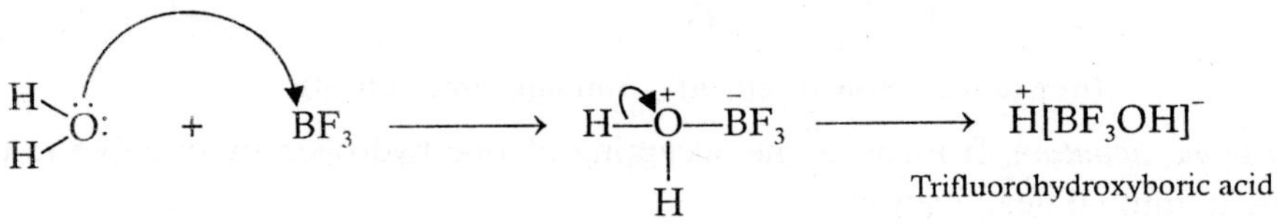

Examples: (*a*) Polymerization of propene

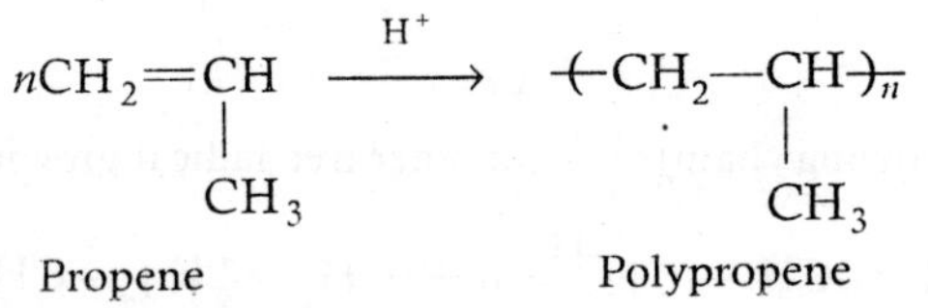

(*b*) Polymerization of isobutylene

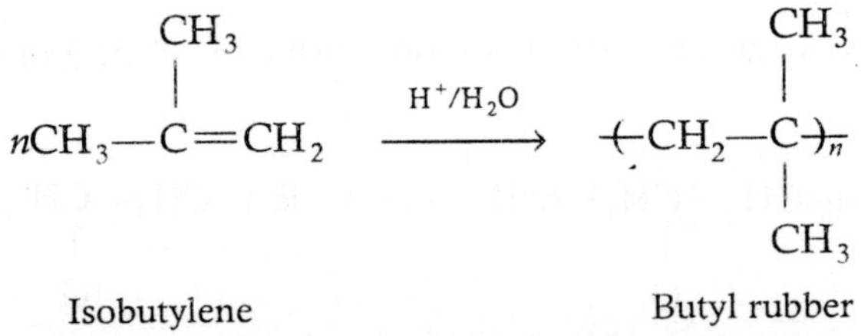

Mechanism. The mechanism involves in the following steps:

(*i*) *Chain initiation step.* Proton is formed in this step:

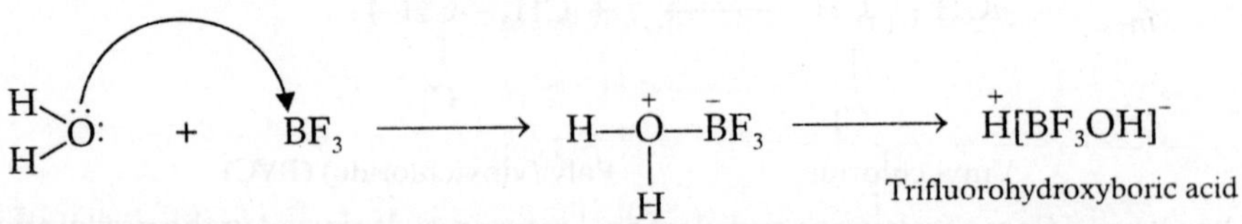

(*ii*) *Chain propagation step.* The proton H^+ attacks on reactant molecule and forms stable carbocation. This step is repeated and forms a long chain.

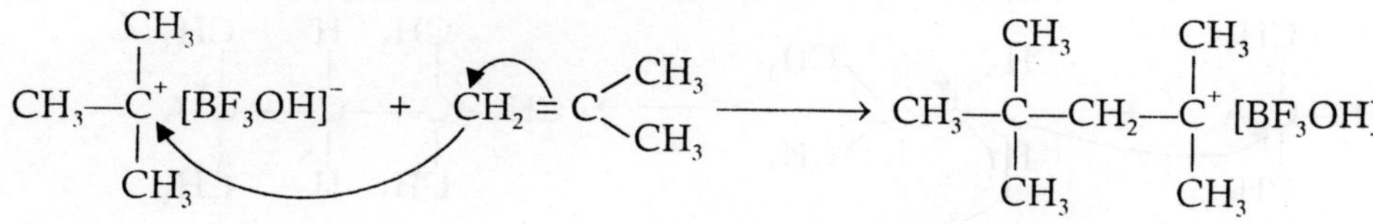

3° carbocation (most stable)

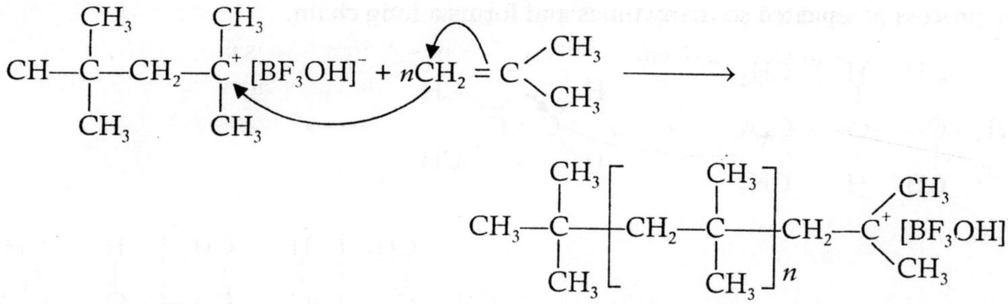

(*iii*) *Chain termination step.* It takes place by the loss of hydrogen ion (H^+) from the growing carbocation chain.

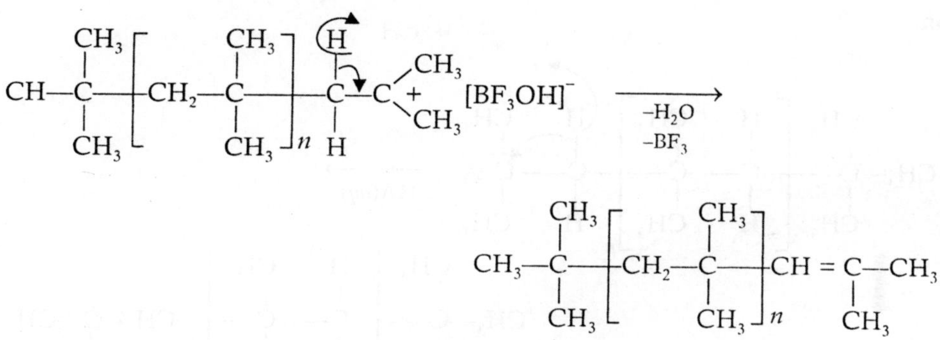

Polyisobutylene (Butyl rubber)

Cationic addition polymerization reactions generally takes place in the presence of lewis acid. If we use acids like HF, H_2SO_4, etc., in place of Lewis acid then the mechanism may be understood as follow:

(*i*) *Chain initiation step.* In this step proton is formed.

$$HA\,(aq) \rightleftharpoons H^+ + A^-$$

(where HA = HF or H_2SO_4 etc.)

(*ii*) *Chain propagation step.* During this step the proton (H^+) attacks on reactant molecule and forms stable carbocation.

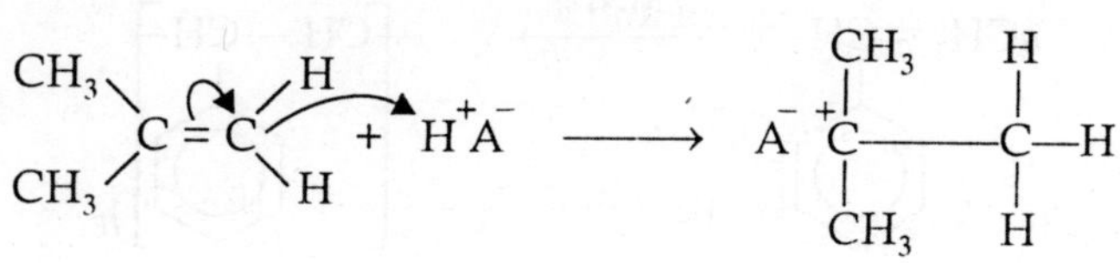

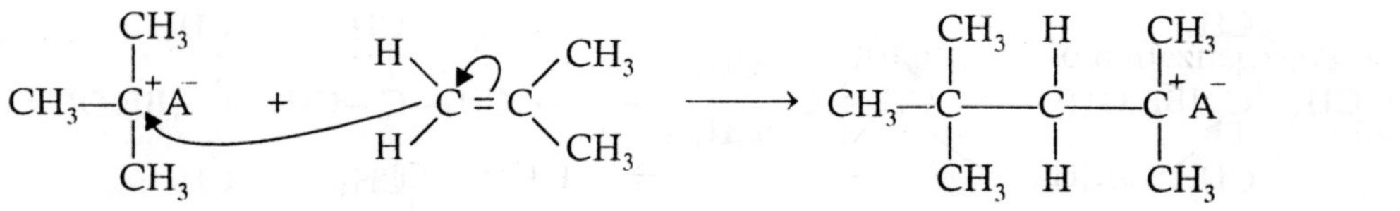

(3° carbocation more stable)

The process is repeated so many times and forms a long chain.

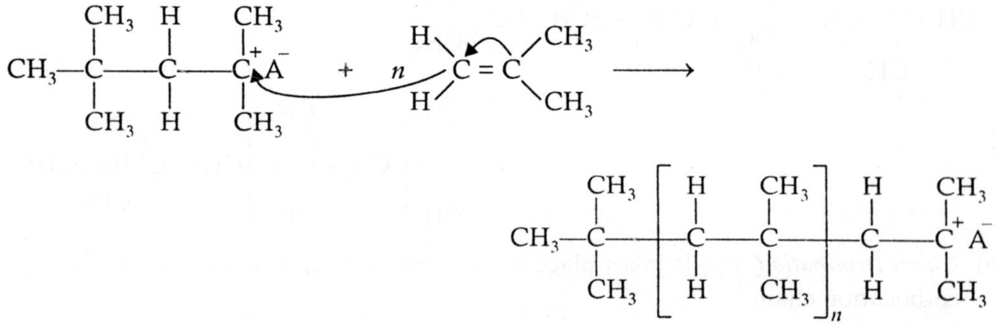

(*iii*) *Chain termination step.* This step takes place by the loss of hydrogen ion (H^+) from the growing carbocation.

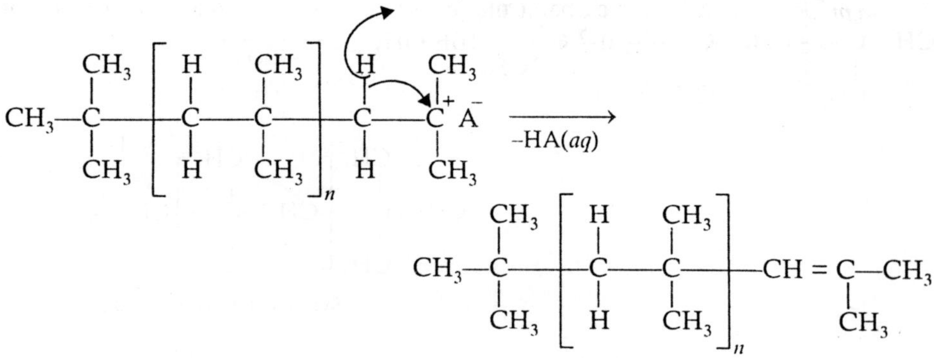

Polyisobutylene (Butyl rubber)

[**Note.** Only alkene carrying electron releasing substituents undergoes such polymerization. Ethylene, vinyl chloride etc., do not shows cationic polymerization.]

(*c*) ***Anionic addition polymerization.*** This type of polymerization is initiated by bases like alkali metals, alkyl metals (*n*-butyllithum), alkali metalamides and Grignard reagents. Alkene carrying electron withdrawing subsituents generally involves in anionic addition polymerization.

Examples:

(*a*) Polymerization of styrene

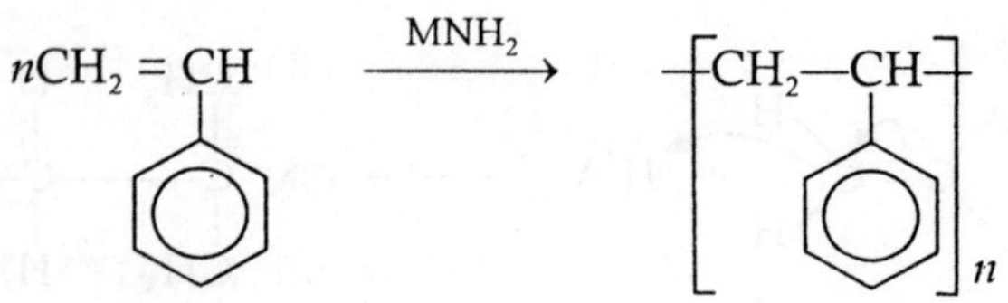

Styrene Polystyrene

(*b*) Polymerization of acrylonitrile

$$n\underset{\substack{|\\ \text{CN}}}{\text{CH}_2\text{=CH}} \xrightarrow{\text{NaNH}_2 \text{ in liq NH}_3} \left(\underset{\substack{|\\ \text{CN}}}{\text{CH}_2\text{—CH}}\right)_n$$

Acrylonitrile — Poly (acrylonitrile) (PAN) or Orlon

Mechanism. The mechanism involves the following steps:

(*i*) *Chain initiation step.* It involves in the formation of carbanion.

$$\text{NaNH}_2 \xrightarrow{\text{liq NH}_3} \text{Na}^+ + \text{NH}_2^-$$

$$\bar{\text{N}}\text{H}_2 + \underset{\substack{|\\ \text{CN}}}{\text{CH}_2\text{=CH}} \longrightarrow \underset{\substack{|\\ \text{CN}}}{\text{NH}_2\text{—CH}_2\text{—}\bar{\text{C}}\text{H}}$$

(2° carbanion)

(*ii*) *Chain propagation step.* This carbanion attacks reactants to form another carbanion. The step is repeated to form a long chain.

$$\underset{\substack{|\\ \text{CN}}}{\text{NH}_2\text{—CH}_2\text{—}\bar{\text{C}}\text{H}} + \underset{\substack{|\\ \text{CN}}}{\text{CH}_2\text{=CH}} \longrightarrow \text{NH}_2\text{—CH}_2\text{—}\underset{\substack{|\\ \text{CN}}}{\text{CH}}\text{—CH}_2\text{—}\underset{\substack{|\\ \text{CN}}}{\bar{\text{C}}\text{H}}$$

$$\xrightarrow{\text{Repetition}} \text{NH}_2\text{—CH}_2\text{—}\underset{\substack{|\\ \text{CN}}}{\text{CH}}\left[\text{CH}_2\text{—}\underset{\substack{|\\ \text{CN}}}{\text{CH}}\right]_n\text{CH}_2\text{—}\underset{\substack{|\\ \text{CN}}}{\bar{\text{C}}\text{H}}$$

(*iii*) *Chain termination step.* It takes place in the presence of H⁺ or some Lewis acid. In the absence of acids or H⁺ the reaction continues till the whole of monomer is polymerized.

$$\text{NH}_2\text{—CH}_2\text{—}\underset{\substack{|\\ \text{CN}}}{\text{CH}}\left[\text{CH}_2\text{—}\underset{\substack{|\\ \text{CN}}}{\text{CH}}\right]_n\text{CH}_2\text{—}\underset{\substack{|\\ \text{CN}}}{\bar{\text{C}}\text{H}} \xrightarrow{\text{NH}_3}$$

$$\text{NH}_2\left[\text{CH}_2\text{—}\underset{\substack{|\\ \text{CN}}}{\text{CH}}\right]_n\text{CH}_2\text{—}\underset{\substack{|\\ \text{CN}}}{\text{CH}_2} + \bar{\text{N}}\text{H}_2$$

Poly (acrylonitrile) (PAN) or Orlon

Orlon (PAN) can be spun into a fibre and can be made to look like silk or wool. It has excellent resistance to sunlight, outdoor exposure, acid etc. It can be woven or knitted and can be blended with wool. It is used in making carpets, sweaters, bathing suits etc.

(*d*) ***Coordination polymerization.*** If the polymerization reaction is catalysed by organo-metallic compounds then it is called coordination polymerization. During the first step of this polymerization a complex of monomer–catalyst is formed which propagates later on.

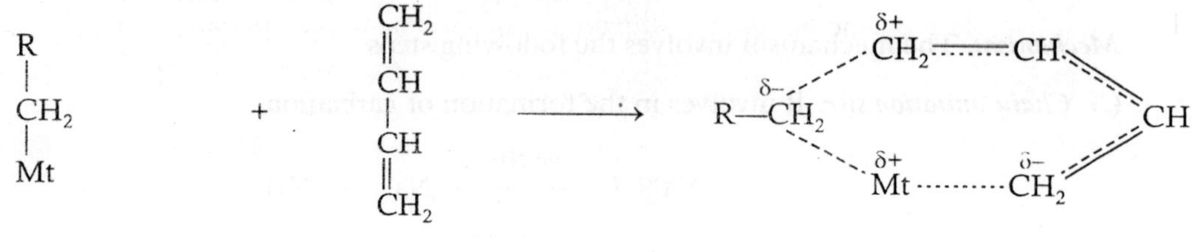

Organo metallic compound (catalyst) | Diene (monomer) | Monomer-catalyst complex

(where Mt represents transition metals like Ti, Mo, Cr, V, Ni & Rh)

Since during the formation of monomer–catalyst complex, a coordination bond is developed between a carbon atom of the monomer and the metal of the catalyst. Hence such type of polymerization is called coordination polymerization. This coordination bond acts as the active centre from where propagation starts. The incoming monomer is incorporated at the active centre of the metal-carbon bond, from where the chain growth starts as and so on.

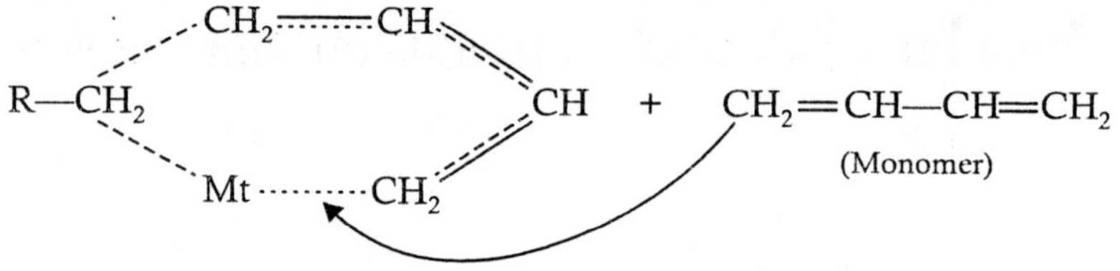

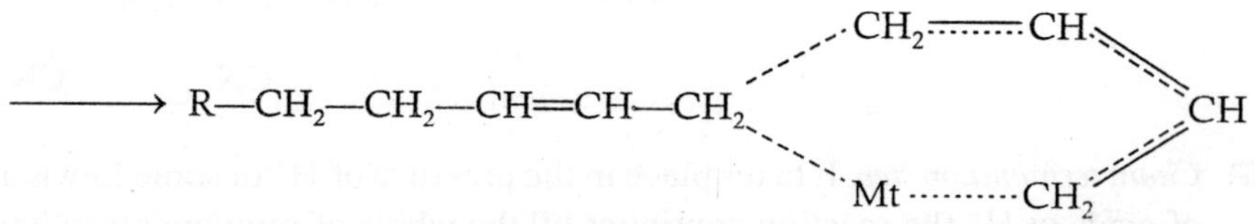

In this polymerization the carbanion of the alkyl group is in the solvent phase and metal ion is in the solid phase, so it is a heterogenous system. Since monomer is inserted between the metal ion and the carbanion so the resulting polymer chain is pushed out from the solid phase and this polymerization is also called insertion polymerization.

Important Features

Following are the important features of coordination polymerization:

1. This type of polymerization takes place especially in olefines and dienes.
2. Such type of polymerization catalysed by organometallic compound.
3. Depending on the polarity of the metal-carbon counter bond and the solvent medium, the metal counter-ion is placed in a particular spatial arrangement with respect to the anion.

4. In certain catalyst systems, the spatial arrangement has a tremendous effect on the spatial orientation and manner of the incoming monomer for insertation to the growing chain, and forms stereo regular polymers.

5. Such type of coordination polymerization can be used to produce highly stereo-regular polymers by choosing a proper catalyst & solvent system.

Ziegler-Natta catalyst is very well known and important which is used for this polymerization. Ziegler-Natta catalysts are formed by the interaction of alkyls or aryls of metals of group I to group III of the periodic table with the halides of transition metals. Generally, the Ziegler-Natta catalysts are made of $TiCl_4$ or $TiCl_3$ and triethylaluminium or diethylaluminium chloride. Here aluminium alkyls act as the electron acceptor and titanium halide act as the electron donor, so after combination coordinate complex is formed. Such type of complex is insoluble in solvent *i.e.*, heterogenous in nature.

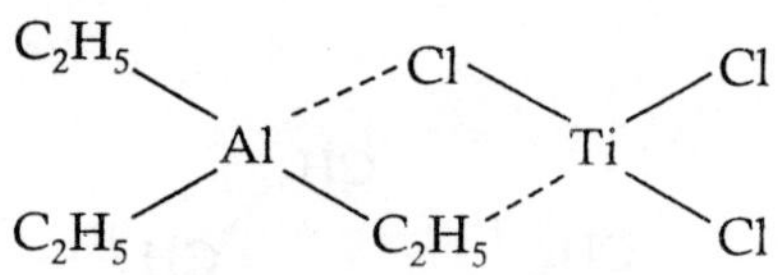

Ziegler-Natta Catalyst
$[TiCl_3 + (C_2H_5)_3Al]$

Here, active centre is formed between the bond of titanium and carbon of alkyl group *i.e.*, Ti—C bond. The monomer is attracted towards this bond. Due to this, the bond between Ti and carbon of alkyl group get opened and produces electron deficient Ti and a carbanion at carbon of alkyl group. The electron deficient Ti ion attracts the π electron pair of the monomer and forms a σ-bond, while the counterion attracts the electron-deficient centre of the monomer as follow:

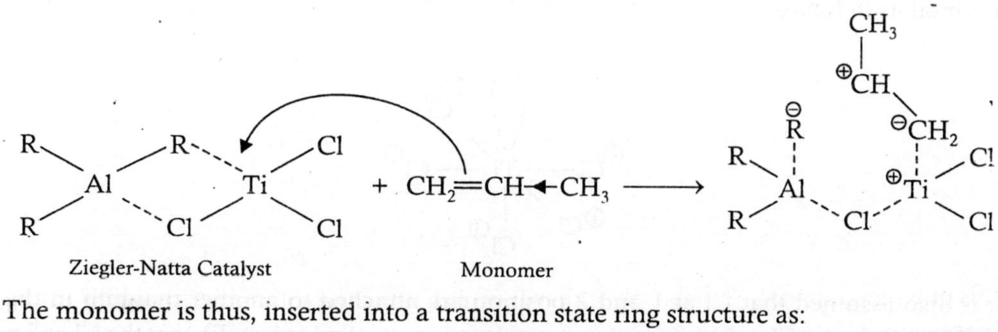

The monomer is thus, inserted into a transition state ring structure as:

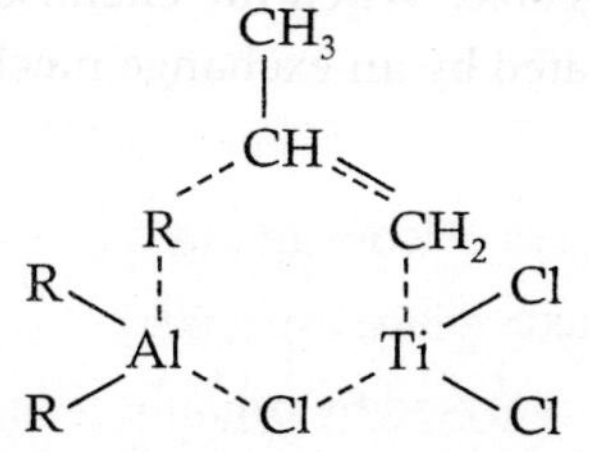

This transition state now regenerates the active centre for chain growth at the metal-carbon bond as follow:

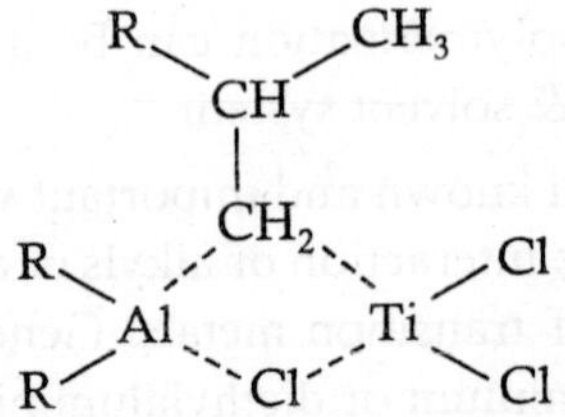

The insertion of monomer is repeated so many times in this manner and the orientation of the substituent group of the monomer is always taken from the metal-ion end. As a result stereo regulated polymer is formed as:

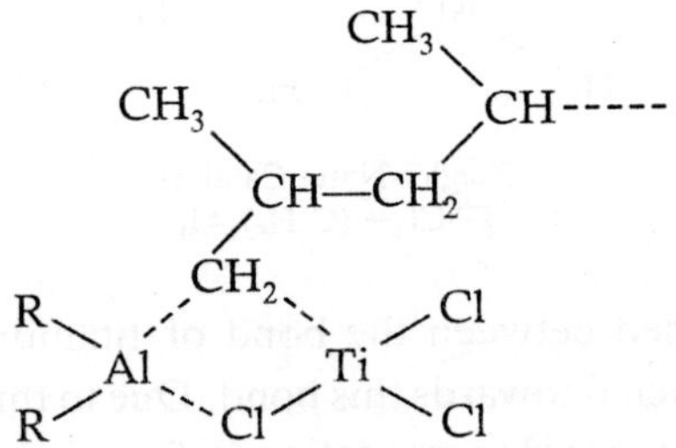

It is also believed that when catalyst and cocatalyst are mixed, the aluminium is chemisorbed on the solid $TiCl_3$ or $TiCl_4$. The five coordinated Ti ion on the surface of the catalyst is assumed to have a vacant *d*-orbital as in figure:

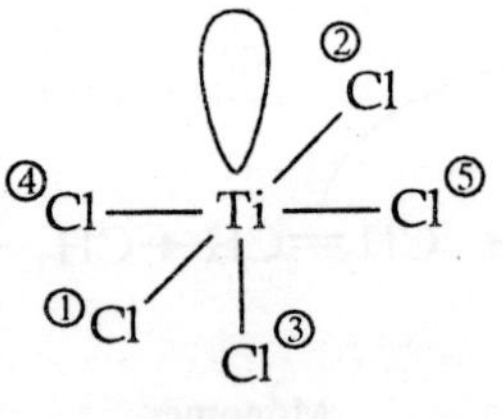

It is also assumed that Cl at 1 and 2 position are attached to another titanium in the crystal lattice of $TiCl_3$. Only the Cl at 5 position can be replaced by an alkyl group. Except the Cl at 5 position, all the other four Cl are non-exchangeable. When the chemisorption of the aluminium alkyl on the $TiCl_3$ crystal takes place, Ti^{3+} is alkylated by an exchange mechanism and forms an active catalyst as follow:

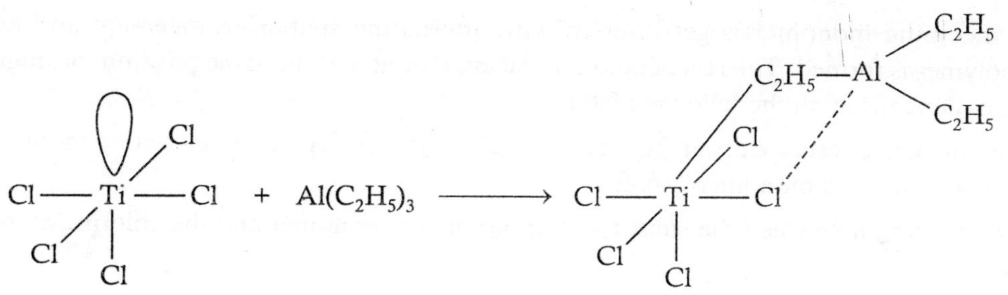

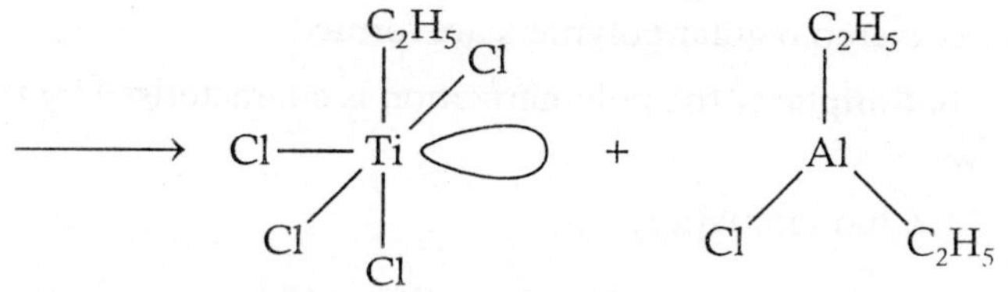

Here the four chlorine atoms are attached to the lattice and the ethyl group is attached to the Ti through a σ-bond alongwith a vacant orbital at the position where originally the fifth chlorine atom was attached and forms an octahedral structure. Once the active catalyst is formed, the monomer is attracted towards the vacant *d*-orbital and forms a transition π complex with the titanium as follow:

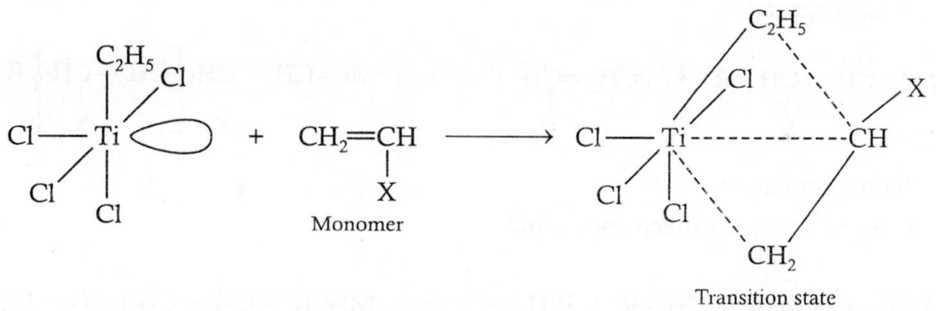

Here the active centre is Ti—C_2H_5 bond, so monomer is inserted at the Ti—C_2H_5 bond. When the monomeric group is inserted, the vacant *d*-orbital can either be regenerated at the same position or at the position where the ethyl group was originally attached as follow:

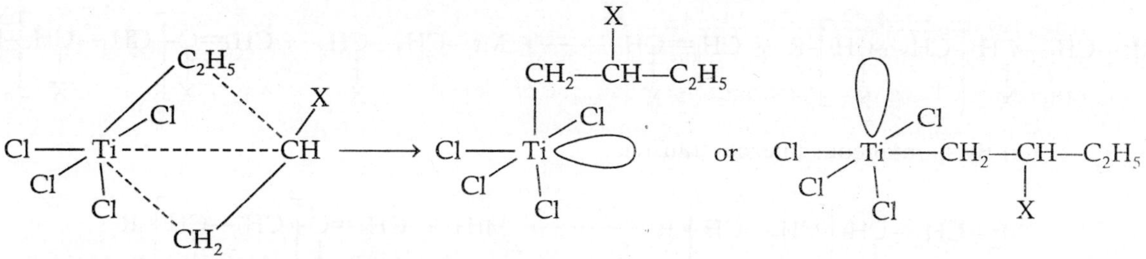

When the vacant *d*-orbital is always regenerated at the same position, the incoming monomeric units will be inserted with the same spatial arrangement and hence isotactic polymer is formed. If the vacant *d*-orbital migrates from one position to the other during the regeneration when the monomer is inserted, the incoming monomer is get inserted with alternating spatial arrangement and hence syndiotactic polymer is formed. The regeneration of vacant *d*-orbital at the same position, or migrates to the other sites depends upon the following factors.

1. The interactive forces existing between the substituent group of the incoming monomeric unit and the already inserted monomeric units.

2. The interaction between the substituent group of the monomer and the chlorine atoms of the active catalyst.

The crystalline structure of catalyst, type of catalyst, temperature of the catalyst complex and molar ratios of the catalyst components have much influence the shift or migration of the vacant *d*-orbital and hence stereo regular polymers are formed.

For Simplicity the polymerization is characterised by the initiation, propagation & termination as follows:

(*i*) Chain initiation

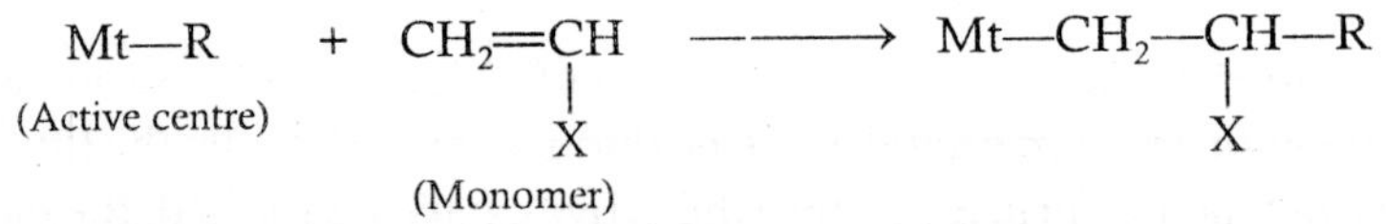

(where Mt represents transition metals like Ti, Mo, Cr, V, Ni, or Rh)

(*ii*) Chain propagation

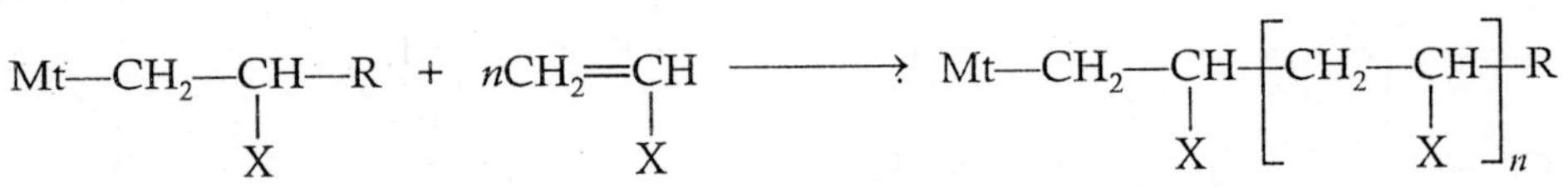

(*iii*) Chain termination

(*a*) By an active hydrogen compound

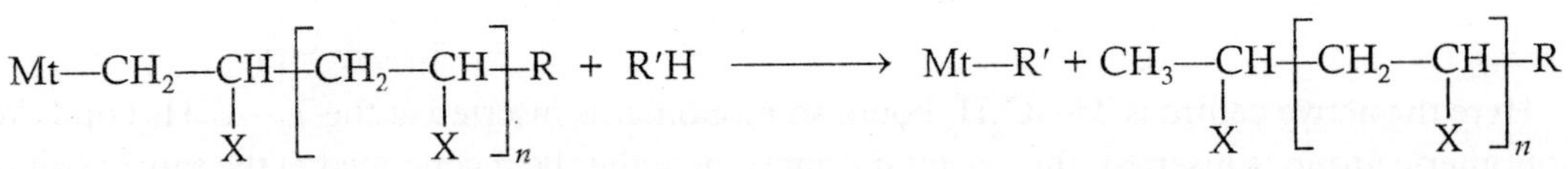

(*b*) By transfer with monomer

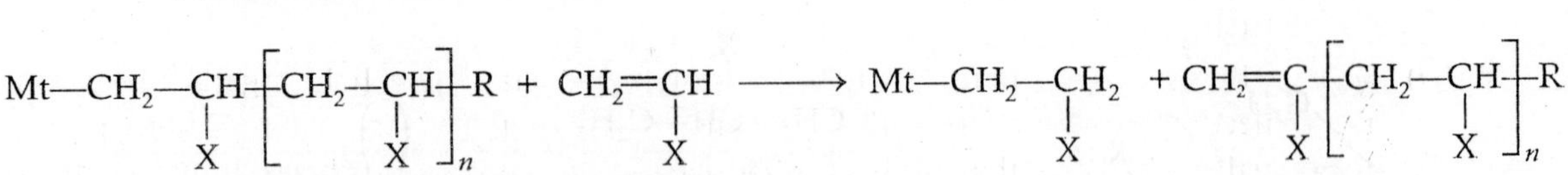

(*c*) By spontaneous internal transfer

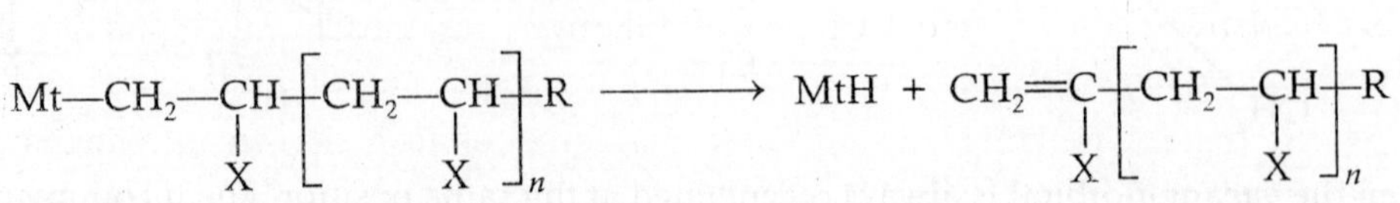

Coordination polymerization is used to prepare stereospecific polypropene, polydiene etc. About 90% or more isotactic polypropene can be prepared by this method.

(B) Step Growth Polymer

Polymers which are formed through a series of stepwise condensation reactions of bifunctional or polyfunctional monomers are called step growth polymer and the process is called step growth polymerization. The monomers must have at least two reactive sites either similar or different functional groups at the end of the chain. These polymers are formed through a series of independent step-wise reactions with or without elimination of smaller molecules. No initiator is required for such type of polymerization. The process begins with the condensation between two monomer units to form dimer and the process continues till a polymer is obtained. For example, polymerization takes place between hexamethylenediamine and adipic acid to form nylon-6 : 6.

$$\underset{\text{Hexamethylenediamine}}{n\mathrm{H_2N{-}(CH_2)_6{-}NH_2}} + \underset{\text{Adipic acid}}{n\mathrm{HOOC{-}CH_2{-}COOH}} \xrightarrow[-\,n\mathrm{H_2O}]{\text{Step growth polymerization}}$$

$$\underset{\text{Nylon 6 : 6}}{\left[\mathrm{HN{-}(CH_2)_6{-}\overset{\displaystyle H}{\overset{|}{N}}{-}\overset{\displaystyle O}{\overset{\|}{C}}{-}(CH_2)_4{-}\overset{\displaystyle O}{\overset{\|}{C}}}\right]_n}$$

Some chain growth polymers and their monomers are following:

Polymer:	Polyethylene	Polybutadiene	Poly (vinylchloride)	Teflon (Poly- -tetrafluroethene)
Monomer:	Ethylene	Butadiene	Vinylchloride	Tetrafluroethene

Following are some step growth polymers and their monomers:

Polymer:	Bakelite	Nylon-6 : 6	Terylene
Monomer:	Phenol and formaldehyde	Hexamethylenediamine and adipic acid	Terepthalic acid and ethyl-ene glycol.

8. On the basis of molecular forces polymers have been classified into the following categories:

(*i*) *Elastomers.* These are polymers in which the polymeric chain are held together by the weakest intermolecular attraction forces. Due to weak intermolecular attraction force elastomers can be stretched by applying an external force. For examples, natural rubber, synthetic rubber, silicon rubber, etc.

(*ii*) *Fibres.* These are polymers in which the polymeric chains are held together by strong intermolecular attraction forces. Due to strong intermoleculer attraction forces, fibres possess high tensile strength, high modulus and close packing of polymeric chains. It may be drawn into long filament-like materials. For examples, nylon, terylene, etc.

(*iii*) *Liquid resins.* These polymers are used as adhesives, sealants etc. in a liquid or colloidal form and are used in industries. For examples, epoxy resins, polysulphide, etc.

(*iv*) *Plastics.* In these polymers the intermolecular forces of attraction are intermediate to those of elastomers and fibres. These are synthetic polymers which are shaped into hard and tough utility articles by applying heat and pressure alongwith a suitable catalyst.

The property by virtue of which a material undergoes permanent deformation under stress is called the 'Plasticity'. The organic compound which are used for softening the thermoplastics at lower temperature and make their moulding easier are called plasticizers. For example, Poly (vinylchloride) (PVC) is a tough thermoplastic but when dibutylphthalate is added to PVC then it becomes soft and easily mouldable. Here dibutylphthalate is plasticizer. Other examples are tricresylphosphate, dioctylphthalate etc. Plastics are different from elastomer. The elastomers can be stretched rapidly at least 150 percent of its original length without breaking and return to their original shape on releasing the stress. For examples, synthetic rubber, etc.

9. Dead polymer and Living polymer

The chain growth (Addition) polymerization of vinyl monomers involve three steps as follow:

(*i*) Chain initiation

(*ii*) Chain propagation and

(*iii*) Chain termination.

During the chain termination the growing chain is terminated by a number of mechanism as

(*i*) Recombination of the initiator free radical

(*ii*) Combination of two activated entities

(*iii*) Elimination

(*iv*) Disproportionation

(*v*) Reactions with inhibitor etc.

As a result the growth of the polymer chain stops and the resulting polymer is called 'Inactive or Dead' polymer.

For example, Inhibitor diphenylpicryl hydrazide (DPPH) exist in the form of stable free radicals which can easily stop the chain growth by direct coupling.

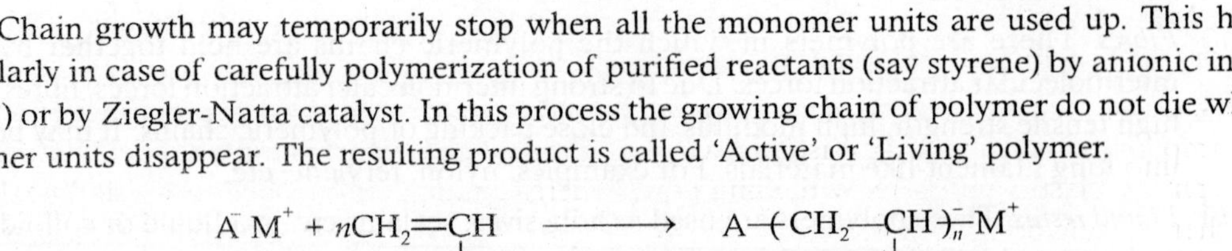

(DPPH) (Growing Chain Polymer) (Dead polymer molecule)

Chain growth may temporarily stop when all the monomer units are used up. This happens particularly in case of carefully polymerization of purified reactants (say styrene) by anionic initiators ($A^- M^+$) or by Ziegler-Natta catalyst. In this process the growing chain of polymer do not die when the monomer units disappear. The resulting product is called 'Active' or 'Living' polymer.

(Living polymer)

The reaction is continued with excess monomer to grow the chain. At the end of the reaction, the polymer may be killed by the addition of any terminating agent like water.

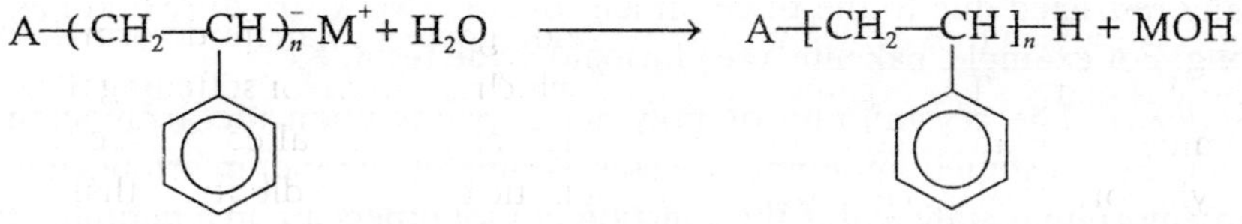

Block copolymers or stereospecific polymers are usually prepared by this technique. In case of step growth or condensation polymerization the growing chain of polymer continues to have one active group at each end and the polymer does not die. However, the length of growing chain of polymer may stop due to increase in the size or molecular weight of the polymer. Cyclization also stops the growing chain of polymer.

7.3 EFFECT OF STRUCTURE ON THE PROPERTIES OF POLYMERS

Since polymers are formed by repeating of monomeric units, the structure of a polymer depends upon the arrangement of its constituent macromolecules (monomers). Following are some of the important factors, which play an important role in deciding the structure and properties of polymers.

1. *Molecular mass.* It is a very important factor which decides the properties of polymers. The number of repeating units of a monomer is called its degree of polymerization. A polymer chain may terminate at any stage during polymerization. Therefore, a polymer may consist of various monomers of different molecular masses. Hence an average molecular mass (number average or weight average) of the polymer is generally used to describe it. Polymers having low molecular masses are usually soft but brittle while polymers having high molecular masses are generally tougher (stronger) and heat resistant. Hence polymers having low degree of polymerization (low molecular mass) does not possess most of the typical characteristics of a polymer.

2. *Nature of monomers.* The chemical properties, mechanical strength and thermal stability of polymer depends upon the nature of monomer. Chemical reactivity of a polymer is very important factor and depends upon the nature of monomer. For example, Teflon is very stable and chemically inert due to the presence of C—C and C—F bonds whereas nylon and polyester undergo hydrolysis due to the presence of a mild ester linkage.

3. *Solubility.* The solubility of a polymer depends very much on its chemical nature and the nature of the solvent. The solubility of a polymer in a particular solvent decreases as the molecular mass of the polymer increases. The solubility of a polymer also decreases by increasing the cross-linking. More crystalline polymers are generally less soluble in solvents due to the chemical resisting power or close packing of monomer molecules.

4. *Geometric arrangement of double bonds in polymers.* It effects the physical properties of a polymer. For example, natural rubber is cis-isomer whereas gutta-percha rubber is trans-isomer. Natural rubber is amorphous, elastic and has low melting point whereas gutta-percha rubber is crystalline, hard and has higher melting point.

5. *Shape of polymer.* The shape of a polymer molecule depend upon the packing of monomer molecules. Molecules of a linear polymer pack themselves closely in a compact manner. Due to this close packing, linear polymers have high densities, high tensile strength and high melting points. In branched chain polymers such close packing does not occur, so they have low densities, low tensile strength and low melting points. In a cross-linked polymer the movement of chains relative to one

another is totally restricted due to the three dimensional network structure and hence they are very tough and strong. For example, bakelite, urea formaldehyde resin, etc.

6. *Crystallinity.* The crystallinity of polymer depends upon the arrangement of monomer molecules. When the constituent monomers posses a completely random arrangement the polymer is said to be in an amorphous state and if the constituent monomers are in a definite crystalline regions then it is said to be in a crystalline form. The degree to which molecules of a polymer are arranged in orderly pattern with respect to each other is termed as crystallinity of the polymer. Crystallinity depends upon two factor: (*a*) Magnitude of attractive intermolecular forces and (*b*) Symmetric structure of the macromolecules. Higher the symmetry of the macromolecules greater is the crystallinity of the polymer. When the polymeric chain is regular and does not contain bulky side group randomly in the main carbon chain, the polymer has a high crystalline structure. For example, polyethene, etc.

But when the polymeric chain is irregular and contain bulky side groups, it has a low degree of symmetry and hence less crystalline or amorphous in nature, for example, polystyrene, poly (vinyl chloride) etc.

7. *Intermolecular forces of attraction.* Intermolecular forces of attraction (Van der Waal's forces, hydrogen bonding, etc.) play an important role on the properties of a polymer. The magnitude of intermolecular attraction forces determine the mechanical strength and thermal stability of polymers. When these forces are strong, the polymer has a very high tensile strength, crystalline nature and sharp melting point. For example, polyesters, etc. on the other hand if the intermolecular forces are weak, the polymer has a low tensile strength, amorphous nature and does not have sharp melting point. It means the polymers may be stretched by applying some external force. For example, elastomers generally show plastic deformation. When a thermoplastic is heated, it gets deformed. This property is known as plastic deformation. Linear polymers get deformed to the maximum extent due to the weak intermolecular forces of attraction. On heating the linear chains slip past over one another. On cooling, they regain their original hardness. Heating and cooling may be repeated many times. They do not affect the nature of polymer.

The polymer is grinded into finely divided state for different use. It facilitates its dissolution and quick responses to most other tests. The grinding of polymer is done in presence of dry ice. During the grinding process, a polymeric material may get over-heated and may even soften. Hence we use dry ice (solid CO_2) for cooling. It prevents softening and makes tough. Thus, elastic materials become brittle so that they can be easily grinded.

Additives in Polymers. Additives are compounding agents which have been used in the processed polymer. The main role of additives is to improve the properties of polymer. Following are the main additives.

(*i*) *Plasticizers.* These are added to the virgin polymers. They have low molecular weight, high boiling point and usually non volatile materials. Plasticizers improve the flexibility of the polymer but lowered the chemical resistance and tensile strength.

e,g., Non-drying vegetable oils, tricrecyl phosphate, dibutyl phthalate, etc.

(*ii*) *Fillers.* Fillers are usually added to reduce shrinkage on setting and cost of the finished product. They also improve the opacity, hardness and tensile strength.

e.g., Saw dust, paper pulp, Carbon black, etc.

Following are some useful fillers which are used for some special characteristics.

(*a*) Heat and corrosion resistance — Asbestos

(*b*) Prevention from X-rays — Barium salt

(*c*) U.V. deactivation — Carbon black

(*d*) Extra hardness — Mica, carborundum, silica.

(*iii*) *Stabilisers.* Stabilisers are those substances which have been added for protecting the polymer product from photodegradation, oxidation, etc.

e.g., Salts of calcium, barium and lead, etc.

(*iv*) *Colouring agents.* These are some suitable organic dyes or opaque inorganic pigments which are added to impart attractive colours to the polymeric product.

Removal of additives. Sometimes it is necessary to remove the additives from the polymer. Removal of additives may be done by following methods.

(*a*) Organic agents like plasticizers, dyestuffs and some stabilizers can be removed by extraction with suitable solvent like ether. Grinded polymer is insoluble in that solvent, so it is removed by soxhlet.

(*b*) Fillers, inorganic stabilizers and opaque inorganic pigments may be removed by dissolving the grinded polymer sample in a suitable solvent. The dissolved components are removed by filtration and the polymer is recovered from the solution by addition of a non solvent.

Crystallite. The crystalline state of a polymer consists of definite crystalline regions called crystallites.

7.4 BIOPOLYMERIZATION AND BIODEGRADABLE POLYMERIZATION

(a) Biopolymers and Biopolymerization

Biopolymers are organic polymers made of biological molecules or microorganisms and the phenomenon of preparation of biopolymer is called biopolymerization. Biopolymers are produced by all living organisms. They are also called natural polymers because they are produced naturally. Most biopolymers are biodegradable. For examples, polysaccharides like cellulose and starch, polyesters like polyhydroxyalkanoates, proteins like silk, poly (γ-glutamic acid), hydrocarbons like rubber etc., are most widespread natural polymers. Generally, natural polymers are not suitable polymers for practical applications due to lack of property as market requirements and very expensive for other than specially high value niche markets such as biomedical applications. That's why blended biopolymers or chemically modified biopolymers like starch, celluloses and proteins are widely used as per requirements. Although biopolymers are not only important for biomedical applications yet they have great importance in industries and engineering branches. They play very important role in the development of advanced construction engineering techniques and building materials. In early time the Romans had already recognised the role of admixture to improve their building materials:

For example, dried blood was used as an air-entraining agent, while biopolymers like proteins served as set retarders for gypsum. Biopolymers were used as admixtures, in the 20th century. In 1920, the first functional biopolymer lignosulphonate was used for concrete plastification. Later on, cellulosics, lignite and micro-biopolymers also became popular. Biopolymers and biodegradable polymers are most recent trend in the ongoing quest for improved functional materials in construction. Some biopolymers used as admixture are listed in the table.

Table 7.1

Year of Introduction	*Admixture*	*Function*	*Type of Admixture*
1920s	Lignosulphonate	Concrete plasticizer	Biopolymer
1940s	Lignite	Bentonite thinner	Biopolymer
1960s	Xanthan gum	Viscosifier	Biopolymer
1962s	Melamine, naphthalene condensates	Concrete super plasticizer	Synthetic polymer
1970s	Cellulose ethers	Water retention agents	Biopolymer
1980s	Vinylsulphonate copolymers	Water retention agent	Synthetic polymer
	Polycarboxylate copolymers	Concrete super plasticizer	Synthetic polymer
1990s	Polyaspartic acid	Biodegradable dispersant, retarder	Biopolymer
2000s	Grafted copolymers	Heavy metal ions retention	Synthetic co-polymer

(b) Biodegradable Polymers and Biodegradable Polymerization

The deterioration in the properties of polymer is called polymer degradation. If the degradation of polymer is carried out by enzymes or microorganisms then the polymer is called biodegradable polymer and the process is called biodegradation (biodepolymerization) of polymer. The degradation is characterised by an uncontrolled change in the molecular weight or constitution of the polymer. Polymer degradation is broadly of two types:

(*i*) Chain end degradation and (*ii*) Random degradation.

The enzymatic degradation of certain natural polymers (biopolymers) follow the chain end degradation mechanism. During the process of biodegradation of polymers, the breakdown of a polymer structure is the primary objective. Scientists are trying.to produce a large number of degradable or biodegradable polymers. A few examples of functional groups contained in degradable polymers are following.

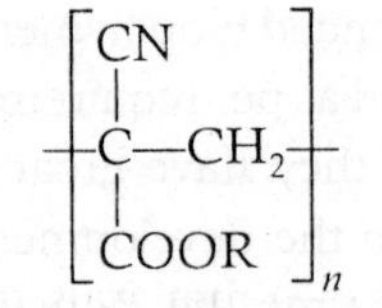

Poly(cyanoacrylates)

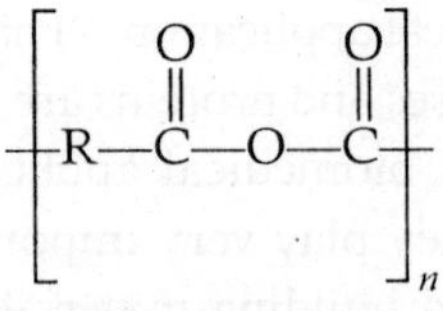

Poly(anhydrides)

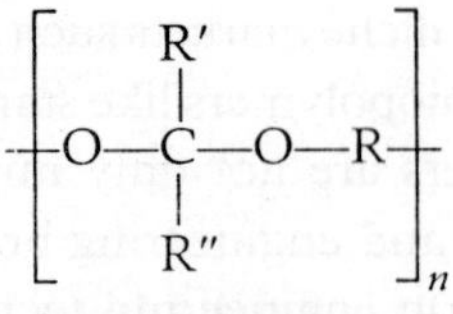

Poly(ketals)

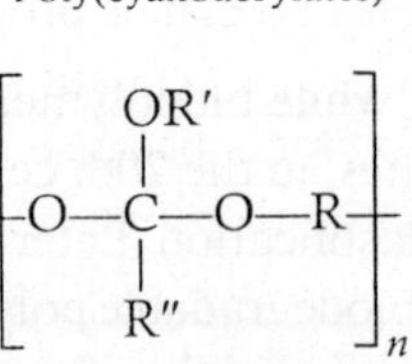

Poly(orthoesters)

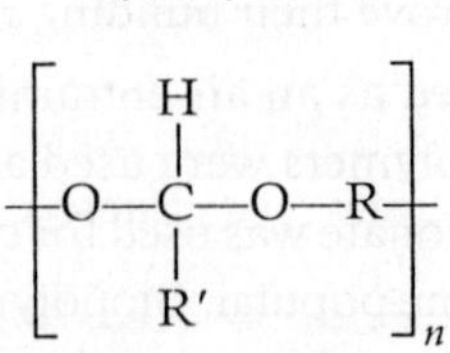

Poly(acetals)

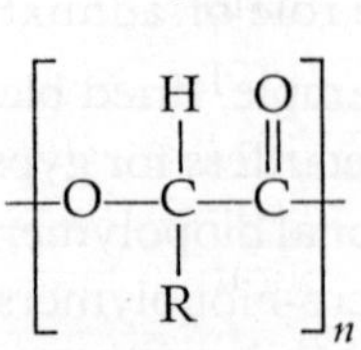

Poly(α-hydroxyesters)

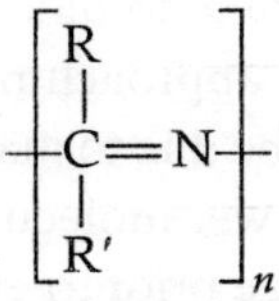

Poly(phosphazenes)

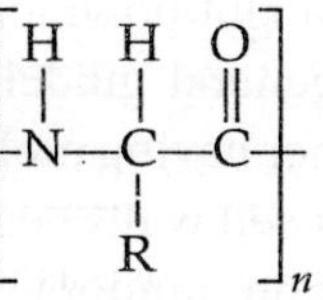

Polypeptides

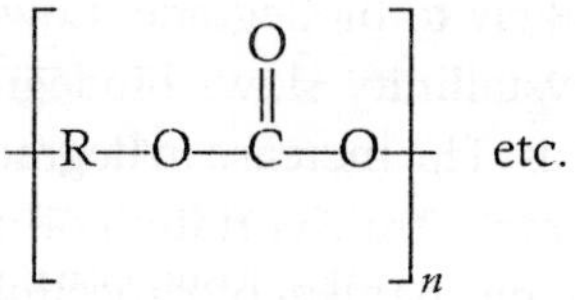

Polycarbonates

Mostly biodegradable polymers can be hydrolysed via enzymatic catalysis. Enzymatic degradation is mainly effective for naturally occurring polymers like polysaccharides, polypeptides etc. Sometimes biodegradation may be very slow. A few examples of biodegradable polymers are following. Silk proteins, proteins, collagen, polyhydroxyalkanoates, microbial polysaccharides, microbial cellulose, hyaluronic acid, alginates etc.

Polysaccharides are largely limited to starch and cellulose derivatives for practical applications either in plastics or as water-soluble polymers. Both these polymers are composed of thousands of D-glycopyranoside repeat units to very high molecular weight. They differ in that starch is poly-1,4-α-D-glucopyranoside and cellulose is poly-1,4-β-D-glucopyranoside as illustrated below.

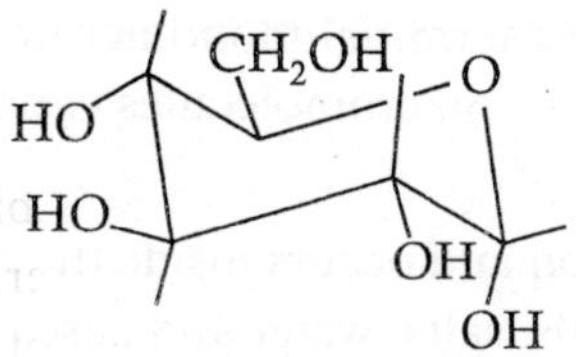

(Starch)

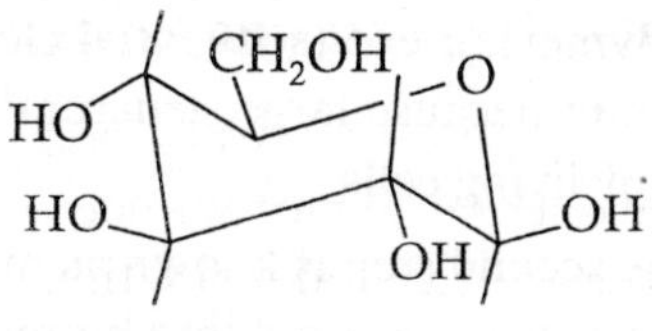

(Cellulose)

Complex carbohydrates such as microbially produced xanthan, curdlan, pullulan, hyaluronic acid, alginates and carageenan are well known biodegradable polymers. Xanthan is the predominant microbial polysaccharide on the market and is used in the food industry and also used as a thickener in many industrial applications. Proteins are not widespread used as plastic materials since they are difficult to process. They are not fusible without decomposition or soluble in practical solvents so they have to be used as found in nature. For examples, silk is used as a fiber. Gelatin (collagen) is used as an encapsulant in the pharmaceutical and food industries, wool is also used as a fiber. The structure of proteins is an extended chain of amino acids joined through amide linkages which are readily degraded by enzymes, particularly proteases. The activity in poly (γ-glutamic acid) with control of stereochemistry by the inclusion of manganese ions may have important for future developments in biodegradable water-soluble polymers with carboxyl functionality.

Polyesters are produced by many bacteria as intracellular reserve materials for use as a food source during periods of environmental stress. They are biodegradable and can be processed as plastic materials. The properties of thermoplastics depends on the structure of the pendant side-chain of the polyester. Plastics have more disposal options than water-soluble polymers because they are usually solid, handleable materials and are generally recoverable after use. Disposal options include landfilling, recycling, incineration and composting. Composting is the type of biodegradation in which probability of oxidation and hydrolysis takes place. Hence, composting is an opportunity for environmentally degradable plastics which are used in food applications such as wrappers and utensils. In such type uses, plastics are contaminated with food residues and the mix is ideally suitable for composting without separation. The general guidelines for predicting the biodegradability of synthetic polymers include hydrophilic/hydrophobic balance; higher ratio is better for biodegradation. Carbon chain polymers are unlikely to biodegrade. Chain branching is deleterious to biodegradation. Condensation polymers are more likely to biodegrade. Lower molecular weight polymers are more susceptible to biodegradation and crystallinity slows biodegradation. The general guidelines are useful for approaching polymer synthesis. The increase in degradation of polymer having molecular weight may be due to many factors, for example, transportation of polymers across cell walls being more likely at lower molecular weight, and the mechanism of exocellular biodegradation, random or chain-end cleavage prior to entering the cell. Chain end, exobiodegradation explains the slower biodegradation at higher molecular weight where there are fewer chain ends than at lower molecular weight. It has been noticed that the terminal groups found in oxidised and photodegraded polyethene are oxygen containing and these should predictably expedite biodegradation via a β-oxidation mechanism.

Mechanism of Biodegradation of Polymers

There are two steps in the biodegradation of polymers:

Step I. The first step is a depolymerisation or chain cleavage step. In this step the long polymer chain is converted into smaller oligomeric fragments due to hydrolysis and/or oxidation. Extracellular enzymes may also be responsible for this step. Enzymes act either endo (random cleavage on internal linkages of the polymer) or exo (sequential cleavage of the terminal monomer unit) process. The first step is very important because large structural material like macromolecules cannot pass through the outer membranes of living cells.

Step II. The second step is known as mineralisation and occurs inside the cell where small size oligomeric fragments are converted into biomass, minerals, salts, water and gases like CO_2 and CH_4.

Most of the methods that measure the extent of biodegradation are respirometric, which is primarily related to the evolution of carbon dioxide. Other methods include assessing the rate of molecular weight loss, measuring the loss of polymer physical properties (*e.g.,* tensile strength as per ASTM standard), measuring the rate of increase of the microbial culture colony size contacting the material, using classical oxygen uptake procedures (*e.g.,* biochemical oxygen demand BOD) and radioactive tracer techniques that use ^{14}C labelling. According to American Society for Testing and Materials (ASTM) standard degradable, biodegradable, hydrolytically degradable and oxidatively degradable plastics are widely accepted either in its pure form or modified form. The ASTM standard decides the extent of degradation.

Biodegradation implies the use of the plastic substrate as the carbon source for the microorganism metabolism. Biodegradation results in the production of CO_2 under aerobic environments or CH_4 under anaerobic environments as well as humic materials. Humic material is an important component of the biodegradation process because it can enhance productivity of agricultural land. Thus composting polymeric materials is a biological recycling of the polymeric carbon. Composting is the process in which degradation of heterogeneous organic matter is accelerated by a mixed microbial population in a moist, warm and aerobic environment under controlled conditions. A typical compost system supports a diverse microbial population in a moist aerobic environment in a temperature range of 40–70°C. Several factors are important in biodegradation, including macromolecule size, structure and chemistry, microbial population and enzyme activity, and various environmental conditions such as darkness, high humidity and adequate mineral and other organic nutrients, as well as temperature, pH and oxygen requirements.

Polyphosphazenes are inorganic biodegradable polymers. They are very important for biomedical applications. Biodegradable polymers have great industrial applications as per market demand. Biodegradable polymers are also used in tissue engineering.

7.5 CLASSIFICATION OF PLASTICS

Plastics have been classified into two classes:

1. Thermoplastics and 2. Thermosetting plastics.

1. **Thermoplastics.** Polymers which soften on heating and become hard upon cooling are known as thermoplastics and they can be converted to any shape by moulding. There is no chemical change during its formation and the mechanism follows the chain growth polymerization process. Due to lack of cross-linking they are soluble in a suitable organic solvent. For examples, Polyethene (LDPE and HDPE), Poly (vinyl chloride) (PVC), Poly (methyl methacrylate) (PMMA) etc.

2. **Thermosetting plastics.** Polymers which change irreversibly into hard and rigid on heating and cannot be reshaped, once they are formed, are called thermosetting plastics. The chemical change occurring during its formation follows step growth polymerization process. Due to the cross-linking between the polymer chain they are insoluble in organic solvents. They have three-dimensional framework. For examples, bakelite, urea formaldehyde resin, epoxy resins, etc.

Differences between Thermoplastics and Thermosetting Plastics

Sr. No.	*Thermoplastics (Two dimensional plastics)*	*Thermosetting Plastics (Three dimensional plastics)*
1.	They are generally long chain linear or branched chain polymers.	They have three dimensional cross-linked network structures joined by covalent bond.
2.	They are generally soft, weak and less brittle.	They are hard, strong and more brittle.
3.	They are formed by addition polymerization only.	They are formed by condensation polymerization.
4.	They can be softened on heating so they can be reshaped and reused.	Due to cross-linking and strong bond strength they do not soften on heating. So they cannot be reshaped and reused.
5.	They are usually soluble in some suitable organic solvents.	Due to cross-linking and strong bond strength they are insoluble in almost all organic solvents.
6.	They can be reclaimed from wastes.	They cannot be reclaimed from wastes.
7.	On strong heating they can be softened very quickly.	On strong and prolonged heating polymers get charred.
8.	Poly (vinyl chloride) (PVC), polyethene, polypropene, etc. are its examples.	Bakelite, epoxy resins, etc., are its examples.

7.5.1. Thermoplastics

(A) Polyethene

It is the polymer of ethene ($CH_2 = CH_2$). It is prepared by addition polymerization of ethene via free radical mechanism at different conditions. On the basis of manufacturing conditions these are of two types:

(*a*) Low Density Polyethene (LDPE) and

(*b*) High Density Polyethene (HDPE).

(*a*) **Low density polyethene (LDPE).** It is prepared by polymerizing ethene at high pressure (1000–5000 atm) and at high temperature 200–250°C in the presence of free radical initiator (oxygen). At first the gas is liquefied at high pressure then it is pumped in autoclaves at 200–250°C in the presence of oxygen. LDPE consists of highly branched structure which does not allow the polymeric chains to pack close together and hence it has a low density.

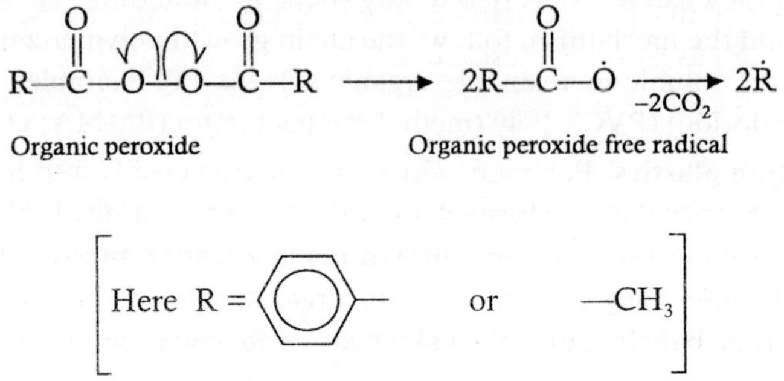

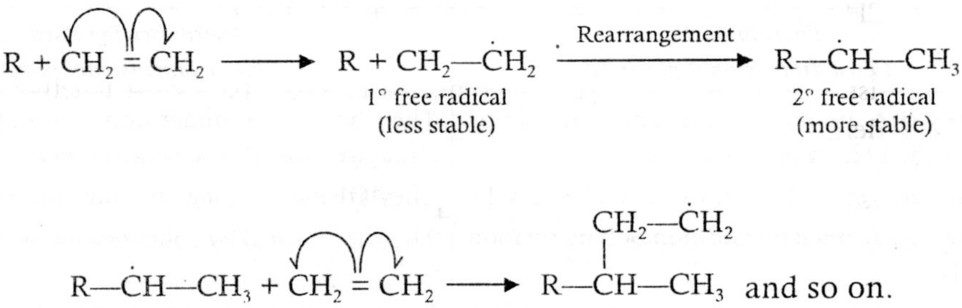

In general it may be represented as

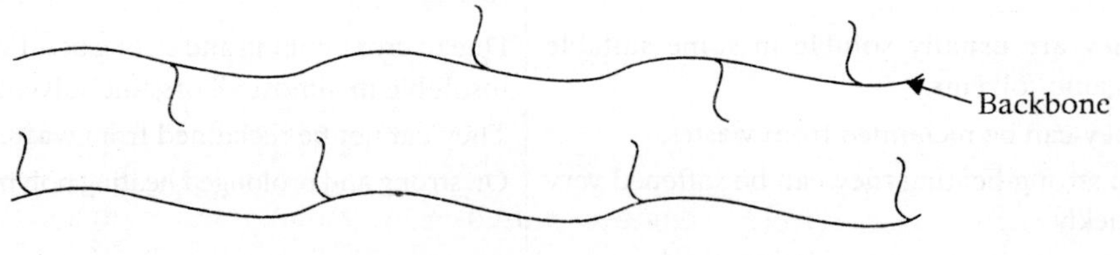

(Branched chain structure of LDPE)

Properties. (*i*) It melts at 110–125°C and is only 40–50% crystalline.

(*ii*) Its density is low (0.91 gm/cm^3).

(*iii*) It is chemically inert and is a poor electrical conductor.

(*iv*) It is moisture resistant and has moderate tensile strength.

(*v*) It is transparent and slightly flexible polymer.

Uses. (*i*) It is used for making pipes for agriculture, irrigation and chemical plants.

(*ii*) Being a poor conductor it is used for insulation of electric wires and cables.

(*iii*) Due to chemical inertness and moisture resistance it is used for packing and wrapping frozen food, textile products etc.

(*iv*) It is also used for the manufacture of squeeze bottles and many attractive containers, etc.

(*v*) It is used for making kitchen and domestic appliances and toys, etc.

Limitations. (*i*) Since it has low density and less tensile strength it is not suitable for load bearing application.

(*ii*) Due to low crystallinity it is not suitable for the manufacture of water pipes.

(*iii*) It is also not suitable for the manufacture of pipes for gas distribution because it is slightly amorphous in nature and gas molecules escape out.

(*b*) **High density polyethene (HDPE).** HDPE is manufactured by the following processes:

1. *Using metal oxide catalyst.* It is prepared by polymerizing ethene at low pressure (1200–1800 atmosphere) at temperature about 60–200°C in the presence of metal oxide like Cr_2O_3.
2. *By using Ziegler-Natta catalyst.* It is also prepared by polymerizing ethene at 60–200°C under a pressure of 6–7 atmospheres in the presence of a Ziegler-Natta catalyst in an suitable inert solvent.

$$nCH_2 = CH_2 \xrightarrow[\text{Ziegler-Natta catalyst}]{60-200^\circ C,\ 6-7\ \text{atm}} \left(CH_2 - CH_2 \right)_n$$

HDPE consists of linear chains of polymer molecules which pack close together resulting in the formation of high density polyethene.

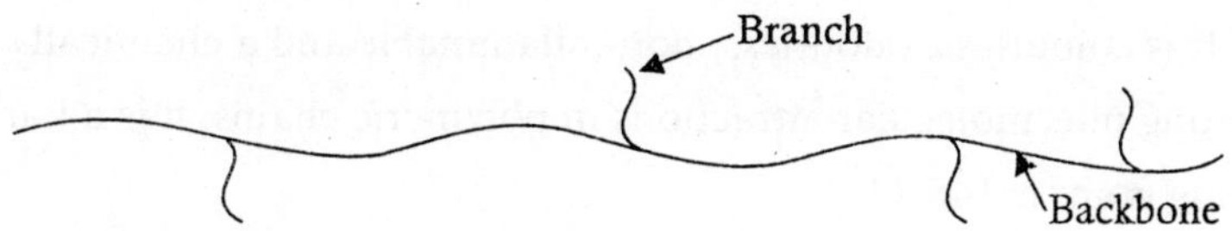

Properties. (*i*) Its density is 0.94–0.96 g/cm^3.

(*ii*) It has high melting range 145–150°C.

(*iii*) It processes chemical and moisture resistant properties.

(*iv*) It is about 90% crystalline and has less permeability for gases.

(*v*) It has higher tensile strength and hardness than LDPE.

Uses. (*i*) Due to highly tensile strength it is used in the manufacture of buckets and overhead tanks for water storage.

(*ii*) Since it is less permeable for gases it is used for domestic gas piping.

(*iii*) Since it has grease proof nature and also inert towards chemicals, so it is suitable for wrapping material for food·products, bottles for milk, household chemicals and drug packaging.

(B) Poly (Vinyl Chloride) (PVC)

It is very important thermoplastic and is known as Koroseal.

Preparation. It is prepared by the free radical addition polymerization of vinyl chloride in the presence of small amount of free radical initiator like benzoyl peroxide. When a water emulsion of vinyl chloride is heated along with benzoyl peroxide in an autoclave under pressure, polyvinylchloride is formed.

$$n\,\underset{\substack{|\\ \text{Cl}}}{CH_2=CH} \xrightarrow[\text{(Benzoyl peroxide, pressure)}]{\Delta} \left[\,CH_2-\underset{\substack{|\\ \text{Cl}}}{CH}\,\right]_n$$

Vinyl chloride — Poly (vinyl chloride)

The monomer vinyl chloride is prepared by the following methods:

(*i*) When HCl gas is passed over ethyne at 70°C in the presence of mercuric sulphate, vinyl chloride is formed

$$\underset{\text{Ethyne}}{CH\equiv CH} + HCl(g) \xrightarrow[HgSO_4]{\Delta} \underset{\text{Vinyl chloride}}{CH_2=CH-Cl}$$

(*ii*) When chlorinated ethylene is treated with alcoholic KOH, vinyl chloride is formed.

$$CH_2=CH_2 \xrightarrow{Cl_2} Cl-CH_2-CH_2-Cl \xrightarrow[-\,HCl]{\text{alc. KOH}} \underset{\text{Vinyl chloride}}{CH_2=CH-Cl}$$

[**Note:** Vinyl monomers should be distilled before polymerization because they usually contain hydroquinone as inhibitor to prevent polymerization during storage. Therefore, the inhibitor should be removed before polymerization. The inhibitor can be removed either by distillation under reduced pressure or by washing with 95% KOH solution in a separating funnel followed by washing with water and then dried over anhydrous $CaCl_2$ or Na_2SO_4.]

Properties. (*i*) It is colourless, odourless, non-inflammable and a chemically inert plastic.

(*ii*) Due to strong intermolecular attractions in polymeric chains, it is a hard polymer.

(*iii*) Its softening point is 148°C.

(*iv*) Its glass transition temperature (T_g) is 81°C.

(*v*) It is soluble in cyclohexanone and tetrahydrofuran (THF).

(*vi*) It has a high strength, lightness and resistance to chemicals, oil and weathering.

(*vii*) When treated with tricresylphosphate or dibutylphthalate it forms plasticized PVC.

Uses. (*i*) Due to excellent resistance to weather it is used in place of wooden windows, rain wears, shoes, etc.

(*ii*) Due to high chemical resistance it is used in acid recovery plants.

(*iii*) It is less brittle, light and flexible and is used for making bottles for storing mineral waters, vinegar, cosmetics, detergents, etc.

(*iv*) It is used for making sheets which are widely used for tanks lining, light fittings, safety helmets, refrigerator parts, cycle and motorcycle mudguards, etc.

(*v*) Plasticized PVC being a good insulator of electricity is used for insulating electrical wires.

(*vi*) Plasticized PVC is used for making sheets, phonograph records, water toys, garden hoses, etc.

(C) Poly (Vinyl Acetate) (PVA)

It is an important thermoplastic and is prepared by the free radical addition polymerization of vinyl acetate in the presence of free radical initiators like benzoyl peroxide. When a water emulsion of vinyl acetate is heated alongwith a small amount of benzoyl peroxide in an autoclave under pressure, poly vinyl acetate is formed.

$$nCH_2=CH{-}OCOCH_3 \xrightarrow[\text{Benzoyl peroxide}]{\Delta} \left[CH_2{-}\underset{\displaystyle O{-}\overset{\displaystyle O}{\overset{\|}{C}}{-}CH_3}{\overset{|}{CH}}\right]_n$$

Vinyl acetate Poly (vinyl acetate)

The monomer vinyl acetate is prepared by treating ethyne with acetic acid in the presence of mercury salt.

$$CH\equiv CH + CH_3COOH \xrightarrow{Hg^{++}} CH_2=CH{-}O{-}COCH_3$$

Properties. (*i*) It is a colourless, transparent, soft and sticky material.

(*ii*) It is soluble in organic solvents.

(*iii*) It is an atactic polymer having glass transition temperature (T_g) 28°C.

(*iv*) Low molecular weight polymers are brittle but become gum like when masticated and hence they are used for making chewing gums.

(*v*) It is resistant to heat but turns slightly yellow on prolonged storage above 120°C.

(*vi*) It can be copolymerized with other polymers.

Uses. (*i*) It is used in adhesives, lacquers and water base emulsion paints (like latex paints).

(*ii*) It is used for bonding papers, leather, textiles, etc.

(*iii*) It is used for making chewing gums, surgical dressings, etc.

(*iv*) Copolymers of vinyl acetate are industrial materials.

(*v*) It is used for the manufacture of poly (vinyl alcohol) etc.

(D) Teflon

Teflon is the trade name of polytetrafluoroethylene. It is very important thermoplastic and is prepared by the free radical addition polymerization of tetrafluoroethylene in the presence of free radical initiator like acetyl peroxide or benzoyl peroxide. The polymerization is carried out by the emulsion method using peroxide initiators.

$$n\,CF_2=CF_2 \xrightarrow[\text{Acetyl peroxide}]{\Delta} \left(CF_2-CF_2\right)_n$$

Tetrafluoroethylene Teflon

During the polymerization of tetrafluoroethylene a large amount of heat is produced, so sufficient precaution should be taken otherwise violent explosions may occur. The monomer tetrafluoroethylene is a gas at room temperature and is prepared by the dechlorination of dichlorotetrafluoroethane or decarboxylation of sodium perfluoropropionate.

Properties:

(*i*) It is a linear polymer having no branching.

(*ii*) It is a highly crystalline polymer.

(*iii*) Its melting point is about around 330°C.

(*iv*) Due to very strong C—F bond the mechanical strength of this polymer remains unchanged over a wide range of temperature from – 100°C to + 350°C.

(*v*) It does not react with strong acids, including hot fuming nitric acid.

(*vi*) It does not react with alkalies. However it has been observed that it reacts with only molten alkali metals to any significant extent.

(*vii*) It has a very low dielectric constant.

(*viii*) It is highly viscous in nature, so the flow rate of its melt is very slow. That's why the conventional techniques used for the processing of other polymers cannot be applied to teflon.

(*ix*) It is very hard and tough.

(*x*) It is a bad conductor of electricity.

Uses:

It has wide applications in industries. Some applications are listed below:

(*i*) It is used for the preparation of pump valves and pipes where chemical resistance is required.

(*ii*) It is used for preparing seals and gaskets which have to tolerate high temperature and pressure.

(*iii*) It is used in non-lubricated bearings.

(*iv*) It is used to provide insulations for high frequency electric installations.

(*v*) It is used for making non-stick surfaces of cooking pans etc., in kitchens.

(*vi*) Its fibre is used to form belts, filter cloth etc., where resistance of acid and alkalies is required.

7.5.2 Thermosets or Thermosetting Plastic

(A) Phenolic Resins or Phenoplasts or Phenol Formaldehyde (PF) Resins

These are condensation polymerization products of phenol with aliphatic aldehydes. Phenol-formaldehyde resin (bakelite) is the most important member of this class. The manufacture of bakelite involves the following steps:

Step I. Phenol and formaldehyde (1 : 1) reacts in the presence of catalyst (acid or base) to form *ortho* and *para* hydroxymethyl phenol.

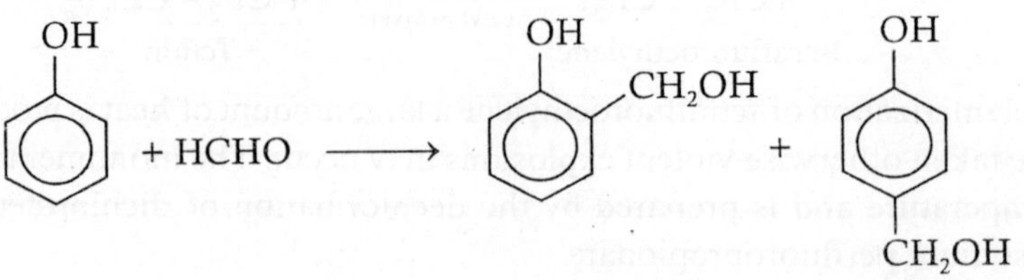

o-hydroxymethylphenol *p*-hydroxymethylphenol

With excess of formaldehyde disubstituted and trisubstituted phenols are formed

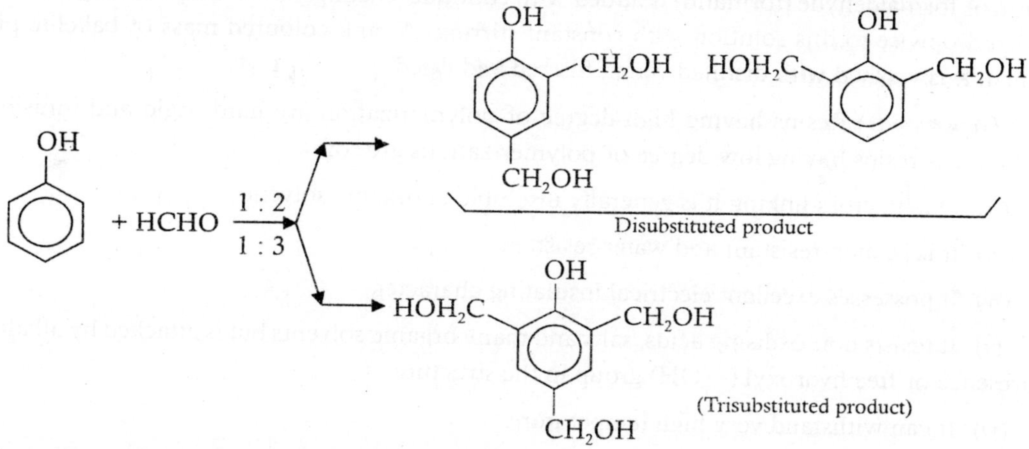

Substituted phenols combines with phenol molecule to form linear polymers known as Novolac.

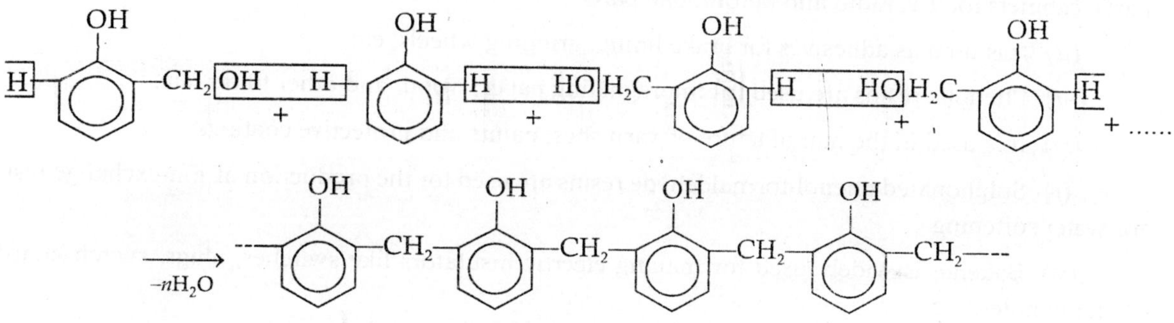

Novolac

Novolac is soluble in organic solvents and is fusible.

Step II. During moulding, hexamethylenetetramine $[(CH_2)_6N_4]$ is used. It provides formaldehyde which converts the soluble and fusible mass of novolac into a hard and insoluble cross-linked structure of bakelite. Here formaldehyde attacks at the vacant para position of the phenol.

$$(CH_2)_6N_4 + 6H_2O \longrightarrow 6HCHO + 4NH_3$$

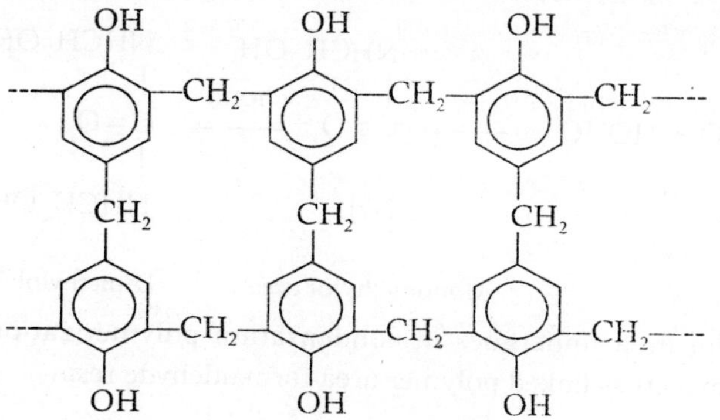

Cross-linked polymer bakelite

Procedure. A mixture of glacial acetic acid and phenol is taken in a beaker. A 40% aqueous solution of formaldehyde (formalin) is added with continue shaking. A few drops of concentrated HCl is added dropwise to this solution with constant stirring. A pink coloured mass of bakelite plastic is obtained. It is washed with distilled water, filtered and dried.

Properties. (*i*) Resins having high degree of polymerization are hard, rigid and infusible solid mass whereas resins having low degree of polymerizations are soft.

(*ii*) Due to cross-linking it is generally insoluble in organic solvents.

(*iii*) It is scratch-resistant and water-resistant.

(*iv*) It possesses excellent electrical insulating character.

(*v*) It resists non-oxdising acids, salts and many organic solvents but is attacked by alkalis due to the presence of free hydroxyl (—OH) group in the structure.

(*vi*) It can withstand very high temperature.

Uses. (*i*) The main use of bakelite is in moulding applications. It is widely used in making telephone parts, cabinets for TV, radio and automobile parts.

(*ii*) It is used as adhesives for brake lining, grinding wheels, etc.

(*iii*) Phenolic resins are used for impregnating paper, wood, and other fillers, etc.

(*iv*) It is used in the manufacture of varnishes, paints and protective coatings.

(*v*) Sulphonated phenol-formaldehyde resins are used for the production of ion-exchange resins for water softening.

(*vi*) Bakelite is widely used for making electric insulators like switches, plugs, switch boards, heater handles, etc.

(*vii*) It is used for making decorative laminates, walls coverings, etc.

(B) Urea-formaldehyde Resins (UF)

Urea-formaldehyde resins are condensation polymerization product of urea and formaldehyde.

Preparation. It is prepared by condensation polymerization of 40% aqueous formaldehyde (formalin) with urea. Urea reacts with formalin to form monomethylol or dimethylol urea as follow:

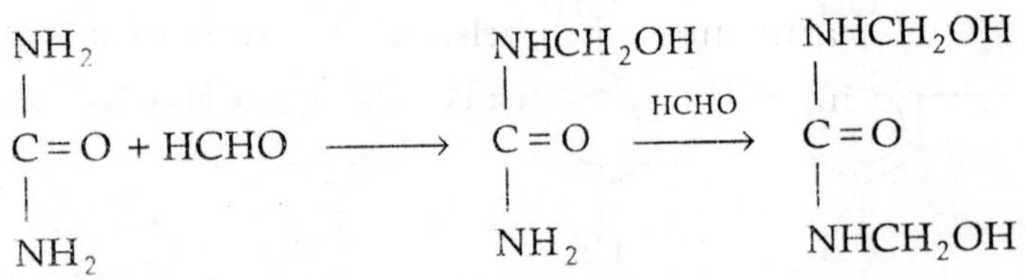

Urea — Monomethylol urea — Dimethylol urea

The dimethylol urea undergoes in condensation polymerization in the presence of acid (conc. H_2SO_4) to form a cross-linked polymer urea-formaldehyde resin.

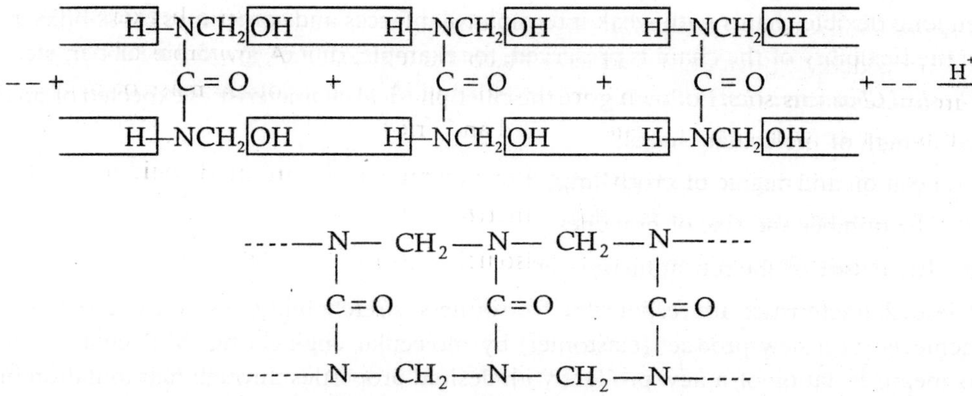

Urea-formaldehyde resin
(Cross-linked polymer)

Urea is dissolved into 40% aqueous solution of formaldehyde (formalin) to get a saturated solution. Concentrated H_2SO_4 is added dropwise with constant stirring. A white solid mass appears which is washed with distilled water and dried. The dried white solid mass is the cross-linked urea formaldehyde polymer.

Properties. (*i*) Resins are clear and colourless hard materials.

(*ii*) They can be synthesized in any desired colours by adding a proper pigment and filler.

(*iii*) They posses good chemical resistance and they are good electrical insulators.

(*iv*) They are abrasion resistant and stable to light.

(*v*) The resins have good adhesive characteristics.

(*vi*) They are resistant to most of the solvents and grease.

(*vii*) They have better tensile strength and hardness in comparison to phenolic resins but their heat and moisture resistance are poor.

Uses. (*i*) The main application is in the manufacture of buttons, bottle caps, surgical items, cosmetic container closures, household appliances, coloured toilet seat, etc.

(*ii*) They are used as adhesive in the plywood industry and furniture.

(*iii*) They are used in the paper industry to improve the wet strength of paper.

(*iv*) They are used in the manufacture of enamels and other surface coatings.

(*v*) They are used for the finishing of cotton textiles making them crease resistant, fire retardant and the control of shrinkage.

7.6 ELASTOMERS

An elastomer is a linear polymer having elastic properties like rubber. Elastomers can be stretched by applying an external force which regains the original size as soon as the external force is removed. They have been long flexible chains with weak intermolecular forces and sometimes cross-links across each other but the flexibility of the chain is preserved, for example, rubber, synthetic rubber, etc.

Physical Characteristics. Following are the chief physical characteristics expected of an elastomer.

(*i*) Length of molecular chains,

(*ii*) Position and degree of cross-links,

(*iii*) The number and size of branches,

(*iv*) The nature of the repeating unit, etc.

These characteristics are responsible for softness, stretchability, resilience, toughness, etc. and can be achieved in a new product (elastomer) by molecular engineering. Molecular engineering of polymers means isolation of a new product with desired properties through manipulation of physical characteristics. An optimum combination of softness, stretchability, resilience and toughness can be achieved in a new elastomer by providing for chain flexibility, free segmental mobility and permanent deformation under prolonged stretching.

While the chain flexibility is achieved by avoiding stiffening units such as alicyclic or aromatic rings in the molecular chains, free segmental mobility can be imparted by

(*i*) Selecting the monomeric units containing C—C and $>C—O—C<$ linkages around which free rotation is possible.

(*ii*) Reducing interchain attractive forces by avoiding polar groups and

(*iii*) Increasing interchain free volume by inserting small side chains which prevent close packing.

A suitable number of cross-links is provided through vulcanization to prevent against permanent deformation. Degree of rigidity can be controlled by varying the degree of cross-linking.

Rubber

Rubber is the best example of an elastomer. It is a polymer of isoprene.

$$\underset{\text{Isoprene (Monomer)}}{CH_2=\overset{\overset{\displaystyle CH_3}{|}}{C}-CH=CH_2} \xrightarrow{\text{Polymerization}} \underset{\text{Polyisoprene (Rubber) (Polymer)}}{\left(CH_2-\overset{\overset{\displaystyle CH_3}{|}}{C}=CH-CH_2\right)}$$

Rubber is of two types:

(*a*) Natural rubber and (*b*) Synthetic rubber.

(*a*) Natural rubber. The main source of natural rubber is the sap of the tree *Hevea braziliensis*. When cuts are made in the bark of the tree, a milky colloidal emulsion is obtained, called rubber latex. Rubber latex contain 25–45% of rubber along with other impurities such as proteins, fatty acids, resins and water. The latex is treated with an acid (acetic acid) to form crude rubber. Crude rubber is composed of 90–95% of rubber hydrocarbons, 2–4% of proteinous matter and 1–4% of resins. The latex (crude rubber) is mixed with appropriate compounding material and then precipitated directly from the solution in the desired shape.

Natural rubber is a polymer of isoprene. By destructive distillation of natural rubber isoprene is formed.

Structure. The X-rays studies suggest that rubber is a cis-isomer whereas the natural rubber Gutta-percha is a trans-isomer. The isoprene units are arranged in the *cis*-form in the natural rubber obtained from the tree *Hevea braziliensis* whereas isoprene units are arranged in the *trans*-form in the natural rubber Gutta-percha obtained from the tree *Dichopsis gutta* and *Palaguim gutta*. It may be represented as in the following figure :

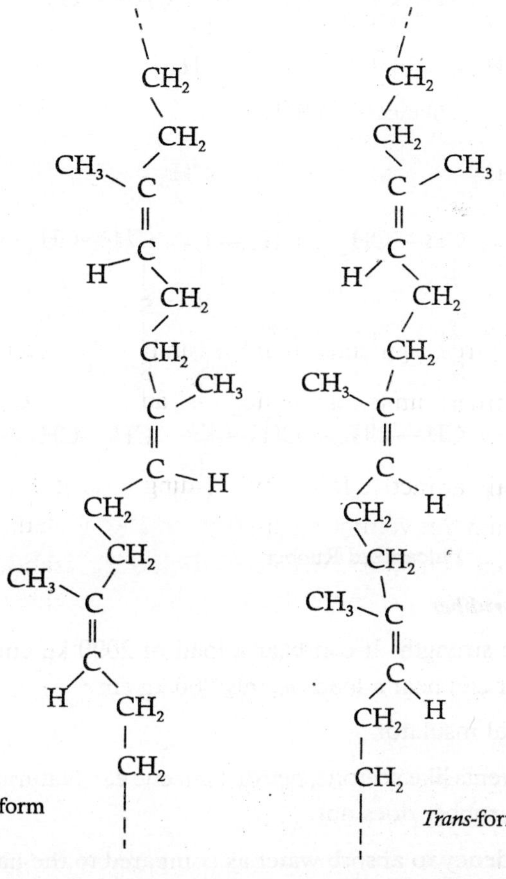

Cis-form *Trans*-form

Vulcanization. Since the natural rubber has no cross-linking, it is a gummy and poorly elastic material. It gets permanently deformed on stretching. So it is necessary to introduce the cross-linking between the chains of natural rubber to improve their utility and importance. Generally polysulphide (—S—S—) linkage are introduced for cross-linking. The process by which the polysulphide (—S—S—) cross-links are introduced in the structure of natural rubber is called vulcanization and the product is called vulcanized rubber. Vulcanization is carried out by heating natural rubber with 2–5% sulphur at 100–110°C for about 30 minutes. This introduces —S—S— cross-links between the adjacent hydrocarbon chains present in the natural rubber. Due to cross linking the vulcanized rubber can be stretched only to a limited extent. It becomes hard with improvement in tensile strength. When the tensile force is removed the chains get coiled up again and it returns to its original shape.

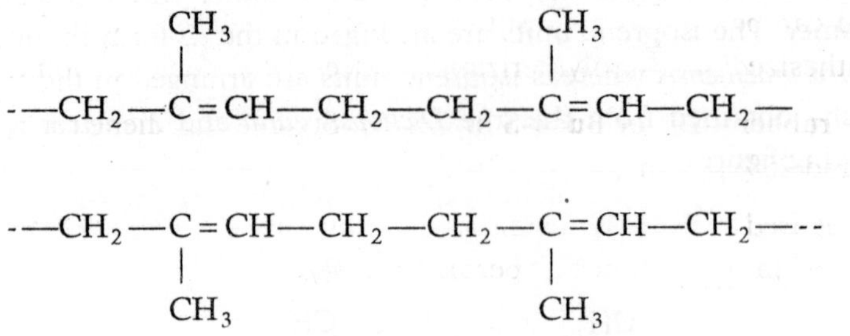

Unvulcanized Rubber

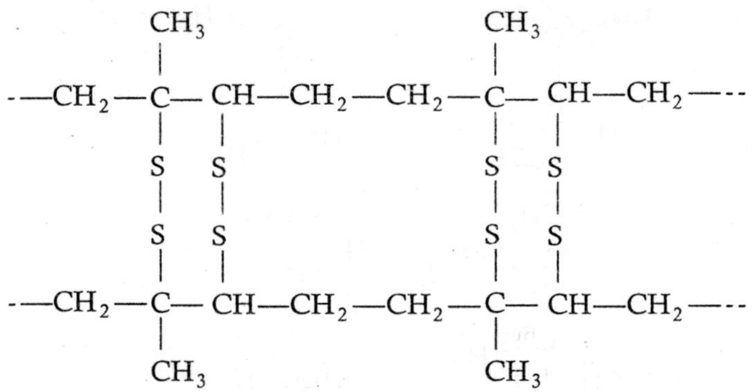

Vulcanized Rubber

Properties of vulcanized rubber

1. It has a high tensile strength. It can bear a load of 2000 kg cm^{-2} before it breaks whereas unvulcanized rubber can bear a load of only 200 kg cm^{-2}.
2. It is a better electrical insulator.
3. It resists organic solvents like acetone, petrol, benzene etc. Natural rubber is soluble in acetone whereas vulcanized rubber does not.
4. It has very little tendency to absorb water as compared to the natural rubber.
5. It comes back to its original shape when the force is removed.
6. Its working temperature range is – 40°C to 100°C whereas unvulcanized rubber works between the range of 10°C to 60°C.

Uses

1. As a lining in tanks used in chemical plants protect them from corrosion.
2. As a rubber mountings to reduce the machine vibrations, noise, etc.
3. As a thread and spongue in shock-absorber.
4. For making toys, sports material, cushions, tyres, gaskets, conveyor belts, etc.
5. For preparing chlorinated rubber, oxidized rubber, etc.

(b) Synthetic rubber. The rubbers or rubber like materials which are prepared in the laboratory by synthetic methods and can be used as a substitute of natural rubber are called synthetic rubbers. Synthetic rubbers are synthesized by the polymerization process of suitable monomers.

Styrene-butadiene rubber (SBR or Buna-S or GR-S). Styrene-butadiene rubber is a copolymer of about 75% butadiene and 25% styrene.

Preparation. It is prepared by the polymerization of a mixture of 75% butadiene and 25% styrene in an emulsion system at 0°C in the presence of peroxide catalyst.

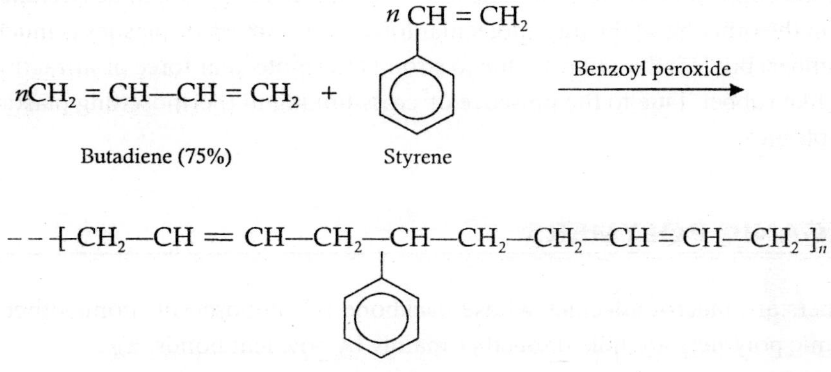

Buna-S (SBR)

Properties. (*a*) It is also vulcanizable like natural rubber.

(*b*) It gets readily oxidized particularly in the presence of traces of ozone present in the atmosphere.

(*c*) It swells in organic solvents upto a limited extent.

(*d*) It has high load bearing capacity, high absrasion resistance and low oxidation resistance.

Uses. It is used in the manufacture of motar tyres, shoes soles, gaskets, floor tiles, cable insulation, adhesives, carpet backing etc.

Nitrile rubber (Buna-N or GR-N or NBR). It is a co-polymer of butadiene and acrylonitrile.

Preparation. It is prepared by the polymerization of 75% butadiene and 25% of acrylonitrile in an emulsion system.

$$nCH_2 = CH—CH = CH_2 + nCH_2 = CH — CN \xrightarrow{\text{Benzoyl Peroxide}}$$

$$\text{---}\left(CH_2—CH = CH—CH_2—\underset{}{\overset{\overset{\large CN}{|}}{CH}}—CH_2\right)_n\text{---}$$

Buna-N (NBR)

Properties. (*a*) It possesses extraordinary resistance to oils, acids, salts, abrasion, heat and sun light.

(*b*) It is less resistant to alkalies.

(*c*) It swells in organic solvents to a certain limit.

Uses. Buna-N is used for making fuel tanks, gasoline hoses, conveyor belts, automobile parts, high altitude air craft components, oil resistant foams, adhesives. As a latex, it is used for impregnating paper, leather and textiles.

Comparison the Chain structure of rubbers with that of Plastics

Rubber is an elastomer. The structure of rubbers is intermediate between thermoplastics and thermosets. The interchain linkages are comparatively low. Due to weak intermolecular force rubber can be stretched without breaking the chemical bonds, by applying an external force. In case of vulcanized rubber the intermolecular force is increases due to cross linking. Hence it can be stretched only upto a limited extent. On the other hand the intermolecular force of attraction of plastics is much greater than the elastomer (rubber) but less than fibres. Due to strong intermolecular force of attraction plastics can not be stretched like rubber. Due to the presence of cross-linking in thermosetting plastics, it is harder than the thermoplastics.

7.7 INORGANIC POLYMERS

Inorganic polymers are macromolecules whose backbone is composed of atoms other than carbon. Atoms in inorganic polymers are linked together mainly by covalent bonds. *e.g.*,

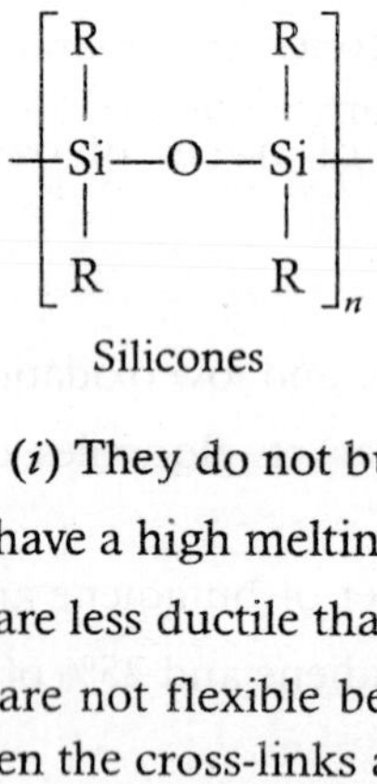

Silicones

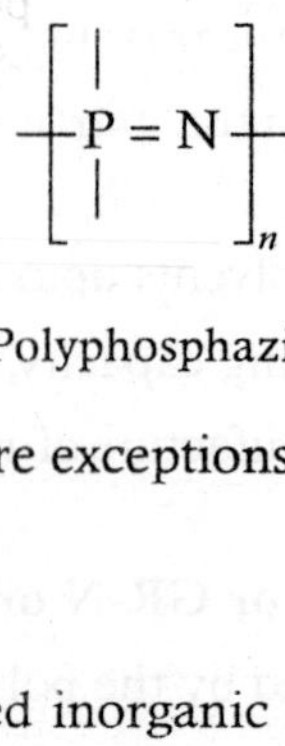

Polyphosphazine

Properties. (*i*) They do not burn (sulphur polymers are exceptions).

(*ii*) They have a high melting point.

(*iii*) They are less ductile than organic polymer.

(*iv*) They are not flexible because in the cross-linked inorganic polymers, the chain segments between the cross-links are short and stiff.

(*v*) They are harder, brittle and stiffer than the organic polymer because these polymers having cross-linked structure with a high density of covalent bond.

(*vi*) They are soluble in polar solvents.

7.7.1 Silicones

Silicone polymers contain Si-O bond in the backbone. $SiCl_4$ on complete hydrolysis give silica SiO_2.

$$SiCl_4 + 4H_2O \longrightarrow Si(OH)_4 \xrightarrow[-2H_2O]{} SiO_2$$

The hydrolysis of alkyl substituted chlorosilanes is not the similar of alkylhalide. *i.e.*, it does not follow this route.

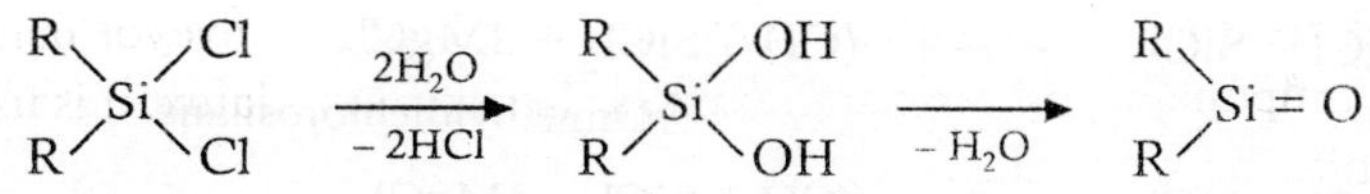

but reacts intermolecularly:

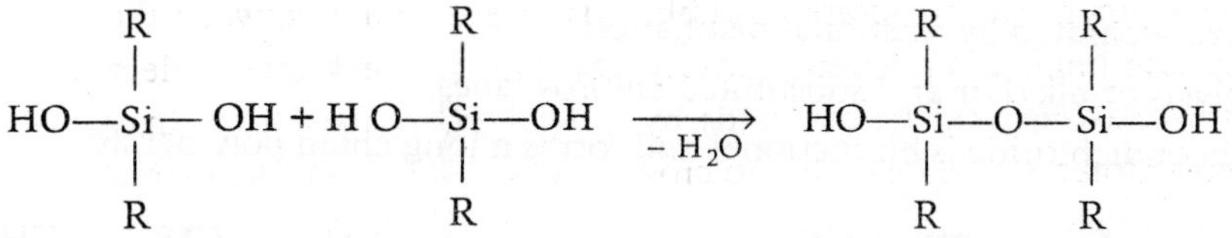

Since reactive OH groups are either side of the chain so polymerization takes place and long chain is formed.

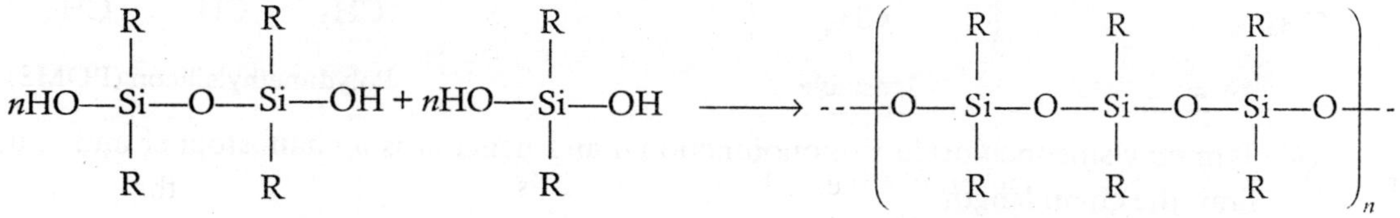

Silicones or Polysilioxanes

The product is called silicones or polysilioxanes. Thus, silicones are organosilicon polymers containing silicon-oxygen bond in the backbone. It resembles inorganic polymers because it has high percentage of ionic character of Si—O bonds and also resembles organic polymers because it has organic groups on silicon atoms.

Preparation. Silicon are generally prepared by the hydrolysis of alkyl or aryl substituted chlorosilanes. It involves in the following steps:

Step I. Preparation of alkyl or aryl substituted chlorosilanes.

(*a*) ***Direct method.*** (*i*) Methyl chloride is reacted with silicon in the presence of copper catalyst at 300°C

$$CH_3Cl + Si \xrightarrow[300°C]{Cu} (CH_3)_2SiCl_2 + CH_3SiCl_3 + (CH_3)_3SiCl$$

The reaction is highly exothermic and the yield of $(CH_3)_2SiCl_2$ is about 50% and can be separated from the mixture through fractional distillation.

(*ii*) When benzene or aromatic hydrocarbon is reacted with the compounds containing Si—H bonds at 250–300°C in the presence of catalysts $AlCl_3$ or BF_3, monoaromatic chlorosilane is formed.

$$C_6H_6 + HSiCl_3 \xrightarrow{250\text{-}300°C} C_6H_5SiCl_3 + H_2$$

(*b*) ***Indirect method (through Grignard reagent.)*** A solution of $SiCl_4$ in dry ether is allowed to react with the Grignard reagent.

$$CH_3MgCl + SiCl_4 \longrightarrow \underset{\text{Methyltrichlorosilane}}{CH_3SiCl_3} + MgCl_2$$

$$2CH_3MgCl + SiCl_4 \longrightarrow (CH_3)_2SiCl_2 + 2MgCl_2$$

Dimethyldichlorosilane

$$3CH_3MgCl + SiCl_4 \longrightarrow (CH_3)_3SiCl + 3MgCl_2$$

Trimethylchlorosilane

The products are separated by fractional distillation.

Step II. Hydrolysis of alkyl or aryl substituted chlorosilanes.

(*a*) Dimethylsilicondichloride is bifunctional and forms a long chain polymer as:

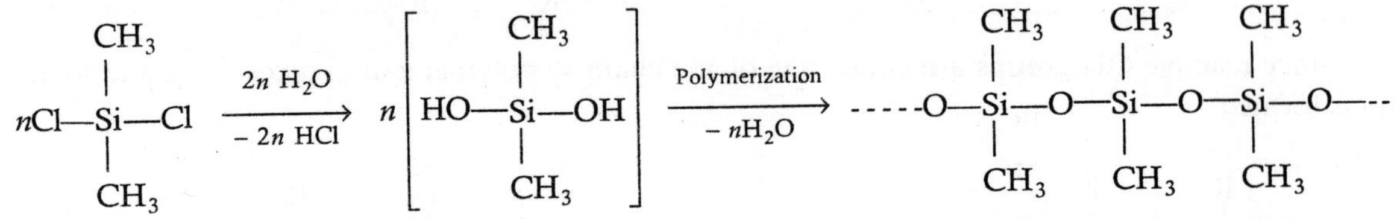

Unstable Polydimethylsilicon (PDMS)

(*b*) Trimethylsiliconchloride is monofunctional and hence it is a chain-stopper and is used to limit the chain length.

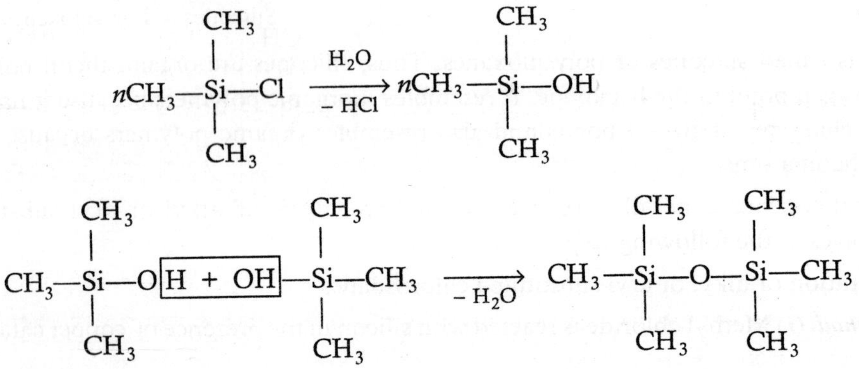

(*c*) Monomethylsilicontrichloride is trifunctional and gives cross-linked polymer.

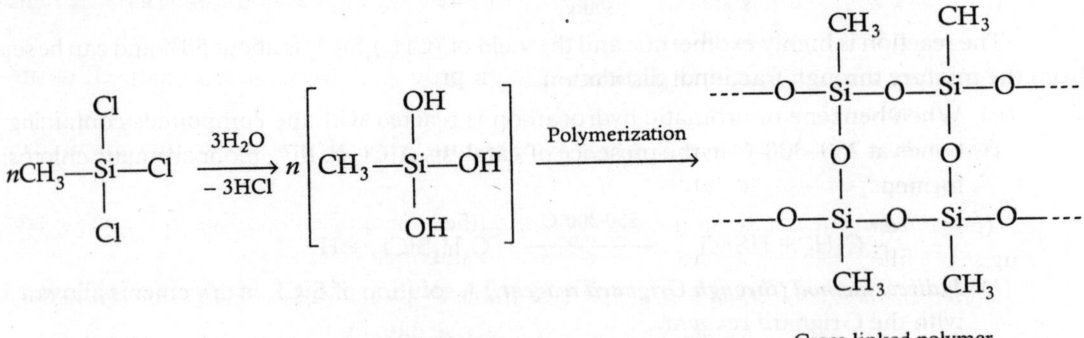

Unstable Cross-linked polymer

Properties. Silicons possess certain unusual but very useful properties.

(*i*) They are stable from – 70°C to 250°C.

(*ii*) They are good water repelling and good resistance to the effect of weathering.

(*iii*) They are chemically and physiologically inert and resist oxidation.

(*iv*) Liquid silicones have only small change in viscosity with change in temperature, hence they are very good lubricant.

(*v*) They have excellent electrical properties even at high temperature.

(*vi*) They have low vapour pressure, high flash points and high degree of compressibility.

(*vii*) They are used in cosmetics and medicines.

(*viii*) They have low surface tension.

(*ix*) They are soluble in organic solvents like toluene and xylenes.

(*x*) They are depolymerized with strong acids and bases.

Types. The silicones are generally classified into the following types:

(*a*) Fluid (*b*) Grease (*c*) Rubber (*d*) Resins.

(*a*) ***Silicon fluid.*** Low molar mass straight chain polymers are obtained by the hydrolysis of dichlorosilanes. It is known as silicon fluid. For its preparation a small amount of trimethylchlorosilane is added to block the chain.

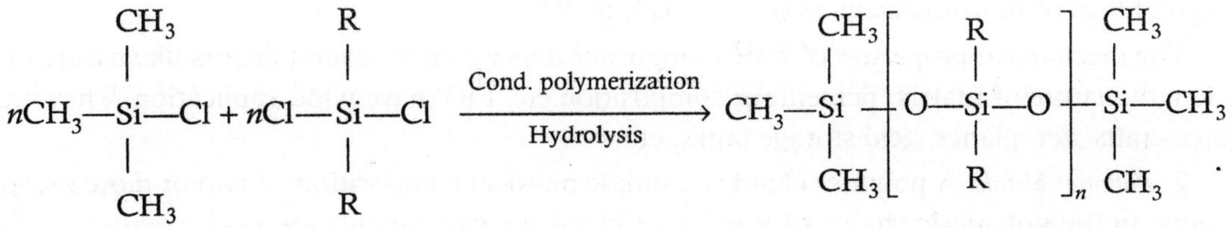

Silicon fluid

Properties and Uses. (*i*) They have high viscosity index so they are used as hydraulic fluids.

(*ii*) They have excellent dielectric properties and are thermally stable and are useful in capacitors and small transformers.

(*iii*) They can withstand high temperature so they are used in high temperature water baths.

(*iv*) They are used in polishes for cars and furniture because dust and dirt particles remove easily from the coated surface.

(*v*) They are nontoxic and possess antifoam properties so they are used to eliminate foam in petroleum oils.

(*vi*) They are good lubricants at high temperature and high pressure so they are used in vacuum pumps, jet turbines, etc.

(*b*) ***Silicon grease.*** Silicon grease is the modified form of silicon fluids. It can be prepared by adding some fillers like silica, carbon black, etc., to silicone oils or fluids.

Properties and Uses. (*i*) Silicon grease are stable to a wide range of temperature from – 100°C to 450°C. So they are used in industries as a very effective lubricant.

(*ii*) When treated with silica it is used as electrical grease in car ignition systems.

(*iii*) Greases with lithium soap filler are used for ball-bearings, etc.

(*iv*) Greases with carbon black filler are used in high temperature conveyers and oven doors, etc.

(*c*) ***Silicon Rubbers.*** These are polymers containing high molecular weight dimethylsilicon mixed with a filler usually finely divided SiO_2 and a peroxide. Organic peroxide helps to form few bridges between linear chains, SiO_2 reinforce the polysilioxane structure.

Properties and Uses. (*i*) They are highly resistant to heat and are used in aircrafts and insulating electrical parts.

(*ii*) They are flexible at 90–250°C and are used in making tyres for fighter aircrafts.

7.8 POLYMERIC COMPOSITES

A composites is a material made up of more than one components. A polymer belonging to a particular class has several properties or characteristics of its own class but may lack some important properties required for a particular application. Hence to improve its properties the polymer may be mixed with some other polymers or non-polymeric materials. The resultant mixture is known as polymer composite. Some polymer composites are following.

1. *Fibre reinforced plastics (FRP).* Fibre reinforced plastics are prepared by bonding a fibre material with a resin matrix (phenolic resins, silicone resins, epoxy resins, polyamides etc.) under heat and pressure. For example, fibre glass is prepared by reinforcing melamine or some other plastic matrix with spun fibres of borosilicate glass (known as E-glass).

The mechanical properties of fibre composites depend upon several factors like nature of fibre, fibre length, nature of matrix, percentage composition etc. FRP have wide application. They are used in space-crafts, aeroplanes, acid storage tanks, etc.

2. *Polymer blends.* A polymer blend is a simple physical composition of two or more incomplete polymers. In the polymeric chains of a polymer blend the constituents are held together by Van der Waal's forces, hydrogen bonding etc. Polymer blending is very useful to improve the properties of a polymer. It may help in protecting the particular polymer from degradation.

For example, (*a*) Polymethylmethacrylate (PMMA) undergoes degradation by γ-rays but the degradation is reduced by blending it with styrene acrylonitrile (copolymer).

(*b*) Blending of nylon-6 with polycarbonate is very tough and is used for making transparent containers and sports equipments.

3. *Polymer alloys.* When two or more compatible polymers chemically interact under a specific set of conditions the resultant mixture is called polymer alloy. The properties like mechanical strength, abrasion resistance etc. of polymer alloy is much better than the constituent polymers.

For example, Acrylonitrile butadiene styrene (ABS) is a copolymer blend of polyacrylonitrile-styrene and butadiene-styrene rubber but ABS-Polycarbonate alloys are typical polymer alloys. These have good mechanical strength and high work ability. They are used for making typewriters, helmets, etc.

EXERCISE

1. Define the following:

   (*a*) Polymer (*b*) Degree of polymerization (*c*) Elastomer (*d*) Thermosetting plastics.

2. How are polymers classified on the basis of (*i*) Structure (*ii*) Synthesis (*iii*) Molecular mass (*iv*) Tacticity and (*v*) Source of origin
3. What do you mean by homopolymer and co-polymer? Explain the term graft co-polymer. ***(K.U.K. June 2006)***
4. Differentiate between chain growth and step growth polymerization. ***(K.U.K. June 2006, Jan. 2006)***
5. (*a*) Short out the chain-growth and step-growth polymers.
   Bakelite, Nylon-6:6, Teflon, Terylene, PVC, Polyethene.
   (*b*) How is PVC prepared? Mention its properties and uses. ***(C.D.L.U. Dec. 2003)***
6. Explain the mechanism of chain-growth polymerization (cationic, anionic and free redical mechanisms). ***(K.U.K. June 2005, Jan. 2006)***
7. What are the differences between thermoplastics and thermosets? ***(K.U.K. Jan. 2007, June 2009, Jan. 2009)***
8. How are the different properties of polymers related to their structure? Explain with examples.
9. What do you understand by the crystallinity of a polymer?
10. What are plasticizers? How does the molecular mass of a polymer affect its properties? How is vulcanization of natural rubber carried out?
11. Define crystallites. Out of linear, branched chain and cross linked polymers, which is the strongest and why? How is PVA prepared? Mention its important properties and uses. ***(K.U.K. Jan. 2006, June 2006)***
12. What is natural rubber? Explain how is the chemical resistance of a plastic influenced by the structure?
13. What are elastomers? Give the preparation, properties and uses of SBR, GR-N, UF, PF resins.
14. What are silicones and how are they prepared? Discuss their important properties and uses. ***(K.U.K. June 2006)***
15. Define polymerization? Differentiate between addition and condensation polymerization. ***(K.U.K. June 2006, June 2009)***
16. Write the structure of natural rubber and gutta-percha rubber? What are the differences between these two?
17. What is plastic deformation? Write a short note on polymer composites.
18. Define polymer blends, reinforced fibreglass and polymer alloys? How is fibreglass obtained?
19. (*a*) Distinguish between addition and condensation polymerisation. ***(K.U.K. Jan. 2006)***
    Give suitable examples. ***(C.D.L.U. Dec. 2003)***
    (*b*) Write short notes on (preparation, properties and uses of)
    (*i*) Styrene-Butadiene Rubber (SBR)
    (*ii*) Poly (vinylchloride) (PVC). ***(K.U.K. Jan. 2005)(C.D.L.U. Dec. 2003)***
    (*iii*) Buna Nitrile rubber (GR-N elastomer) ***(K.U.K. June 2005)***
    (*iv*) One thermoplastic material ***(K.U.K. June 2005, Jan. 2007)***
20. What do you understand by molecular weight of a polymer? Explain weight-average and number-average molecular weights and the methods for their determination.
21. Write a short note on the following:
    (*a*) Glass transition temperature
    (*b*) Polydispersity index
    (*c*) Plasticizers
    (*d*) Silicon grease

(*e*) Vulcanization of rubber

(*f*) E-glass.

22. Write an essay on 'The effect of structure on properties of polymers'. ***(K.U.K. Jan. 2005, June 2006)***
23. Discuss the free radical chain growth polymerization. ***(K.U.K. Jan. 2005, Jan. 2006, June 2006, Jan. 2007, Jan. 2009)***
24. Write an essay on following:
    (*i*) Polymeric composites. ***(K.U.K. June 2006)***
    (*ii*) Industrial applications of silicones. ***(K.U.K. Jan. 2005, June 2007)***
    (*iii*) Biodegradable polymers.
25. PVC is soft and flexible whereas Bakelite is hard and brittle, explain. ***(K.U.K. June 2004)***
26. What are the chief physical characteristics expected of an elastomer? How are they achieved in a new product? ***(K.U.K. June 2004)***
27. What is meant by copolymerization? Describe the preparation and technical applications of silicon fluids? ***(K.U.K. June 2004)***
28. How does the chain structure of rubbers compare with that of plastics? ***(K.U.K. Jan. 2004)***
29. Why cannot thermosetting plastics be reshaped and reused? ***(K.U.K. Jan. 2004, June 2008)***
30. Why do all simple organic molecules not produce polymers? Write a short note on glass-resin forced plastics. ***(K.U.K. Jan. 2004)***
31. Describe the preparation and technical applications of silicone rubbers. ***(K.U.K. Jan. 2004)***
32. How inorganic polymer is differ from organic polymer? Explain Why is the softening temperature of polystyrene is lower than that of polyethene although the former contains polar group.
33. Define virgin polymer. What do you mean by 'living' and 'dead' polymers?
34. Why should the vinyl monomers be distilled before polymerization?
35. What are the factors governing the solubility of a polymeric material?
36. Why and how is a polymeric material brought into finely divided state?
37. What is dry ice? Why is it added during grinding of a polymeric material?
38. What are the main types of additives present in the processed polymers? How are these removed?
39. How teflon is prepared? Write its properties and applications.
40. What are biopolymers? How they are differ from biodegradable polymers?
41. Define biodegradable polymerisation. Explain its importance.
42. Explain the free radical polymerisation mechanism of a vinyl monomer. ***(K.U.K. Jan. 2008)***
43. What are silicones? Draw the structure of silioxane polymer obtained by hydrolysing dichlorodimethylsilane. ***(K.U.K. Jan. 2008)***
44. Distinguish between nylon-6 and nylon-6:6. ***(K.U.K. Jan. 2008)***
45. What are polymers and what do you understand by polymerisation? Explain with examples the different type of polymerisation. ***(K.U.K. Jan. 2009, June 2005)***
46. Differentiate isotactic and syndiotactic polymer. ***(K.U.K. June 2008)***
47. Define composites and write its applications over polymer. ***(K.U.K. June 2008)***
48. Write preparation, properties and uses of PVC. ***(K.U.K. June 2008, Jan. 2006, June 2006)***
49. Give the mechanism of cationic chain growth polymerisation. ***(K.U.K. June 2009)***
50. Discuss the preparation, properties and uses of a thermosetting polymer. ***(K.U.K. June 2009, June 2005)***

51. What are silicones? Give some of their important applications. *(K.U.K. June 2009)*

52. What do you mean by functionality of monomer? *(K.U.K. June 2009)*

53. Write a short note on glass reinforced plastics. *(K.U.K. June 2009)*

54. What is an elastomer? Name one synthetic elastomer. Mention the raw material for its manufacture. *(K.U.K. Jan. 2006)*

55. Why are silicones called inorganic polymers? *(K.U.K. Jan. 2006)*

56. Define various types of polymerisation. Explain the mechanism of ziegler-natta coordination polymerisation. *(K.U.K. Jan. 2006)*

57. Write the industrial applications of silicones. *(K.U.K. Jan. 2006)*

58. Discuss the preparation, properties and the important applications of phenol-formaldehyde polymer. *(K.U.K. Jan. 2006)*

59. Define tacticity. Discuss the various types of tacticity present in the polymers. *(K.U.K. Jan. 2006)*

60. Write a short note on silicones. *(K.U.K. June 2005)*

61. Describe the method of preparation, properties and some important applications of polyvinylacetate. *(K.U.K. Jan. 2006)*

62. Explain method of preparation, properties and uses of bakelite. *(K.U.K. Jan. 2007)*

63. How ziegler-natta coordination polymerisation is better than free radical polymerisation? *(K.U.K. Jan. 2009)*

64. What is an elastomer ? Live the preparation, properties and uses of a synthetic elastomer. *(K.U.K. Jan. 2009)*

65. Give the preparation of a dimethylsilicone polymer. *(K.U.K. Jan. 2009)*

66. Describe the method of preparation, properties and applications of Urea-formaldehyde resin.

67. What is silicon fluid? Write its structural formula. Give some important industrial applications and properties.

68. What do you understand by vulcanisation of rubber? How is it carried out? Write the structure of vulcanised rubber. Write its some important properties and applications.

69. What are synthetic rubbers? Write the preparation, properties and uses of SBR and NBR.

70. What are silicon rubbers? How is it prepared? Write the properties and uses of silicon rubbers.

71. How silicon rubber is differ from the natural or vulcanised rubber?

72. What is silicon grease? How is it prepared? Write its important properties and applications.

73. What do you understand by coordination polymerisation? Why is it called insertion polymerisation ? Explain its mechanism. Write its important characteristics.

74. What are ziegler-natta catalysts? Explain the role of ziegler-natta catalyst in coordination polymerisation.

75. Define tacticity. Explain the mechanism for the preparation of stereo regular polymers. Write its important features.

76. Write a short note on following:

    (*i*) Fibre reinforced plastics — (*ii*) Polymer blends

    (*iii*) Polymer alloys — (*iv*) Silicon fluid

    (*v*) Silicon grease — (*vi*) Silicon rubber

77. What do you mean by polymeric composites? Explain it by taking some suitable examples. Write its important applications in industries.

78. What is borosilicate glass (E-glass)? How is it differ from ordinary glass? Give its some applications.

Chapter 8 ANALYTICAL METHODS

8.1 INTRODUCTION

The branch of chemistry which deals with the experimental techniques for analyzing the compounds or molecules qualitatively or quantitatively and methods of expressing the results with maximum precision/accuracy is called the analytical chemistry. The analytical chemistry is classified into two classes: qualitative and quantitative. The qualitative analysis deals with the identification of the elements, ions or compounds present in an unknown sample. For example, mixture analysis, detection of compounds, chromatography, etc. On the other hand, quantitative analysis deals with the determination of quantity of one or more components presents in the sample. For example, gravimetric analysis, volumetric analysis (iodometry, iodimetry, precipitation titrations, etc.), instrumental based like conductometry, colorimetry, thermal gravimetry, spectrophotometry, etc. The quantitative calculations are based on experimental results obtained with measurements of different physical parameters.

Needs of Analytical Methods/Techniques

The analytical techniques play very important role in the identification and characterization of various compounds. Some important needs are as follows:

(*i*) In research laboratories, the various analytical methods have great importance in the isolation, identification, characterization and synthesis of compounds, qualitatively as well as quantitatively. For example, chromatography is useful for the separation of compounds.

(*ii*) The vulcanization of rubber, cross-linking of polymers, grafting and functionalization of polymers, etc. can be checked by various analytical techniques.

(*iii*) The purity of substances or compounds is easily tested by different analytical techniques.

(*iv*) These techniques are also useful in the determination of structure of compounds.

(v) The reaction mechanism, rate of reaction and the extent of reaction can be determined by various analytical techniques.

(vi) The formation of intermediates during reactions is also studied by these techniques.

(vii) The adulterants in various food products, medicines, cold drinks, water, milk, etc. can be checked smoothly.

(viii) These methods are also useful for checking the air pollution, soil pollution, water pollution, etc.

(ix) In industries and engineering branches various analytical methods have been employed in the testing of thermal and mechanical stabilities of materials, lubricants, etc.

(x) The quality of products may be controlled by these methods.

(xi) These techniques are also useful in the determination of the hardness of water which is used in different type of boilers.

(xii) In biotechnology these techniques give the fruitful results to analyse and control the diseases.

(xiii) The DNA finger print can be tested by these techniques.

(xiv) In medical science, urine, blood, stool, etc. can be examined by the various analytical techniques.

(xv) These techniques are very useful in the diagnosing the diseases for their proper treatment.

(xvi) In agriculture field, these techniques are useful for testing the soil fertility.

In summary, we can say that the analytical methods are the backbone of every branch in different fields and industries. There are several methods employed for analyzing the compounds qualitatively or quantitatively. In this chapter, we shall focus on some important quantitative analytical methods.

8.2 THERMOGRAVIMETRIC ANALYSIS (TGA)

It is the quantitative measurement of any change in weight of a substance under investigation or examination with the increase in temperature. The usual temperature range is from ambient to 1200°C in inert or reactive atmospheres. The result is expressed as a graph of mass (m) versus temperature (T) and is known as thermogram. A plot of $\frac{dm}{dT}$ versus temperature (T) is called differential thermogram. Thermogravimetry (TG) can directly record the loss in weight with time or temperature due to dehydration or decomposition on heating the sample. Thermograms are characteristic for a given compound or system because of the unique sequence of physicochemical reactions which occur over definite temperature ranges and at rates that are a function of the molecular structure. The change in weight of the compound or system occurs as a result of the rupture and/or formation of various bonds at elevated temperature.

Instrumentation

The sample is heated to elevate temperature and the weight of sample is continuously weighed. The samples are placed in a crucible that is attached to an automatic-recording balance. The automatic null type balance incorporates a sensing element which detects a deviation of the balance beam from its null position. One transducer is a pair of photocells, a slotted flag connected to the balance arm and a lamp (light source). Once the balance has been adjusted; any changes in sample weight cause the rotation of the balance. This moves the flag so that the light falling on each photocell is no longer equal. The resulting non-zero signal is amplified and feedback as a current to a taut-band torque motor (the pivot-point of the balance) to restore the balance to equilibrium. It may be represented as in Fig. 8.1.

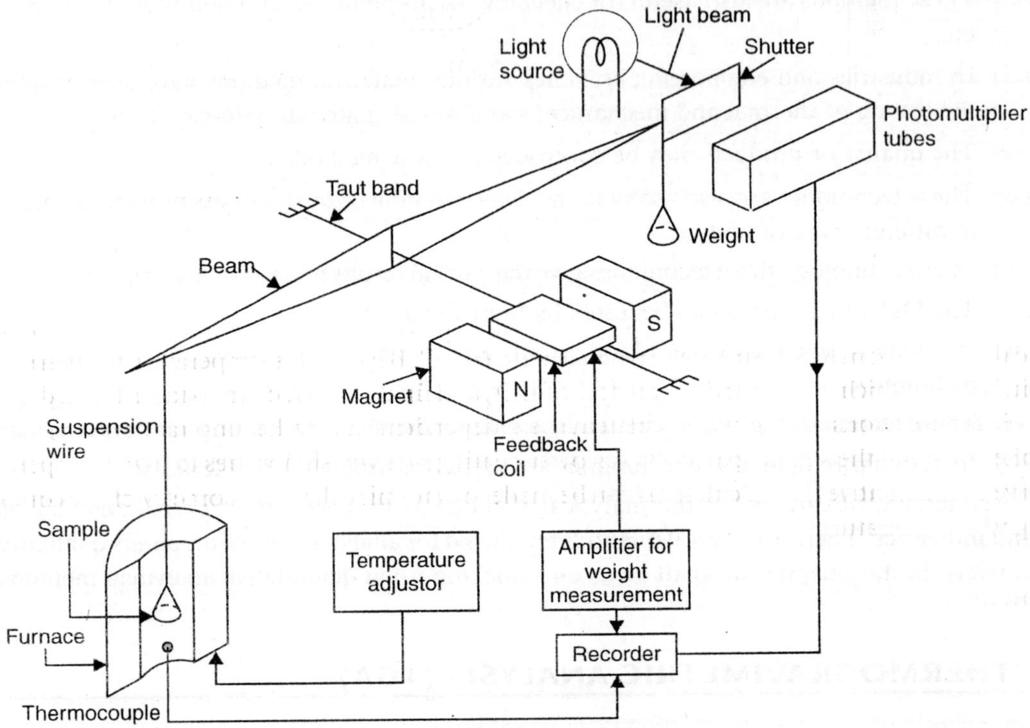

FIGURE 8.1 **Schematic diagram of TG equipment with optical sensor (Courtesy of Shimadzu Seisakusho, Ltd.)**

This current is proportional to the change in weight and is recorded on the y-axis of the recorder. Furnace temperature is continuously monitored by a thermocouple whose signal is applied to the x-axis of the recorder. Generally the linear heating rates 5°C to 10°C per minutes are employed. In differential thermogravimetry the actual measurement signal is derived from a solid state resistance-capacitance circuit which uses the direct output of change in weight as electrical signal from the thermo-balance for the primary signal input. The resulting output is the derivative $\frac{\Delta W}{\Delta t}$, which is used in kinetic in interpretations. The amount of sample under investigation is taken in the range from 1 mg to 300 mg and sensitivities down to a few micrograms of weight change are common. For example, we consider about the thermogram of calcium oxalate. It is represented as in Fig. 8.2.

The thermogram contains horizontal portions and curve portions. Horizontal portions represent no weight change and curve portions represent loss in weight due to dehydration, escaping of volatile products etc. In Fig. 8.2 of thermogram monohydrated calcium oxalate undergoes decomposition by increasing the temperature 5°C per minute. Calcium oxalate is stable upto 100°C. After that water molecule escapes out. At 250°C a plateau is obtained. Anhydrous calcium oxalate is stable from 250°C

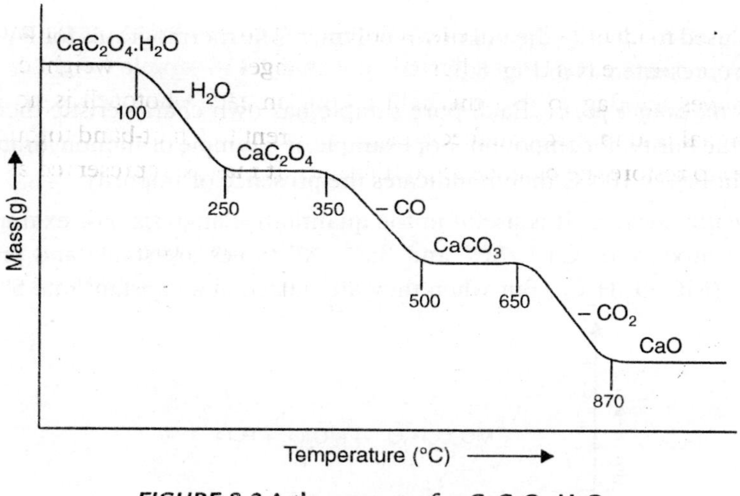

FIGURE 8.2 **A thermogram for $CaC_2O_4.H_2O$**

to 350°C. After that calcium oxalate undergoes decomposition and carbon monoxide comes out. At 500°C calcium carbonate is formed and is stable upto 650°C. Beyond this temperature the thermogram shows inflection which corresponds to the loss of CO_2 and finally the calcium oxide is formed at above 870°C. Exact locations of the weight plateaus are dependent on the heating rate and the ambient atmosphere around the sample particles. A slower heating rate will shift values to lower temperatures. The curve is quantitative and calculations can be made to determine the stoichiometry of the compound at any given temperature.

Applications

TGA has wide applications in analytical chemistry. Some of them are as follows:

(*i*) *In the study of polymers.* The TGA thermogram gives the important information in the study of decomposition of polymers. Each kind of polymer has a characteristic particular thermogram. The

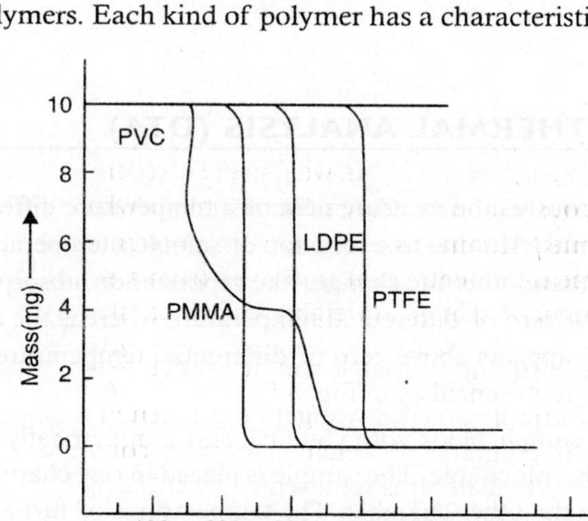

FIGURE 8.3 **Thermograms of PVC, PMMA, LDPE and PTFE**

thermogram is also used to identify the unknown polymer. The thermograms for PVC, PMMA, LDPE and PTFE may be represented as in Fig. 8.3.

(*ii*) *In testing the sample purity.* Each pure sample has own characteristic thermogram which is useful for checking the purity of compound. For example, if a sample of monohydrated calcium oxalate losses its own weight below 100°C then it indicates the presence of impurity.

(*iii*) *In quantitative analysis.* It is useful in the quantitative analysis. For example, it is applicable to the analysis of a mixture of Ca^{2+}, Sr^{2+} and Ba^{2+}. All threes are stable and precipitated as their monohydrate forms ($MC_2O_4.H_2O$). But when they are heated at a constant rate 5°C per minute they

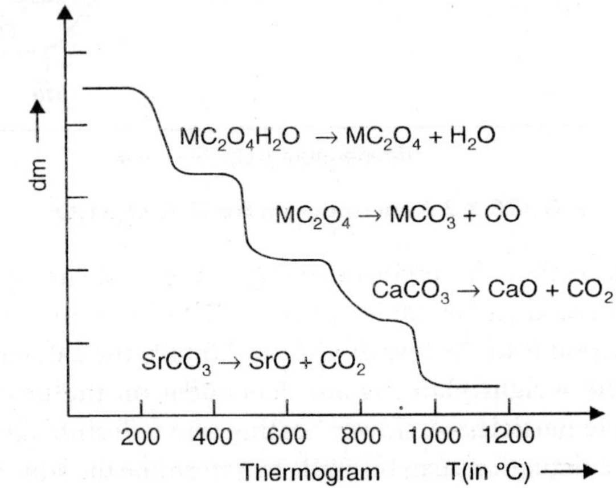

FIGURE 8.4

forms anhydrous oxalates between 300°C–400°C. After that they get converted into carbonate by losing carbon monoxide. The carbonates of these metals are stable between 580°C to 630°C. Beyond this temperature the loss of CO_2 occurs by $CaCO_3$ and $SrCO_3$ to form CaO and SrO respectively. From the above data, it is easy to calculate the mass of each element present in the sample. It may be represented as in Fig. 8.4.

8.3 DIFFERENTIAL THERMAL ANALYSIS (DTA)

Differential thermal analysis involves the measurement of a temperature difference of a sample and a thermally inert reference material (Alumia) as a function of sample's temperature. The thermogram of DTA shows peaks or bands. The endothermic changes like vaporization, absorption, desorption, fusion, sublimation, etc. appears below zero of differential temperature whereas the exothermic changes like oxidation, decomposition, etc. appears above zero of differential temperature. A rough line sketched diagram of DTA equipment is represented as in Fig. 8.5.

The furnace contains a sample block with identical and symmetrically located chambers. Each chamber contains a centered thermocouple. The sample is placed in one chamber (S) and the reference material (α-Al_2O_3) is placed in the other chamber. The temperatures of furnace and sample block are then increased at a linear rate (5°C to 10°C/per minute), either by increasing the voltage through the heater element by a meter-driven variable transformer or by a thermocouple-actuated feedback type of controller. The difference in temperature between sample and reference thermocouples, connected in

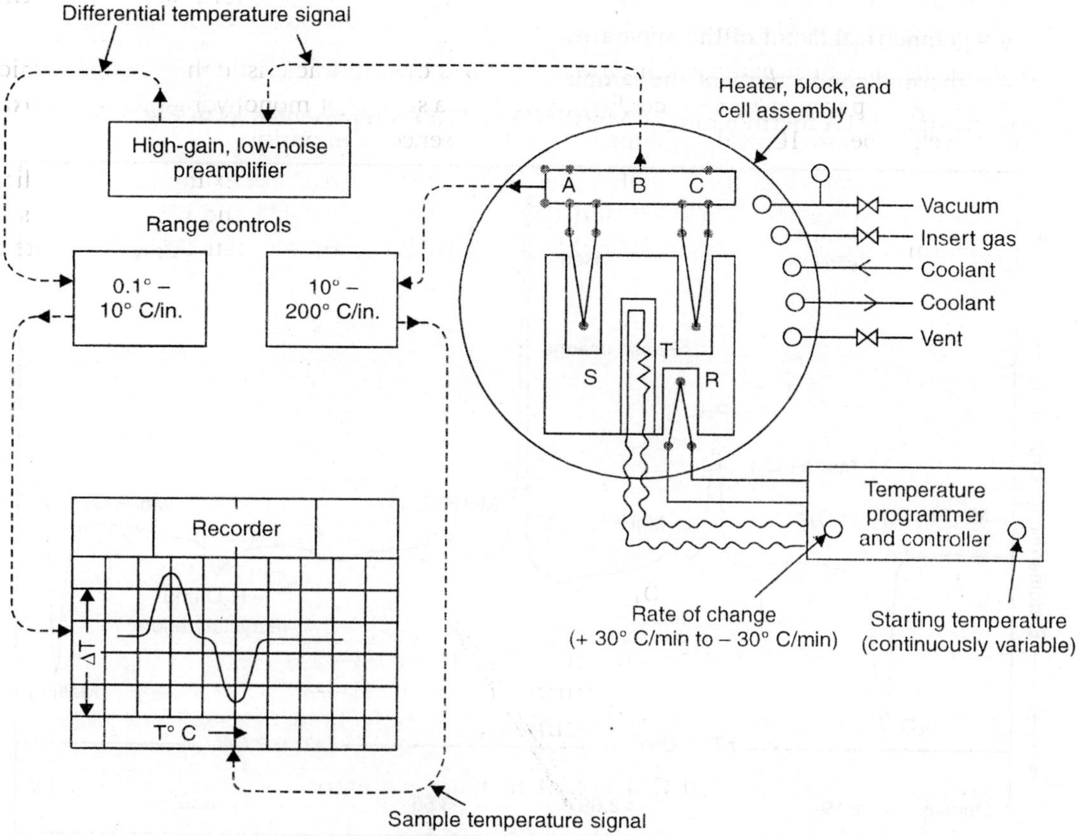

FIGURE 8.5 **Schematic diagram of the DuPont differential thermal analyzer (Courtesy of E. I. DuPont de Nemours, Inc.)**

'series opposition, is continuously measured. Differential temperature signal is amplified about 100 times by a high-gain, low noise, DC amplifier for the microvolt-level signals. The difference signal is recorded on the *y*-axis of a millivolt recorder. The temperature of the furnace is measured, by a separate thermocouple which is connected to the *x*-axis of the recorder, frequently through a reference ice junction or room-temperature compensator. Since the thermocouple is placed directly in the sample, this technique gives the maximum accuracy among all the thermal methods. The area of the output curve is not necessarily proportional to the amount of energy transferred in or out of the sample. For more accuracy the sample and reference thermocouple are removed from direct contact with the sample. The temperature range is between – 190°C and 1600°C and the sample size range from 0.1 mg to 100 mg. Quantitative heats of transition require area integration to determine the total amount of energy transferred into or out of the sample. The area (ΔT × time) (enclosed under a peak) is given by

$$\text{Peak area (A)} = \int_{t_1}^{t_2} \Delta T \, . \, dt = \frac{m \, . \, \Delta H}{g \, . \, k}$$

where m = mass of the sample

ΔT = differential temperature

ΔH = enthalpy of involved reaction

g = geometrical factor of the apparatus

and k = thermal conductivity of the sample.

For example, DTA thermogram of $Mn(PH_2O_2)_2 . H_2O$ is represented as in Fig. 8.6.

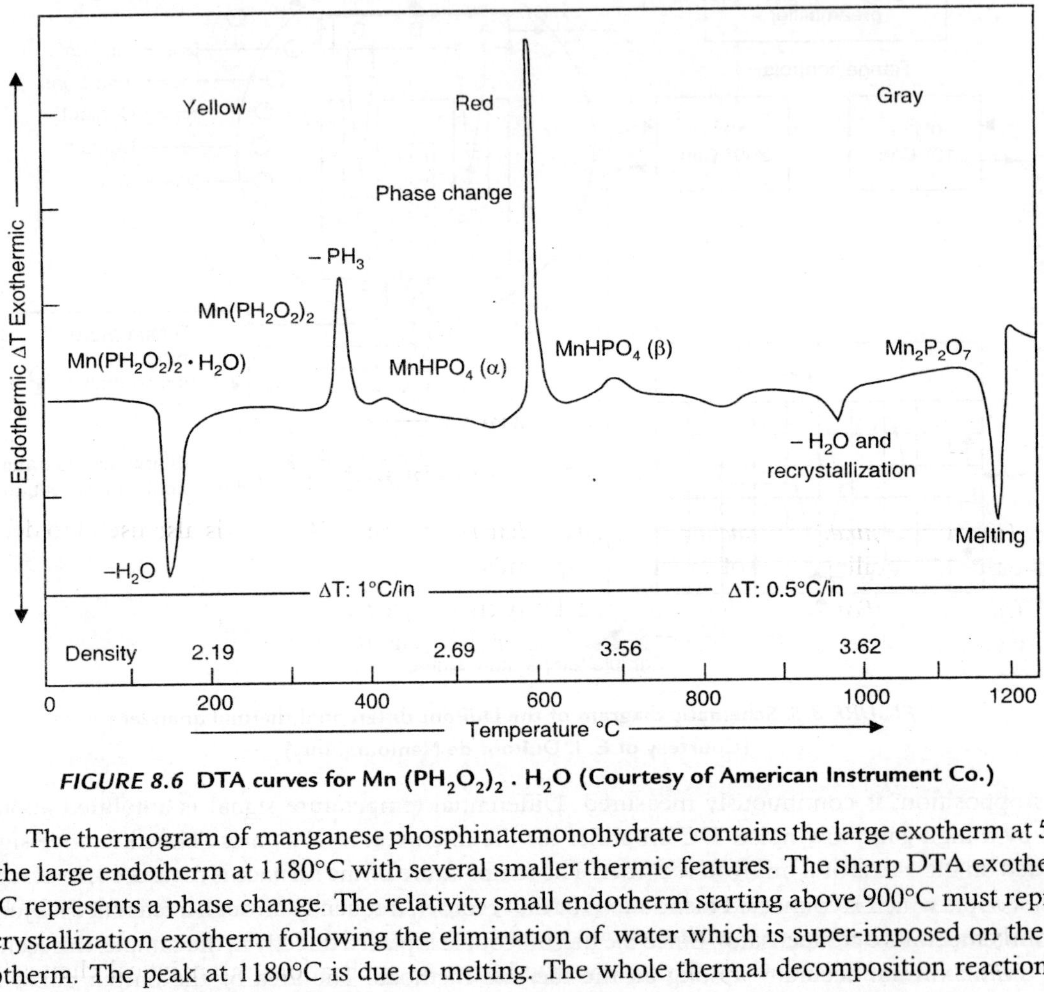

FIGURE 8.6 **DTA curves for Mn $(PH_2O_2)_2 . H_2O$ (Courtesy of American Instrument Co.)**

The thermogram of manganese phosphinatemonohydrate contains the large exotherm at 590°C and the large endotherm at 1180°C with several smaller thermic features. The sharp DTA exotherm at 590°C represents a phase change. The relativity small endotherm starting above 900°C must represent a recrystallization exotherm following the elimination of water which is super-imposed on the latter endotherm. The peak at 1180°C is due to melting. The whole thermal decomposition reactions and phase changes are represented as

$$Mn(PH_2O_2)_2 . H_2O(s) \longrightarrow Mn(PH_2O_2)_2(s) + H_2O(g)$$

$$Mn(PH_2O_2)_2(s) \longrightarrow MnHPO_4(s) + PH_3(g)$$

$$\alpha\text{-}MnHPO_4(s) \longrightarrow \beta\text{-}MnHPO_4(s)$$

$$2MnHPO_2(s) \longrightarrow Mn_2P_2O_7(s) + H_2O(g) \quad \text{(and re-crystallization)}$$

$$Mn_2P_2O_7(s) \longrightarrow Mn_2P_2O_7(l)$$

Applications

It is the complimentary of TGA, *i.e.*, the information obtained by TGA is often enhanced by the application of DTA. Some important applications are as follows:

(*i*) *In the study of phase transitions.* It is helpful in the phase diagrams and the study of phase transitions. The DTA thermogram of pure sulphur may be represented as in Fig. 8.7. The thermogram consists of four peaks at 113°C, 124°C, 179°C and 446°C corresponds to the different phase transitions, 446°C is the boiling point of sulphur.

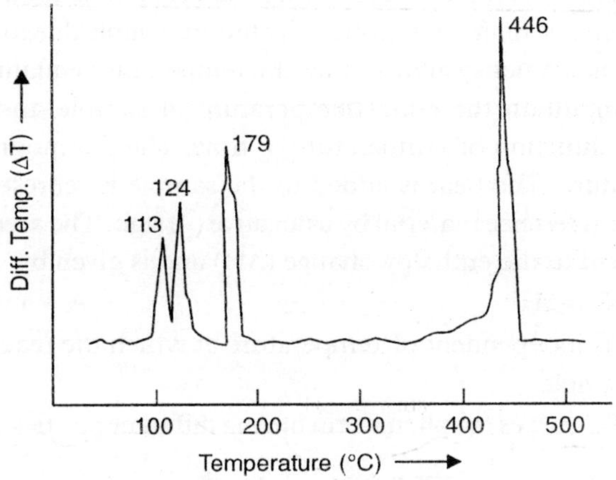

FIGURE 8.7 **DTA curve for sulphur**

(*ii*) *In determination of melting point and boiling point.* The DTA curve is also useful to determine the melting and boiling point of organic compounds.

(*iii*) *In identification of substances.* Since DTA curves are not identical for two substances. Hence the substance can be identified by the comparing of thermogram with the original sample.

(*iv*) *In industries.* This technique has a great importance in industries for the quality control of materials like cement, glass, etc. For example, in cement industry the characterization of limestone used in the production of Portland cement has been done by DTA. The amount of $MgCO_3$ in cement is calculated or determined quantitatively by DTA curve.

(*v*) *In determining the thermal stabilities of compounds.* The thermal stability of compounds or complexes are studied by DTA thermogram.

(*vi*) *In polymer chemistry.* This technique is very helpful for qualitative analysis of polymer mixture by analysing the characteristic melting point of each polymer.

(*vii*) *In analytical chemistry.* This technique has wide applications in analytical chemistry. One important application is the testing of purity of substances.

(*viii*) *In physical chemistry.* It is useful for the study of specific heat and heat of reaction, phase transformation, etc.

8.4 DIFFERENTIAL SCANNING CALORIMETRY (DSC)

Differential Scanning Calorimetry (DSC) is very closely related to DTA. In this method the sample and reference material are also subjected to a closely controlled programmed temperature. The heat energy is added to or subtracted from the sample or reference containers in order to maintain both sample and reference at the same temperature. This heat energy is recorded as a function of temperature or time.

Because this energy input is precisely equivalent in magnitude to the energy absorbed or evolved in the particular transition, a recording of this balancing energy yields a direct calorimetric measurement of the transition energy.

The apparatus of DSC is identical to that of DTA. The main difference is that the DSC apparatus consists of individual heaters for sample and reference material. Whenever the temperature difference between the sample and reference material is noticed by thermocouple due to exothermic or endothermic reaction, heat is added to the cooler system so that the temperature equality may be maintained. The added required heat for maintain the equal temperature of sample and the reference material is continuously recorded as a function of temperature or time. The thermogram is expressed in calories per second versus temperature. The heat is added to the sample is represented by a positive (+) sign whereas that supplied to the reference material by a negative (–) sign. The area (A) under a peak (maxima or minima) is directly related to the enthalpy change (ΔH) and is given by

$$A = K' m\Delta H \quad \Rightarrow \quad A \propto m\Delta H$$

where K' = constant and is independent of temperature at which the reaction takes place,

and m = mass of the sample

For example, the DSC curves for phenacetin having different purities is represented as in Fig. 8.8.

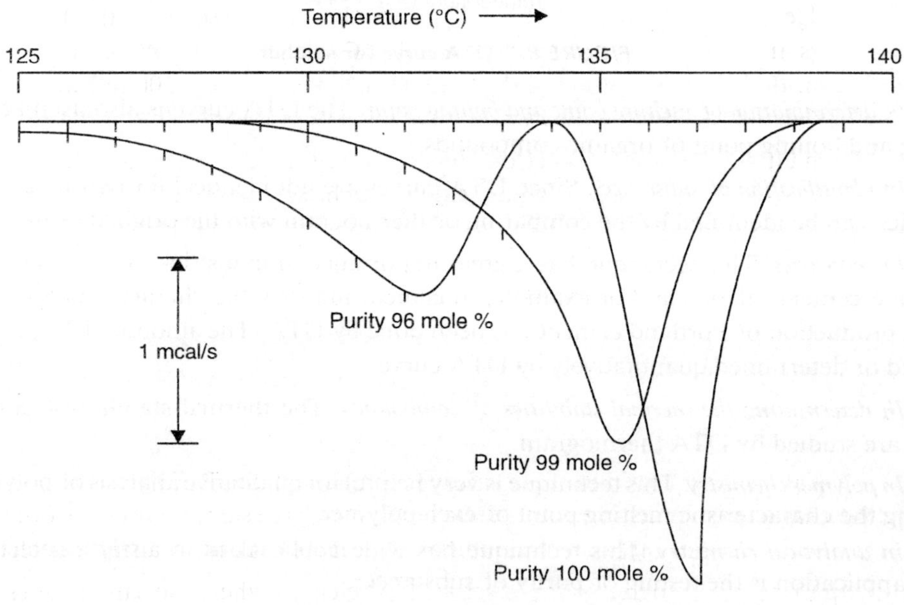

FIGURE 8.8 **DSC curve of phenacetin**

Applications

DSC has wide applications in industries and laboratories. Some of the important applications are follows:

(*i*) *In the determination of melting/boiling point.* It is used to determine the melting or boiling point, decomposition points of organic compounds.

(*ii*) *In checking the purity of compounds.* Each pure compound has own characteristic thermogram. Hence it is easy to check the purity of compounds (especially drugs) by comparing the thermogram with the pure sample. The curve provides purity data within 1% error. Fig. 8.8 shows the purity of phenacetin.

(*iii*) *In the characterization of organic compound.* This technique is useful for the characterization of organic compound by knowing the melting point, boiling point, decomposition point, etc. with the help of DSC thermogram.

(*iv*) *In determination of enthalpies of transitions.* It is widely used in calculating enthalpy of transitions such as enthalpy of fusion, enthalpy of melting, enthalpy of crystallization, etc. of polymeric materials.

8.5 SPECTROPHOTOMETER

A spectrophotometer is an instrument for measuring the transmittance or absorbance of a sample as a function of wavelength. The instrument may be manual or recording. Generally spectrophotometer is classified into two types: (*i*) single beam spectrophotometer and (*ii*) double beam spectrophotometer. On the basis of spectral region and techniques it is classified into different types like UV spectrophotometer, IR spectrophotometer, etc. A complete understanding of spectrophotometers requires a detailed knowledge of optics and electronics which is far beyond the scope of this book. Most of the spectrophotometers are double beam spectrophotometer. Ordinary spectrophotometer covers a range of 200 nm–800 nm. But if the whole path length is evacuated then below 200 nm wavelengths region can be studied.

The modern ultra-violet-visible spectrometers consist of light source, monochromator, detector, amplifier and the recording devices. The most suitable light sources are tungsten filament lamp or hydrogen-deuterium discharge lamp which covers the whole of the UV-visible region. The schematic line diagram of a possible double-beam optical-null recording spectrophotometer is represented as in Fig. 8.9.

The produced light from the light source is passed through a monochromator and encounters a rotating chopper. The chopper is a rotating mirror of such type of shape and placement that it permits the beam to pass straight through during half of its period of rotation. During the other half the beam encounters a reflective surface that turns it through a right angle as in Fig. 8.9. The direction may be changed again by other stationary mirrors as desired. Thus the monochromatic beam is divided into two beams having equal intensities. One beam is passed through the sample while the other encounters a reference solution. The beams are then recombined so that they fall upon a single detector.

Let us suppose the monochromator is set at a wavelength where the sample does not absorbs then the sample beam and the reference beam are of equal power and the detector observes same thing. In such cases the electrical output of the detector as a voltage does not vary with time, in other words a DC voltage. The signal from the detector is fed to an AC amplifier. The DC voltage from the detector is not amplified and the output of the amplifier is adequate to do anything in the instrument. Now we change the wavelength of radiation so that the sample absorbs. In this case the radiant power in the reference beam is greater than in the sample beam. The detector, observes first at one and then the other, generates an electrical signal which reflects this pulsation in radiant power. This signal is an AC voltage which actually superimposed upon the DC signal. The amplifier amplifies this AC voltage, which is then fed to a motor called servomotor. The servomotor drives an 'optical wedge' into the reference beam. The wedge is a special device which blocks part of beam and diminishes the radiant power in a smoothly progressive fashion as it moves. When the wedge has moved so as to attenuate the reference beam to the same extent as the sample absorption has attenuated the sample beam, then once again the detector sees the same thing regardless of where the chopper is in its rotational period. Thus the electrical signal is again DC and the amplifier does not amplify hence the motor stops. The position of the optical wedge at this point is reflected mechanically by the position of a pen which moves up and down on a piece of a chart paper. If we derive the monochromator with a motor which also moves the chart paper at right angles to the pen motion then a plot of wavelength (λ) vs transmittance or absorbance if the wedge is shaped properly and the chart paper is approximately calibrated.

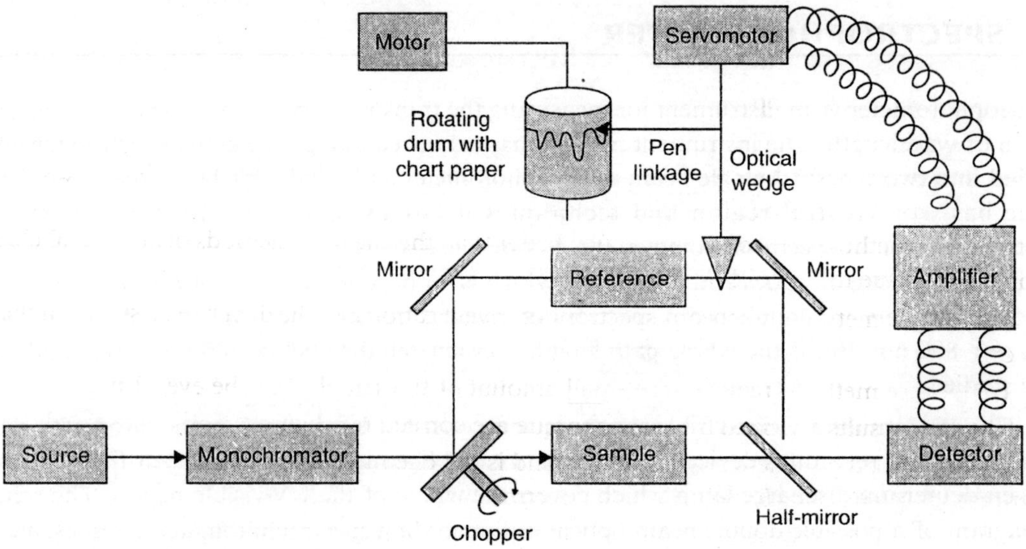

FIGURE 8.9 **Schematic diagram of a possible double-beam optical-null recording spectrophotometer**

Salient Features of Spectrophotometer

Different types of spectrophotometers have different salient features. Some of them are as follows :

(*i*) It is useful for plotting the absorption spectra (percentage transmittance *vs* wavelength).

(*ii*) It has special features in the identification of chemical substances. Spectra-structure correlations in both UV-visible and infrared regions are frequently very useful in the identification of different functional groups in unknown compounds.

(*iii*) The stability of a compound is determined by spectrophotometer (Thermogravimetric Analysis).

(*iv*) Spectrophotometer is useful to determine the rate of reaction, study of the reaction mechanism and formation of intermediates.

(*v*) UV-visible spectrophotometer is useful to determine the configurations of geometrical isomers.

(*vi*) This technique also distinguishes between equatorial and axial conformations.

8.6 SPECTROSCOPY

Spectroscopy may be defined as the interaction between the matter and electromagnetic radiation. In spectroscopy some remarkable physical methods known as spectroscopic methods have been developed which serve as important tools for the structural analysis of organic compounds. These are based upon the study of the spectra of the compounds even if available in small amount. These physical methods are known as spectroscopic methods of analysis. Some important methods are following:

(*i*) Ultra-Violet (UV) spectroscopy

(*ii*) Infra-Red (IR) spectroscopy

(*iii*) Nuclear Magnetic Resonance (NMR) spectroscopy

(*iv*) Mass spectroscopy (MS)

The mass spectroscopy helps in the determination of molecular mass and molecular formula. UV and IR used to study the nature of functional groups while NMR is used to determine the nature of proton and hence tells about the complete structure of the compound as well as its stereochemistry. Spectroscopic methods are considered much better than the classical methods of structural elucidation of compounds due to the following reasons:

(*i*) These methods take very little time for investigation and the data can be stored in the tabular form or a chart.

(*ii*) These methods require very small amount of the sample, may be even 1 mg.

(*iii*) The results obtained by this technique are correct, reliable and may be repeated.

(*iv*) During the process organic compound is not chemically affected and can be recovered after a particular investigation. Exception mass spectroscopy.

(*v*) These methods also help in the study of the reaction mechanism as well as the rates of chemical reactions.

(*vi*) These methods are useful in the study of stereochemistry of organic compounds

[**Note.** For the best results, the organic compound under investigation must be in pure form.]

8.6.1 Electromagnetic Radiations

Visible light is a form of energy which has both particle (corpuscular) and wave nature. Neither of these theories alone can completely account for the properties of light. Some of the properties like interference, polarization and diffraction are best explained by the wave theory while the properties like black body radiations and photoelectric effects are best explained by corpuscular theory. According to corpuscular theory, a beam of light is a stream of particles called photons and according to wave theory it is an electric and magnetic field moving through space like wave. Since the light radiation is associated with electric and magnetic fields both, therefore, it is known as electromagnetic radiation. All the electromagnetic radiations travel with the velocity of light *i.e.*, 3×10^8 m/sec.

Some Useful Terms in Electromagnetic Radiation

(*i*) *Wavelength* (λ). It may be defined as the distance between the two consecutive crests or troughs in a particular wave. The height of the particular crest or depth of the trough is called amplitude. It may be represented as in Fig. 8.10.

It is expressed in angstrom (Å), micrometer (mm) or (nm).

$$1 \text{ Å} = 10^{-10} \text{ m}$$

$$1 \text{ μm} = 10^{-6} \text{ m, also called micron, μ}$$

$$1 \text{ nm} = 10^{-9} \text{ m} = 1 \text{ mμm or simply milli micron mμ}$$

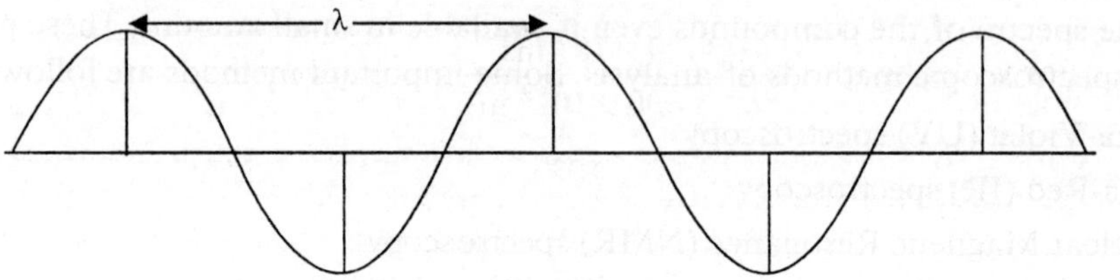

FIGURE 8.10

(*ii*) *Frequency* (ν). It may be defined as the number of waves which pass through a point in one second. It is expressed in cycles per seconds (cps) or Hertz (Hz). It is proportional to the reciprocal of the wavelength

i.e.,

$$\nu \propto \frac{1}{\lambda}$$

(*iii*) *Wave number* ($\overline{\upsilon}$). It may be defined as the total number of waves that are present in unit length. It is the reciprocal of wavelength.

$$\overline{\upsilon} = \frac{1}{\lambda}$$

It is expressed in cm^{-1} or m^{-1}.

(*iv*) *Velocity (c).* It may be defined as the linear distance travelled by a wave in one second. It is expressed in meters per seconds.

Mathematically,

$$c = \nu \times \lambda = \frac{\nu}{\overline{\upsilon}} \quad \text{because} \quad \lambda = \frac{1}{\overline{\upsilon}}$$

$$\therefore \quad c = \frac{\nu}{\overline{\upsilon}}$$

[**Note.** All electromagnetic waves (in vacuum) travel with the same speed ($3 \times 10^8 \text{ ms}^{-1}$) whatever may be their frequency or wavelength.]

SOLVED NUMERICALS

1. *Express the wavelength 20 microns in angstroms, nanometers, centimeters and as wave number.*

Solution. (*i*) $20 \text{ μ} = 20 \times 10^{-6} \text{ m} = \dfrac{20 \times 10^{-6} \text{ m} \times 1 \text{ Å}}{10^{-10} \text{ m}} = 2 \times 10^5 \text{ Å}$

(*ii*) $20 \text{ μ} = 20 \times 10^{-6} \text{ m} = \dfrac{20 \times 10^{-6} \text{ m} \times 10^9 \text{ nm}}{1 \text{ m}} = 2 \times 10^4 \text{ nm}$

(*iii*) $20\ \mu = 20 \times 10^{-6}\ \text{m} = \dfrac{20 \times 10^{-6}\ \text{m} \times 100}{1\ \text{m}} = 2 \times 10^{-3}\ \text{cm}$

(*iv*) $\bar{\upsilon} = \dfrac{1}{\lambda} = \dfrac{1}{20} = \dfrac{1}{20 \times 10^{-6}\ \text{m}} = \dfrac{1}{2} \times 10^{-5}\ \text{m}^{-1} = 0.5 \times 10^{-5}\ \text{m}^{-1}.$

2. *Calculate the frequencies of violet and red light of wavelength 400 nm and 750 nm respectively.*

Solution. For violet light $\nu = \dfrac{c}{\lambda} = \dfrac{3.0 \times 10^{8}\ \text{ms}^{-1}}{400 \times 10^{-9}\ \text{m}} = 7.5 \times 10^{14}\ \text{s}^{-1}$

For red light $\nu = \dfrac{c}{\lambda} = \dfrac{3.0 \times 10^{8}\ \text{ms}^{-1}}{750 \times 10^{-9}\ \text{m}} = 4.0 \times 10^{14}\ \text{s}^{-1}$

Characteristics. Following are the characteristics of electromagnetic radiations:

(*i*) Electromagnetic radiations consist of electric and magnetic fields that oscillate in the directions perpendicular to each other and perpendicular to the direction of propagation.

(*ii*) They do not require any medium for propagation and can travel through vacuum.

(*iii*) All electromagnetic waves travel in vacuum with the light velocity ($3 \times 10^{8}\ \text{ms}^{-1}$), however different electromagnetic waves have different wavelengths and frequencies.

According to Max Planck, the energy (E) associated with a photon is given by the relation

$$E = h\nu$$

where υ = frequency of the electromagnetic wave,

and h = Planck's constant (6.6×10^{-34} Js)

According to Einstein the energy of photon (E) is related to its momentum, p by the relation

$$E = pc$$

where c = velocity of light

By combining the above two equations, we get

So $$p = \frac{E}{c}$$

$$p = \frac{h\nu}{c}$$

$$p = \frac{h}{\lambda} \quad \text{or} \quad \lambda = \frac{h}{p}.$$

3. *Calculate the frequency of a radiation of visible light having wavelength 600 nm.*

Solution. $c = \nu\lambda$ or $\nu = \dfrac{c}{\lambda}$

Here, $c = 3 \times 10^{8}\ \text{ms}^{-1}$ and $\lambda = 600\ \text{nm} = 600 \times 10^{-9}\ \text{m} = 6 \times 10^{-7}\ \text{m}$

$\therefore \quad \nu = \dfrac{3 \times 10^{8}\ \text{ms}^{-1}}{6 \times 10^{-7}\ \text{m}} = 4 \times 10^{14}\ \text{s}^{-1} = 4 \times 10^{14}\ \text{Hz}.$

4. *A ratio operator broadcasts at a frequency of 14.0 MHz (Megahertz). Calculate the wavelength of the radio waves put out by the transmitter ?*

Solution. $c = \nu\lambda$ or $\lambda = \frac{c}{\nu}$

Here, $c = 3 \times 10^8\ ms^{-1}$ and $\nu = 14.0\ MHz = 14 \times 10^6\ Hz = 14 \times 10^6\ s^{-1}$

$$\therefore \quad \lambda = \frac{3 \times 10^8\ ms^{-1}}{14.0 \times 10^6\ s^{-1}} = 21.428\ m.$$

8.6.2 Types of Spectroscopy

The spectroscopy is broadly classified into two types:

(*i*) Emission spectroscopy

(*ii*) Absorption spectroscopy

Emission spectroscopy is used in case of the study of atomic structure whereas the *absorption spectroscopy* is quite useful to study the structure of organic compounds.

When a beam of electromagnetic radiation having a certain frequency interacts an atom then its electron can jump from one energy level to another by emission or absorption of energy. Like atoms, molecules also exhibit spectra when they absorb or emit energy only in discrete amounts or packets called quanta. The basic fact about spectroscopy is that absorption or emission by an atom or molecule can take place only if the energy associated with the radiation is equal to the energy difference (ΔE) between two energy levels in the atom or molecule.

i.e., $\Delta E = E_2 - E_1$ should be equal to $h\nu$

where h = Planck's constant

ν = frequency of the emitted or absorbed radiation.

Each type of spectra is observed in a characteristics region of electromagnetic spectrum. In molecular spectra there are a series of closely spaced lines (band) and are termed as band spectra. These lines correspond to rotational, vibrational and electronic transitions. Molecular spectra are generally observed as absorption spectra not as emission spectra because the conditions required for recording of an emission spectrum, the molecules or compounds generally break. They do not acquire that temperature which is required for emission spectra.

8.6.3 Molecular Energy Levels

For the clarity of rotational, vibration-rotation and electronic band spectra, it is essential to understand the energy levels in a molecule. Consider a diatomic molecule AB. The molecule AB consists of two nuclei of the atoms A and B. The electrons in these atoms can exist in a number of energy levels like in all atoms. When the molecule AB is excited, an electron may shift from one energy level to another. If electronic transition was the only mode by which the molecules could absorb energy or emit energy, molecular spectrum would also have a line spectra like that of atoms. However, in case of molecules, the absorbed energy is utilized not only for electronic transitions like atoms, but also to cause a vibration of the nuclei with respect to each other and to set the molecule rotating as a whole. The rotational and vibrational motions are also quantized like electronic transitions. The quantized vibrational and rotational motion coupled with the electronic levels to give a complicated picture of the energy levels of the molecule as in Fig. 8.11.

The group of levels represented by n. The lowest energy level is represented by $n = 1$ and the next higher electronic energy level is represented by $n = 2$ and so on. Each of the electronic levels is sub-divided into a number of vibrational sublevels represented by the vibrational quantum number (υ). $\upsilon = 0, 1, 2, 3, \ldots$ give various vibrational quantum numbers belonging to the same electronic level $n = 1, 2, 3, \ldots$, etc. Each of the vibrational quantum number (sublevel) is associated with a number of rotational quantum levels represented by the rotated quantum numbers J. The differential values of J (0, 1, 2, ... so on) give the values of various rotational quantum numbers belonging to the same vibrational level.

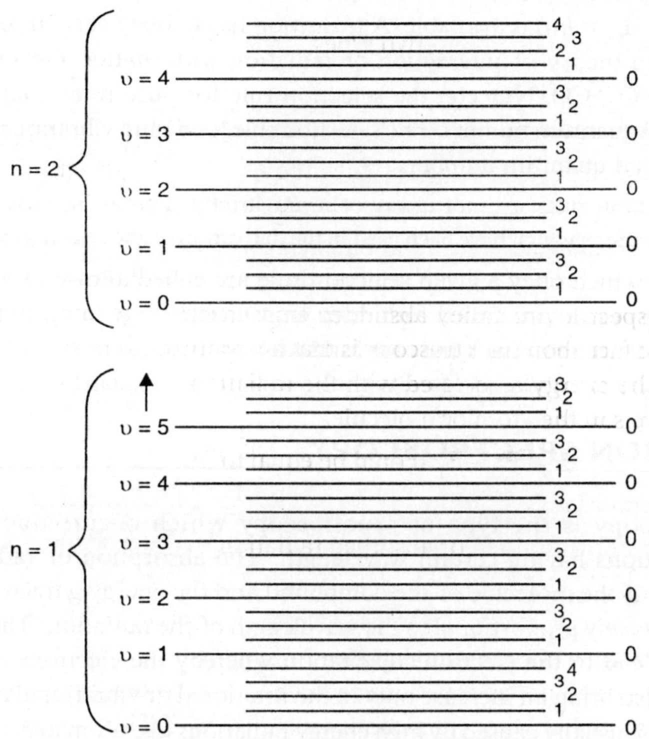

FIGURE 8.11 **Molecular energy levels**

From the above discussion, it is clear that the energy required for exciting emissions is the least for rotational transitions, higher for vibrational transitions and much higher for electronic transitions. For a small energy only rotational transitions are possible (from one rotational level to another rotational level) within a given vibrational level and the observed emission spectra is known as rotational spectrum. Since the energies required for excitation are small, the rotational spectra are observed in the far infrared region or microwave region of the electromagnetic spectrum.

Similarly the exciting energies required for producing vibrational spectra are sufficiently large. Such spectra are observed in the ultraviolet region. Since change in vibrational levels also involve changes in rotational levels, the overall result is a vibrational spectrum in which each line is accompanied by a rotational fine structure. Thus, such type of spectrum is known as vibrational-rotational spectrum.

If much higher excitation energies are employed, electronic transitions may take place which are accompanied by vibrational level changes and each of these in turn by rotational fine structure. The spectrum obtained in such a case would be quite complex and will consist of lines due to electronic, vibrational and rotational transitions and is known as electronic band spectrum. Each electronic band accompanied by its own vibrational and rotational levels. Since electronic transitions require much higher excitation energies hence such type of spectra are found in the visible and ultraviolet regions.

8.6.4 Selection Rule

The molecular spectra are governed by certain selection rules which are the backbone of spectroscopy. Transitions may not take place between any two levels in an atom or a molecule even if the required amount of energy ($E_2 - E_1 = h\upsilon$) is available. A transition has to obey certain selection rules which are obtained from quantum theory of interaction of radiation with matter. For example, for a diatomic molecules such as H_2, O_2, CO, NO, etc. the selection rule for pure rotational transition is $\Delta J = \pm 1$, where J is the rotational quantum number. The selection rule for a pure vibrational transition is $\Delta \upsilon = \pm 1$, where υ is the vibrational quantum number.

[**Note.** The selection rules are not always obeyed strictly. This is because of the fact that certain approximations which have been used in the derivation of the selection rules are not valid strictly.]

The transitions which obey a given selection rule are called allowed transitions whereas those which violate a selection rule are called forbidden transitions. In general, the bands due to allowed transitions are more intense than the bands for forbidden transitions.

8.7 ABSORPTION SPECTROSCOPY

Absorption spectroscopy is the type of spectroscopy which occurs due to the absorption of electromagnetic radiations having certain wavelength. The absorption of radiation brings about an increase in the energy of the molecule of the compound and the energy gained is directly proportional to the frequency or inversely proportional to the wavelength of the radiation. The increase in the energy of the molecule may lead to the electronic excitation whereby the electrons are raised to the higher energy levels. It may also bring an increase only in the rotational or vibrational energy of the atom. The electronic excitations are usually caused by high energy radiations like ultraviolet (UV) radiations whereas the vibrational or rotational energy levels can be raised by less energy radiations like infrared (IR) radiations.

The absorption of radiations is observed instrumentally by passing the radiations through the sample of the substance under investigation and the intensity of the transmitted radiations is studied. The spectrum is made up of bands or peaks of maximum intensity which gives the information of functional group and structure of the compound. The apparatus which is used for detecting the spectra is known as spectrophotometer.

Laws of Photochemistry

When a beam of monochromatic radiation passes through a homogeneous medium, a part of the incident light is reflected, a part of it is absorbed within the medium and the remainder is transmitted. Let the intensity of incident light be represented by I_0, that of the absorbed light by I_a, that of transmitted light by I_t and that of reflected light by I_r, then

$$I_0 = I_a + I_t + I_r$$

Lambert and Beer explained the laws of photochemistry as follows:

1. *Lambert's Law.* Lambert proposed a relation between I_0 and I_t. He found that intensity of incident light is greatly reduced on passing through a medium. He proposed that "the rate at which the intensity of light decreases with thickness of the medium is proportional to the intensity of the incident light."

Mathematically

$$-\frac{dI}{dx} = kI \qquad \text{or} \qquad -\frac{dI}{I} = k \, . \, dx$$

where I is the intensity of incident light, x is the thickness of the absorbing medium traversed by the radiation and k is a constant called absorption coefficient of the medium for the light of a particular wavelength. The negative sign indicates that I diminishes as x increases.

On integrating, we get

$$\int_{I_0}^{I_x} \frac{dI}{I} = -\int_{x=0}^{x=x} k \, dx$$

$$\ln \frac{I_x}{I_0} = -kx \qquad \text{or} \qquad I_x = I_0 e^{-kx}$$

The intensity of light absorbed can be calculated by putting the value of I_t as

$$I_a = I_0 - I_t \qquad (\because \; I_r \text{ is negligible})$$

$$= I_0 (1 - e^{-kt}) \; ; \; \text{'}t\text{' being the thickness of the medium.}$$

Lambert's law is valid for media other than solutions.

2. *Lambert-Beer's law.* When the absorbing medium is a solution, the relationship between the intensities of incident and transmitted radiations is given by Lambert-Beer's law.

The law states *"on passing through a solution, the rate of decrease of intensity of the beam of monochromatic radiation with the thickness of the solution is proportional to the intensity of incident radiation as well as concentration of the solution."*

Mathematically

$$\frac{dI}{I} = -k' \, . \, C \, . \, dx$$

where C = molar concentration of the solution

K′ = molar absorption coefficient

dx = thickness of the layer of solution.

On integrating, we get

$$\int_{I_0}^{I_x} \frac{dI}{I} = -k'C \int_{x=0}^{x=x} dx$$

$$\ln \frac{I_x}{I_0} = -k'Cx \qquad \text{or} \qquad I_x = I_0 e^{-k'Cx} \qquad \ldots(1)$$

The intensity of light absorbed is given by

$$I_a = I_0 - I_t; \text{ '}t\text{' being the total thickness of the medium}$$
$$= I_0(1 - e^{-k'Ct})$$

Transmittance is given by $\frac{I_t}{I_0} = T$

Taking log on both sides, we get $\ln \frac{I_t}{I_0} = \ln T$ ...(2)

Using equation (1)

$$\ln \frac{I_t}{I_0} = \ln e^{-k'Ct}$$

$$\ln \frac{I_t}{I_0} = -k'Ct$$

On putting the value of $\ln \frac{I_t}{I_0}$ in equation (2), we get

$$-k'Ct = \ln T \quad \text{or} \quad A = -\ln T$$

where $A = k'Ct$ and is known as absorbance or optical density of the solution. The plots of T versus C and of A versus C are shown in Figs. 8.12 (*a*) and 8.12 (*b*). The plot of A versus C is a straight line passing through the origin. Hence, this plot of absorbance A versus C is more convenient to interpret than the plot of T versus C.

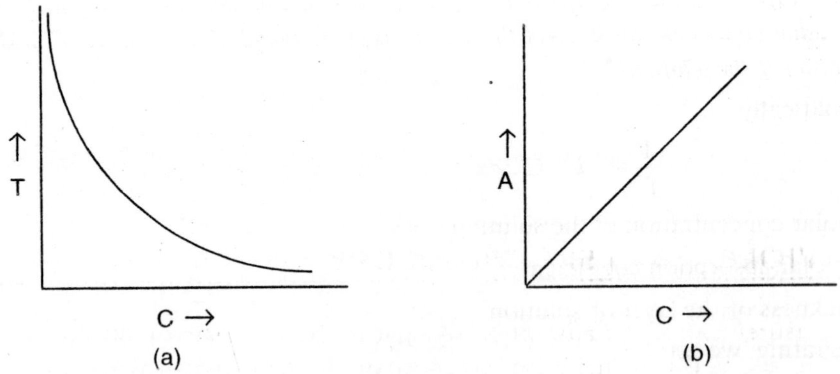

FIGURE 8.12

Limitations of Lambert-Beer's Law

1. The law is not applicable for non monochromatic radiation.
2. The law governs the absorption behaviour of dilute solutions only. At high solute concentrations, the ions of a solute, if it is an electrolyte, are close enough to disturb the charge distribution of their neighbours. The interionic interaction can drastically alter the

ability of the solute to absorb a given wavelength of the incident radiation. Thus, the relationship between A and C is no longer linear. They are distinct deviations from linearity as shown in Fig. 8.13.

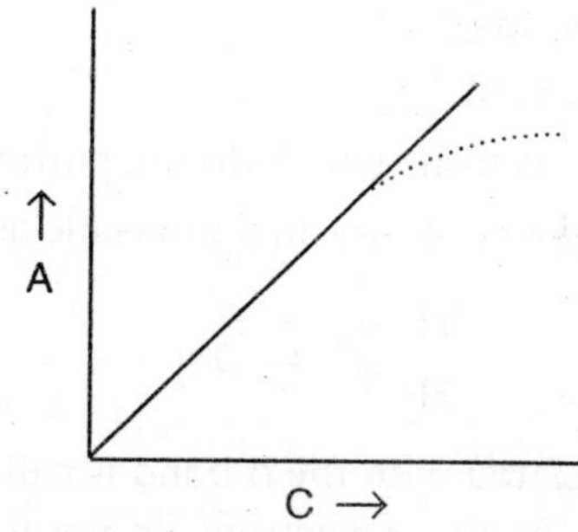

FIGURE 8.13 **Deviation from the Lambert-Beer's law**

Types of Absorption Spectroscopy

The absorption spectroscopy may be divided into different types. Some important types are as follows:

(*i*) Ultraviolet (UV) spectroscopy

(*ii*) Infrared (IR) spectroscopy

(*iii*) Microwave spectroscopy

(*iv*) Nuclear magnetic resonance (NMR) spectroscopy

(*v*) Electron spin resonance (ESR) spectroscopy

(*vi*) Mass spectroscopy, etc.

The mass spectroscopy helps in the determination of molecular mass and molecular formula. UV and IR spectra are used to study the nature of functional groups while NMR is used to determine the nature of proton and hence tells about the complete structure of the compound as well as its stereochemistry. The absorption of radiations may be understood with the help of law of photo-chemistry. In this chapter, we shall only discuss ultraviolet (UV) and infrared spectroscopy (IR) in brief.

8.8 ULTRAVIOLET (UV) SPECTROSCOPY

It involves the measurement of the absorption of light in the UV region (200–400 nm) and visible region (400–800 nm) by the compound under investigation. It is also known as electronic spectroscopy because it occurs due to the transition between the electronic energy levels within a molecule.

Principle

When a substance is placed in UV or visible region of electromagnetic radiations, the electrons get excited and prompt from a lower to a higher energy level. Due to this the electronic states of the molecules get changed. Each electronic level in a molecule is associated with a number of vibrational sub-level with a small energy difference. Similarly each vibrational level is associated with a number of rotational sub-level with small energy difference. Therefore, during transition an electron can go from any of the sub-levels (various vibrational and rotational states) in the ground state to any of the sub-level orbitals present in the excited state. These orbitals are usually vacant in the ground state of the molecule. Three types of electrons present in the ground state of the molecule are:

(*i*) σ-electrons (present in σ-bonds)

(*ii*) π-electrons (present in π-bonds)

(*iii*) *n*-electrons (non-bonding electrons which do not participate in bond formation)

All three types of electrons may be shown in a molecule as follows:

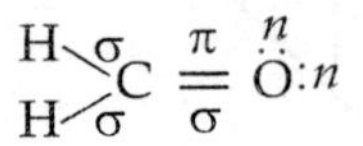

The antibonding orbital associated with the σ bond is called the σ* orbital and that associated with the π bond is called π* orbital. As the *n* electrons do not form bonds, there are no antibonding orbitals associated with them. The electronic transitions (→) that are involved in the ultraviolet and visible region are of the following types σ → σ*, *n* → σ*, *n* → π* and π → π*. The energy which is required for the excitation of electrons during electronic transitions may be represented in Fig. 8.14.

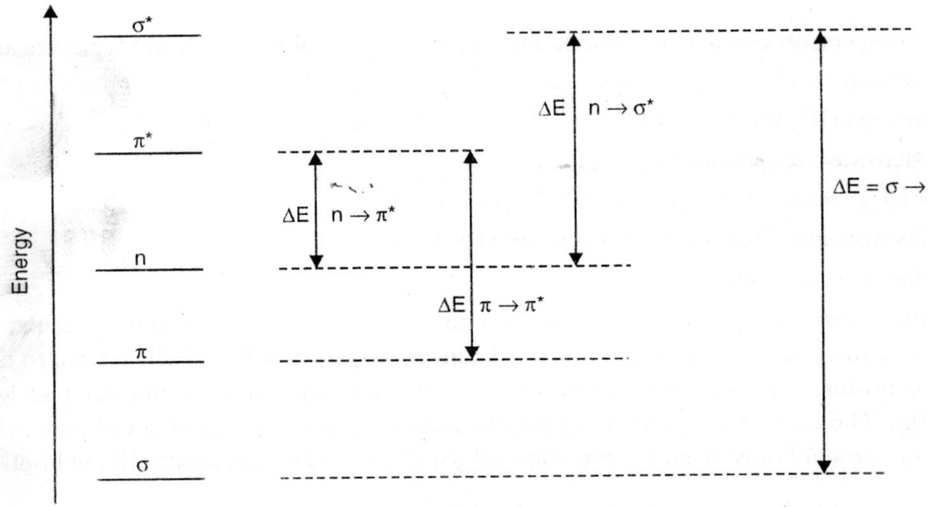

FIGURE 8.14

The energy required for the σ → σ* transitions is very high, so they do not show absorption in the ordinary ultraviolet region. An exception is cyclo-propane, which shows λ_{max} about 190 nm. The order of energy is σ → σ*> *n* → σ*> π → π*> *n* → π*.

(*i*) σ → σ* *transitions.* These are the type of transitions in which the electron promotes from π-bonding orbital to σ* antibonding molecular orbital. Such type of transitions occurs by those compounds in which all electrons are involved in the formation of σ-bonds, for example, alkanes and cycloalkanes. It requires very high energy hence the transition occurs at very short wavelength. For example, methane absorbs at 125 nm and propane absorbs at 135 nm. Cycloalkanes generally absorbs at higher wavelength. Cyclopropane absorbs at 195 nm.

[**Note.** Cyclopropane is a strained ring. Due to angle strain, the ground state energy level of cyclopropane is higher than the respective open chain isomer or alkane. That is why cyclopropane absorbs at high wavelength in compare to alkane.]

(*ii*) $n \rightarrow \sigma^*$ *transitions.* These are the type of transitions in which non-bonding electrons (n) promoted to σ^* anti-bonding molecular orbital. Such type of transitions occur by those compounds which contain atleast one heteroatom like nitrogen, halogen, oxygen, sulphur, etc. These heteroatoms contain atleast one lone pair of electron. For example,

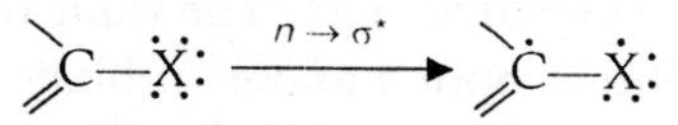

These transitions require lower energy to $\sigma \rightarrow \sigma^*$ transitions and hence occur at longer wavelengths.

[**Note.** Lower the electronegativity of the heteroatom, higher will be the wavelength for absorption. For example, methyl chloride absorbs at λ_{max} 173 nm while methyl iodide absorbs at λ_{max} 258 nm.]

(*iii*) $\pi \rightarrow \pi^*$ *transitions.* These are the type of transitions in which π electrons promoted to an antibonding π^* molecular orbital. Such type of transitions occurs by those compounds which contain atleast one multiple bonds. For example,

>C = C<, >C = O, —N = O

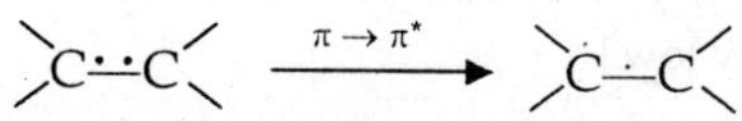

These transitions require high energy and occur at short wavelength. The $\pi \rightarrow \pi^*$ transitions are of intermediate energy, absorptions owing to these transitions are usually between those owing to $n \rightarrow \pi^*$ and $n \rightarrow \sigma^*$ transitions. Absorptions usually occur within the region of UV and visible region. For example, in carbonyl compounds, the $\pi \rightarrow \pi^*$ transitions occurs at about 180 nm.

(*iv*) $n \rightarrow \pi^*$ *transitions.* These are the type of transitions in which an electron promoted from a non-bonded atomic orbital to the antibonding π^* molecular orbital. Such type of transitions occur by those compounds in which carbon is attached with hetero atom by multiple bond like >C = O, >C = S, —C ≡ N, etc.

These types of transitions require the least energy and thus appear at maximum wavelength (λ). For example, saturated aldehyde (H(R)>C = O) shows two type of transitions, one at lower energy due to $n \rightarrow \pi^*$ and another at higher energy due to $\pi \rightarrow \pi^*$ excitations.

[**Note.** The absorption occurs at lower wavelength is generally more intense.]

SOLVED NUMERICAL

5. *The UV spectrum for a compound shows two peaks at* λ_{max} *280 nm,* $\epsilon_{max} = 15$ *and* λ_{max} *190,* $\epsilon_{max} =$ *100. Identify the electronic transition for each and indicate which is more intense ?*

Solution. It is clear fact that longer wavelength is associated with smaller energy and vice-versa.

Here, λ_{max} = 280 nm represents $n \rightarrow \pi^*$ transition

λ_{max} = 190 nm represents $\pi \rightarrow \pi^*$ transition

Since $\pi \rightarrow \pi^*$ transition has larger value of $\in_{max}$ ($\in_{max}$ = 100). Hence it is more intense peak.

8.8.1 Allowed and Forbidden Transitions

Allowed transitions. The transitions which obey a given selection rule are called allowed transitions. Allowed transitions usually take place due to $\pi \rightarrow \pi^*$. The value of intensity $\in_{max}$ ranging between 10^4 to 10^6. Actually in the $\pi \rightarrow \pi^*$ transitions, both π and π^* orbitals lie in the same plane. Therefore, the overlap between the orbitals in the excited state is very large. The $\pi \rightarrow \pi^*$ transitions are highly probable. For example, the $\in_{max}$ value for $\pi \rightarrow \pi^*$ electronic transition in 1, 3-butadiene system is 20000. The peaks in allowed transitions are more intense than the peaks in forbidden transitions.

Forbidden transitions. The transitions which violate a selection rule are called forbidden transitions. These are usually take place due to $n \rightarrow \pi^*$ transitions. The value of $\in_{max}$ ranging between 10–1000. Actually in $n \rightarrow \pi^*$ transitions, the non-bonded (n) electrons present on the hetero atom lie in a plane perpendicular to the plane containing π^* molecular orbital and belonging to the other atom involved in the bond formation. Therefore, the overlap between the orbitals in the excited state is very low. The probability of the excitation of the n electrons to the π^* orbital is very low. For example, for $n \rightarrow \pi^*$ transition in case of saturated aldehydes and ketones, $\in_{max}$ is less than 100.

[**Note.** Forbidden transition does not mean that it cannot take place at all, but it implies that the probability of the transitions is very low.]

8.8.2 Role of Solvent in the Electronic Transitions

The nature (polarity) of solvents play an important role in the electronic transitions in the UV spectroscopy. The solvent may cause a shift in the position of absorption band either toward longer or shorter wavelength. It may be explained as follows:

(*i*) *Effect of solvent on $n \rightarrow \pi^*$ and $n \rightarrow \sigma^*$ transitions.* Consider a particular group which is more a polar in the ground state than in the excited state. When such type of compound is dissolved in polar solvents, the electrons in the ground state are more stabilized due to hydrogen bonding or dipole-dipole interactions with the molecules of the polar solvents like H_2O, in comparison to the excited state. As a result the energy difference between the ground state and the excited state increases as in Fig. 8.15. Thus the absorption shifts towards the smaller wavelength region. For example, a carbonyl compound ($>$C=O) of acetone is more polar in ground state than in excited state

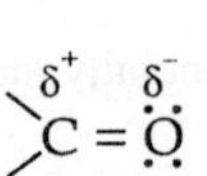

(More polar in ground state)

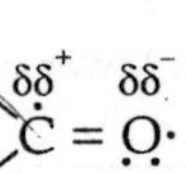

(Less polar in excited state)

[**Note.** $\delta\delta^+$ and $\delta\delta^-$ represents the less polar character.]

Hence the absorption (λ) for acetone occurs at 280 nm in non-polar solvent like hexane whereas it becomes 264 nm in polar solvent like water.

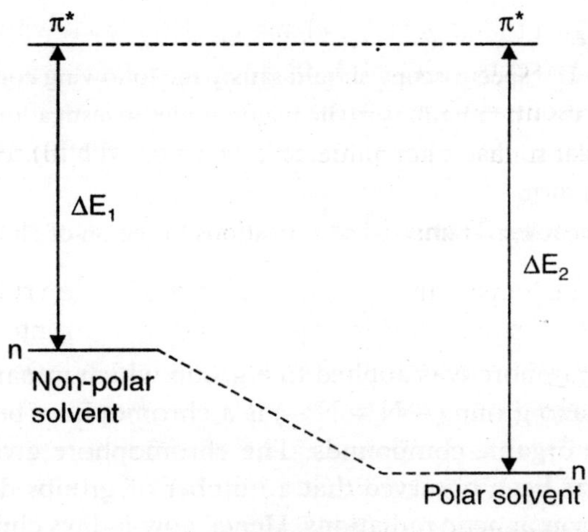

FIGURE 8.15

(*ii*) *Effect of solvent on π → π* transitions*. Consider a particular group which is less polar in the ground state and more polar in the excited state. When such type of compound is dissolved in polar solvent, the absorption is shifted towards longer wavelength region. The electrons in the excited state are more stabilized in comparison to the ground state because of hydrogen bonding or dipole-dipole interactions with the molecules of the polar solvents. In other words, the excited state will be relatively more stabilized than the ground state or the energy of the excited state will decrease. As a result the energy difference between the ground state and the excited state decreases as in Fig. 8.16 and hence the absorption will shift towards longer wavelength. For example, the ethylene molecule is more polar in the excited state than the ground state due to $\pi \rightarrow \pi^*$ transitions.

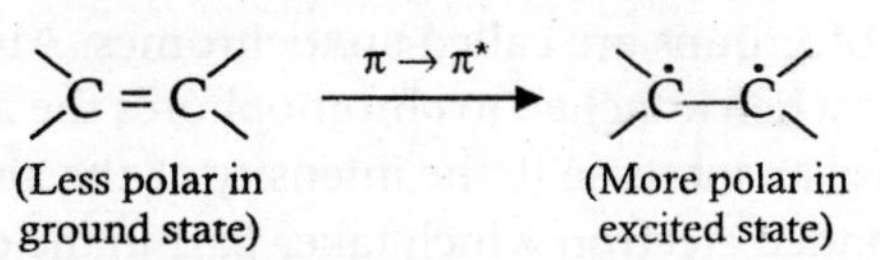

In the polar solvent π^* molecular orbital is more stabilized than in the ground state. Hence the absorption is shifted to a longer wavelength region.

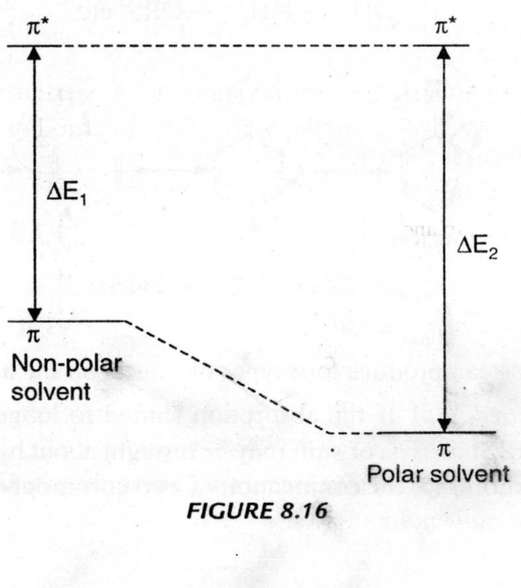

FIGURE 8.16

[**Note.** ***Choice of solvent.***

The solvent used in the UV spectroscopy should satisfy the following conditions:

(*i*) It should not itself absorb radiations in the region under investigation.

(*ii*) It should be less polar so that it has minimum interaction with the solute molecules.

(*iii*) It should be in pure form.

The polarity of the solvent also shifts $n \rightarrow \sigma^*$ transitions to region of shorter wavelength.]

8.8.3 Chromophore

In early days the term chromophore was applied to a group which imparts characteristics colour to a compound. For example, diazo group (—N = N—) is a chromophore because it gives characteristics orange or red colour to the organic compounds. The chromophore gives the absorption in the UV region. But now-a-days it has been observed that a number of groups do not give any colour to the compound but absorbs electromagnetic radiations. Hence now-a-days chromophore may be defined as the functional group that absorbs electromagnetic radiations irrespective of the fact whether it imparts a particular colour to the compound or not. For example, chromophore carboxylic acid (—COOH) absorbs electromagnetic radiations at λ_{max} 208 nm in the UV region and has ϵ_{max} = 15 but does not give any colour to the compound. Examples of some chromophores are:

$$>C=O, \quad —C\equiv N, \quad —COOH, \quad —N=N—, \quad >C=C<, \quad \text{etc.}$$

There are many chromophores which absorb the electromagnetic radiations below than 220 nm. Such type of regions cannot be properly studied with the help of the commercially available spectrophotometers. To solve this problem certain groups are attached to the basic chromophores in place of in their hydrogen atoms. The subsituent groups can shift the absorption towards the longer wavelength region. Such types of groups are called auxochromes. Auxochromes are groups which do not act as the chromophores but when attached to chromophores the absorption shifted towards longer wavelength region and also cause an increase in the intensity of the absorption band. All auxochromes have at least one pair of non-bonded electron which takes part in the conjugation in the chromophoric group. For example,

$$—\ddot{O}H, \quad —\ddot{N}H_2, \quad —\ddot{O}R, \quad \text{etc.}$$

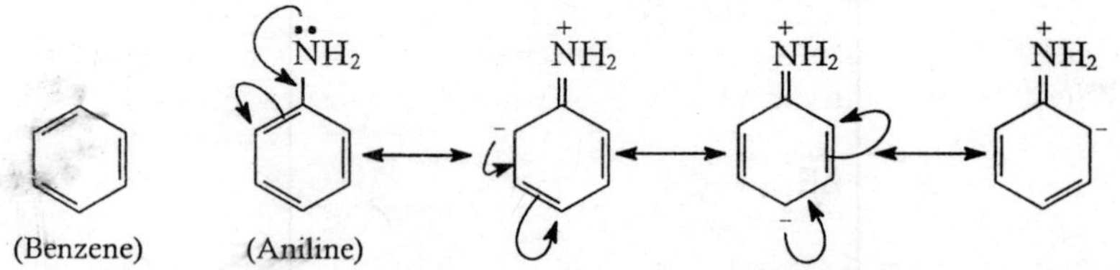

(Benzene) (Aniline)

λ_{max} = 225 nm λ_{max} = 280 nm

ϵ_{max} = 230 ϵ_{max} = 9350

The substituent groups can produce four types of effects on the absorption.

(*i*) *Bathochromic shift (red shift).* If the absorption shifted to longer wavelength region is called bathochromic shift. Such type of shift may be brought about by (*a*) attaching auxochromes to the basic chromophores (*b*) the conjugation of two chromophores.

(*ii*) *Hypsochromic shift (blue shift).* If the absorption shifted to smaller wavelength region, is called hypsochromic shift.

(*iii*) *Hyperchromic shift.* Such type of shifted cause an increase in intensity.

(*iv*) *Hypochromic shift.* Such type of shifted cause a decrease in intensity.

Conjugated Chromophores

When two chromophores are in conjugation ($CH_2 = CH - CH = CH_2$) the absorption shifted to a longer wavelength region up to a maximum extent due to bathochromic shift. There is also an increase in the intensity of absorption due to hyperchromic shift. This is due to the delocalization of the electron charge the energy of the excited state decreases. The energy difference between the excited state and the ground state tends to decrease. As a result, the absorption shifted to a longer wavelength region. For example,

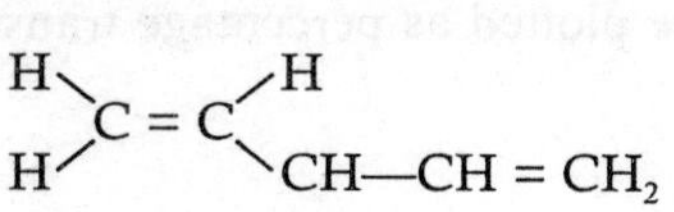

$\lambda_{max} = 176$ nm

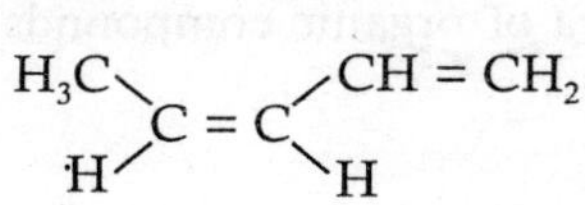

$\lambda_{max} = 211$ nm

Applications

Ultraviolet spectroscopy has wide applications. Some important applications are as follows:

(*i*) *Detection of functional group.* It is very helpful to detect the presence of particular functional groups. For example, if a compound does not show λ_{max} above 200 nm, it shows the absence of conjugation, carbonyl group (—CHO or $>$C$=$O), benzene and other polynuclear hydrocarbons.

(*ii*) *Extent of conjugation.* Greater the number of conjugated double bonds always absorbs at higher wavelength region. In case of eight or more double bond conjugation the absorption may occur even in the visible region of light. Each extent double bond gives the addition of 30 nm. For example,

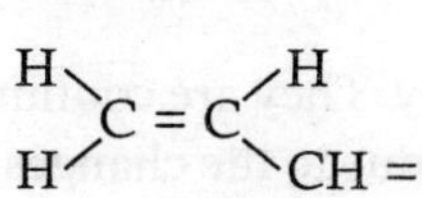

$\lambda_{max} = 217$ nm

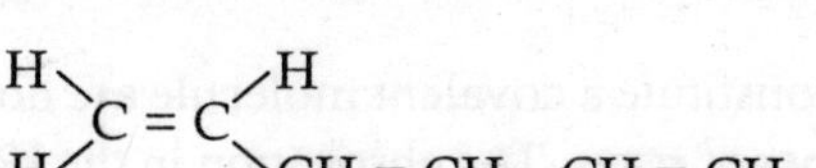

$\lambda_{max} = 247$ nm

(*iii*) ***Distinction between conjugated and non-conjugated compounds.*** The conjugated compounds absorb λ_{max} at higher wavelength region in comparison to non-conjugated compounds. For example,

$CH_2 = CH—CH_2—CH = CH_2$ $\lambda_{max} = 176$ nm

$CH_3—CH = CH—CH = CH_2$ $\lambda_{max} = 215$ nm

(*iv*) *Identification of geometrical isomers.* The trans isomer absorbs electromagnetic radiations at a longer wavelength with longer extinction coefficient than the corresponding cis-isomer. For example,

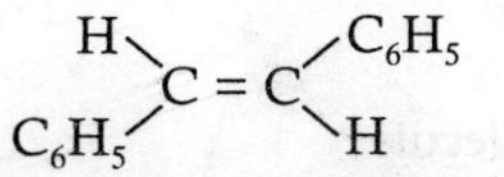

$\lambda_{max} = 290$ nm

$\epsilon_{max} = 24000$

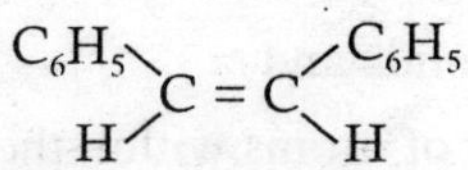

$\lambda_{max} = 278$ nm

$\epsilon_{max} = 9350$

8.9 INFRA-RED SPECTROSCOPY

It is the most important technique for the structure elucidation of organic compounds. In contrast to the relatively few absorption peaks observed in the ultraviolet region for most organic compounds, the infra-red spectrum provides a large number of absorption bands which are very useful to determine the structure of the organic compound under investigation. Many of the absorption bands cannot be assigned accurately; those that can, however, provide a wealth of structural information about a molecule. The ordinary infrared region extends from 2.5 μ to 15 μ (4000–667 cm^{-1}). The region from 0.8 μ to 2.5 μ (12500–4000 cm^{-1}) is called the near infrared region and the region from 15 to 200 μ (667-50 cm^{-1}) is called the far infrared region.

The absorption of IR radiations can be expressed either in wavelength (λ) or in wave number ($\bar{\upsilon}$). Mostly IR spectra of organic compounds are plotted as percentage transmittance against wave number ($\bar{\upsilon}$)

$$\text{Wave number } (\bar{\upsilon}) = \frac{1}{\text{Wavelength}}$$

For wavelength (λ) 2.5 μ

$$\text{Wave number } (\bar{\upsilon}) = \frac{1}{2.5 \times 10^{-4} \text{ cm}} = 4000 \text{ cm}^{-1}$$

For wavelength (λ) 15 μ

$$\text{Wave number } (\bar{\upsilon}) = \frac{1}{15 \times 10^{-4} \text{ cm}} = 667 \text{ cm}^{-1}$$

Thus, in terms of wave number the ordinary infrared region is 4000 cm^{-1} to 667 cm^{-1}.

Principle

The atoms which constitute a covalent molecule are not stationary. They are continuously rotating and vibrating in a number of ways. The absorption in the IR region is due to the changes in dipole moments or vibrational and rotational levels. When radiations with frequency range less than 100 cm^{-1} are absorbed, molecular rotation takes place in the substance and hence discrete lines are formed in the spectrum as the absorption is quantized. When more energetic radiations with frequency 10^2 to 10^4 cm^{-1} are passed through the sample of the substance, molecular vibrations get set up. Such type of absorption is also quantized. Clearly, a single vibrational energy change is accompanied by a large number of rotational energy changes. Thus, the vibrational spectra appear as vibrational-rotational bands. In the IR spectroscopy, the absorbed energy brings about predominant changes in the vibrational energy which depends upon:

(*i*) Masses of the atoms present in a molecule,

(*ii*) Strength of the bonds and

(*iii*) The arrangement of atoms within the molecule.

[**Note.** No two compounds can have similar IR spectra, except the enantiomers.]

The atoms in a molecule are not tightly held together. When they absorb IR radiations, the covalent bonds get either stretch or bend. The stretching or bending depends upon the stretch of the bonds. For simplicity, the stretching and bending may be understood as follow. The molecule may be visualized as consisting of balls of different sizes tied with springs of varying strengths. Here balls and springs correspond to atoms and chemical bonds respectively. When it is passed through the sample, the vibrational and the rotational energies of the molecules are increased. Fundamental vibrations are of two types: (*i*) Stretching and (*ii*) Bending.

(*i*) *Stretching.* In stretching, the distance between the two atoms increases or decreases but the atoms remain in the same and axis.

(*ii*) *Bending.* In bending, the positions of the atoms change with respect to the original bond axis.

Stretching and bending may be represented as in Fig. 8.17.

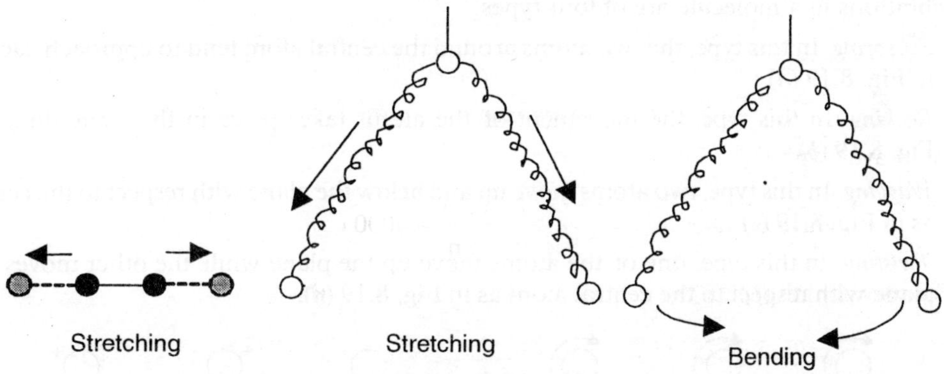

FIGURE 8.17

It is well known fact that more energy is required to stretch a spring than that required to bend it. Thus, the stretching absorptions of a bond appear at higher frequencies as compared to the bending absorptions of the same bond. The various stretching and bending vibrations of a bond occur at certain quantized frequencies. When IR radiation is passed through the substance, energy is absorbed and the amplitude of the vibration is increased. When the molecule returns to the ground state from the excited state, it releases the extra amount of energy by rotational, collision or translational processes. As a result, the temperature of the sample under investigation increases.

8.9.1 Types of Stretching Vibrations

The diatomic molecule like H—H, Cl—Cl, H—Cl etc. vibrate only in one way but the triatomic molecules such as CO_2, NO_2, etc. vibrate in different way. Hence stretching vibrations are of two types :

(*i*) *Symmetric stretching.* In this type of vibration, the movement of the atoms with respect to a particular atom in a molecule is in the same direction as in Fig. 8.18 (*a*).

(*ii*) *Asymmetric stretching.* In this type of vibration, the movement of the atoms with respect to a particular atom in a molecule is in opposite direction. One atom approaches the central atom while the other departs from it as in Fig. 8.18 (*b*).

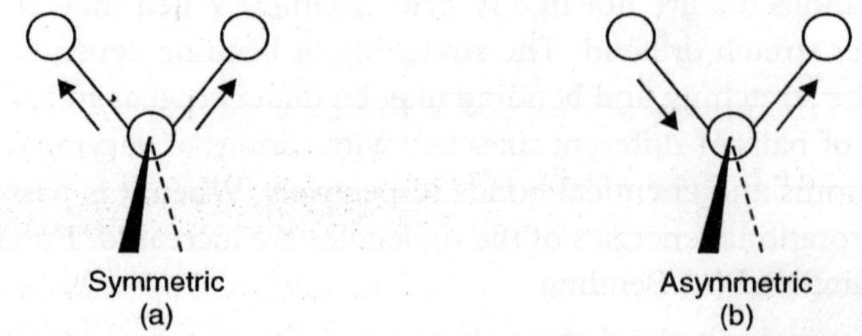

FIGURE 8.18 **Stretching vibrations**

8.9.2 Types of Bending Vibrations

Bending vibrations in a molecule are of four types:

(*i*) *Scissoring.* In this type, the two atoms around the central atom tend to approach each other as in Fig. 8.19 (*a*).

(*ii*) *Rocking.* In this type, the movement of the atoms takes place in the same direction as in Fig. 8.19 (*b*).

(*iii*) *Wagging.* In this type, two atoms move up and below the plane with respect to the central atom as in Fig. 8.19 (*c*).

(*iv*) *Twisting.* In this type, one of the atoms move up the plane while the other moves down the plane with respect to the central atom as in Fig. 8.19 (*d*).

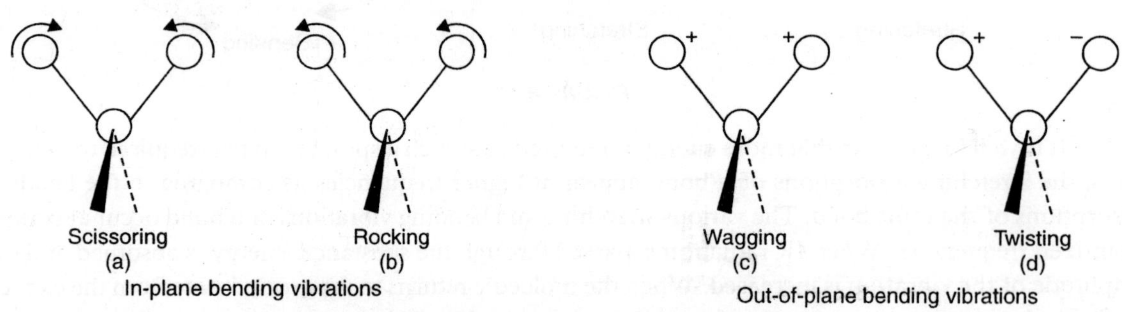

FIGURE 8.19 **Vibrations of a group of atoms (+ and – signify vibrations perpendicular to the plane of the paper)**

Bending vibrations always occur at lower wave numbers than stretching vibrations.

8.9.3 Vibrational Frequencies

The stretching vibrational frequency of a bond can be calculated by Hook's law as:

$$\bar{\upsilon} = \frac{\upsilon}{c} = \frac{1}{2\pi}\left[\frac{K}{m_1 m_2/(m_1 + m_2)}\right]^{1/2} = \frac{1}{2\pi}\left[\frac{K}{\mu}\right]^{1/2}$$

where $\mu = \dfrac{m_1 m_2}{m_1 + m_2}$ and is called reduced mass

K = Force constant of the bond and is related to the strength of the bond,

m_1 and m_2 are masses of atoms in a molecule.

Vibrational frequency or wave number depends upon:

(*i*) Bond strength and (*ii*) Reduced mass.

8.9.4 Number of Fundamental Vibrations and Fundamental Frequency

For infrared absorption, the vibrations should not be centro-symmetric. Only those vibrations are active in IR spectra which are not centro-symmetric. This technique is most informative to organic chemists. The symmetry properties of a molecule in a solid can be different from those of an isolated molecule. The IR spectrum of a molecule results due to the transitions between two different vibrational energy levels. The vibrational motion resembles the motion observed for a ball attached to a spring *i.e.*, harmonic oscillator. The vibrational energy of a chemical bond is quantized and can have the value.

$$E_{Vib} = \left(V + \frac{1}{2}\right) h\upsilon$$

where V = 0, 1, 2, 3, ... and is known as vibrational level

υ = vibrational frequency of the bond.

At ordinary temperature, where molecules are in their lowest vibrational energy levels, the potential energy diagram approximates that of a harmonic oscillator. But at higher temperature the deviations occurs as in Fig. 8.20.

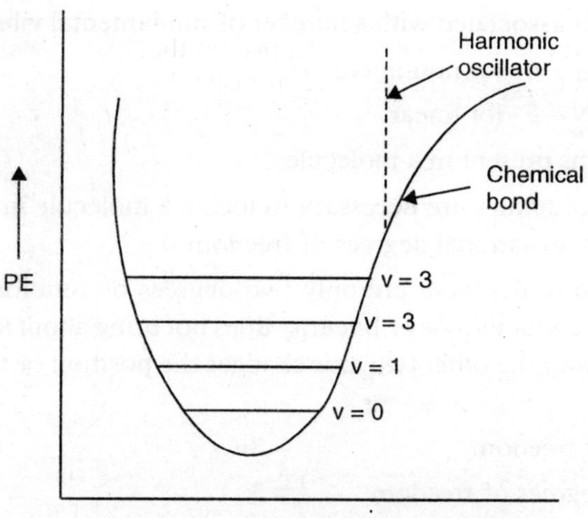

FIGURE 8.20 **Potential energy diagram**

Absorption of radiation with energy equal to the difference between two vibrational energy levels (ΔE_{Vib}) will cause a vibrational transition to occur. Transitions from the ground state (V = 0) to the first excited state (V = 1) absorb radiation strongly and give rise to intense bands called the fundamental bands. The energy difference (ΔE_{Vib}) between the lowest possible energy level and the next higher energy level is given by

$$\Delta E_{Vib} = E_{Vib(V=1)} - E_{Vib(V=2)} = \left(1 + \frac{1}{2}\right) h\upsilon - \left(0 + \frac{1}{2}\right) h\upsilon = h\upsilon$$

This gives the frequency of a fundamental band. Transitions from the ground state (V = 0) to the second excited state (V = 2) with the absorption of infrared radiation give rise to weak bands called overtones. The energy of the first overtone is given by $\Delta E_{Vib} = 2h\upsilon$. The polyatomic molecules may exhibit more than one fundamental vibrational absorption bands. The number of these fundamental bands is related to the degree of freedom in a molecule which is equal to the sum of the co-ordinates necessary to locate all the atoms of a molecule in space. Each atom has three degrees of freedom corresponding to the three cartesian co-ordinates (x, y, z) necessary to describe its position relative to other atoms in a molecule. An insolated atom which is considered as a point mass has only translational degree of freedom. It cannot have vibrational or rotational degree of freedom. When atoms combine to form a molecule, no degrees of freedom are lost, *i.e.*, the total number of degrees of freedom of a molecule will be equal to $3n$ where n is the number of atoms in a molecule. A molecule having finite dimensions is made up of rotational, vibrational and translational degree of freedom *i.e.*,

$3n$ degree of freedom = translational + rotational + vibrational

Rotational degrees of freedom results from the rotation of a molecule about an axis through the centre of gravity. Since we are concerned only with the number of fundamental vibrational modes of a molecule, so we calculate only the number of vibrational degrees of freedom of a molecule.

Fundamental Frequency

The frequency associated with the fundamental vibration of the molecule is called fundamental frequency. Every molecule is associated with a number of fundamental vibrations given by the formula

$3N - 6$ for non-linear

and $3N - 5$ for linear

where N = Number of atoms present in a molecule.

Since only three co-ordinates are necessary to locate a molecule in space, so we can say that a molecule has always three translational degrees of freedom.

In case of linear molecule, there are only two degrees of rotation. It is due to fact that the rotation of such a molecule about its axis of linearity does not bring about any change in the position of the atoms while rotation about the other two axis changes the position of the atoms. Thus, for a linear molecule of n atoms.

Total degrees of freedom $= 3n$

Translational degrees of freedom $= 3$

Rotational degrees of freedom $= 2$

$\therefore$ Vibrational degrees of freedom $= 3n - (3 + 2)$

$= 3n - 5$

Each vibrational degree of freedom corresponds to the fundamental mode of vibration and each fundamental mode corresponds to a band. Hence, theoretically there will be $3n - 5$ possible fundamental bands for the linear molecules. For example, CO_2 is a linear triatomic molecule.

Here number of atoms (n) $= 3$

$\therefore$ Total degrees of freedom $= 3n$

$= 3 \times 3 = 9$

and Vibrational degrees of freedom $= 3n - 5$

$= 9 - 5$

$= 4$

Therefore, there are four vibrational degrees of freedom should occur in case of CO_2 molecule.

In the case non-linear molecule, there are three degrees of rotation, as the rotation about all the three axes will result in a change in the position of the atoms. Thus, for a non-linear molecule having n atoms

Total degrees of freedom $= 3n$

Translational degrees of freedom $= 3$

Rotational degrees of freedom $= 3$

$\therefore$ Vibrational degrees of freedom $= 3n - (3 + 3)$

$= 3n - 6$

For example, C_6H_6 is a non-linear molecule

Here number of atoms $= 12$

Total degrees of freedom $= 3n$

$= 3 \times 12 = 36$

Translational degrees of freedom $= 3$

Rotational degrees of freedom $= 3$

$\therefore$ Vibrational degrees of freedom $= 3n - 6$

$= 36 - 6$

$= 30$

Hence, theoretically there should be 30 fundamental bands in the IR spectrum of benzene.

8.9.5 Factors Influencing Vibrational Frequency

It has been noticed that the calculated value of the frequency or wave number is not observed in experiments. It is due to the fact that the vibration of each atom or group is influenced by the structure of the neighbouring atoms or groups in a molecule. Some of the important factors are given below :

1. Electronic Effect. Electronic effects like inductive effects and resonance effects influence the stretching vibrational frequencies of both ($>C=C<$) bond and ($>C=O$) group. It may be illustrated as follows :

(*i*) *Inductive effect.* The groups having + I effect ($—CH_3$, $—C_2H_5$, etc.) present in the molecule always tend to reduce the value of force constant. Therefore, the absorption shifts to lower wave number. For example,

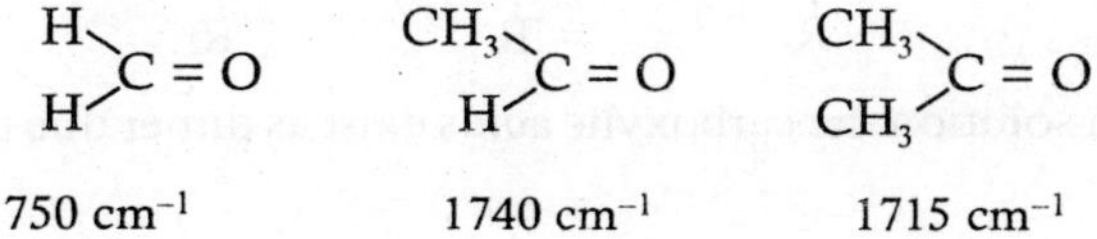

$1750\ cm^{-1}$ $1740\ cm^{-1}$ $1715\ cm^{-1}$

Thus, greater the + I effect ; smaller will be the magnitude of wave number.

Similarly, the atoms or groups having –I effect (—CN, —Br, etc.) present in the molecule always tend to increase the value of force constant. Therefore, the absorption shifts to longer wave number. For example,

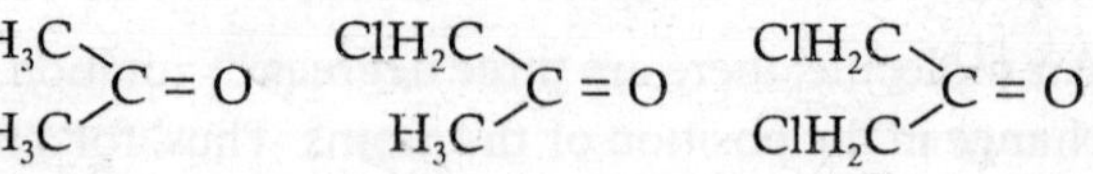

1715 cm^{-1} 1725 cm^{-1} 1740 cm^{-1}

Thus, greater the – I effect, higher will be the magnitude of wave number.

(*ii*) *Resonance effect.* The resonance whether in the unsaturated hydrocarbons or in the aromatic ring tends to decrease the magnitude of the force constant. Therefore, the absorption shifts to lower wave number. For example,

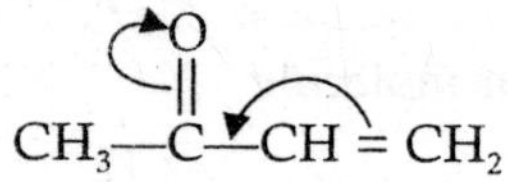

Here absorption shifts from 1715 cm^{-1} to 1690 cm^{-1}.

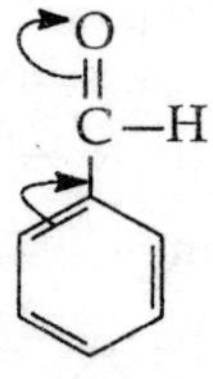

Here absorption shifts from 1715 cm^{-1} to 1700 cm^{-1}.

2. Hydrogen Bonding. The hydrogen bonding whether intermolecular or intramolecular decreases the value of force constant. Hence, the absorption shifts to lower wave number. Intermolecular hydrogen bonding gives broader bands while intramolecular hydrogen bonding gives sharp and well defined bands. The absorption bands in the intermolecular hydrogen bonding are dependent upon concentration whereas the absorption bands in the intramolecular hydrogen bonding are independent of the concentration since it takes place within the molecule. For example,

(*i*) *Alcohols and phenols.* The free O—H group in both the cases (alcohols and phenols) absorbs strongly in the region 3650–3600 cm^{-1}. But if the O—H group is involved in the H-bonding, then the broad absorption band appears around 3300 cm^{-1}. Actually due to H-bonding the bond length increases hence the bond becomes weak. Therefore, the magnitude of the force constant decreases and hence the value of absorption frequency is shifted towards lower wave number.

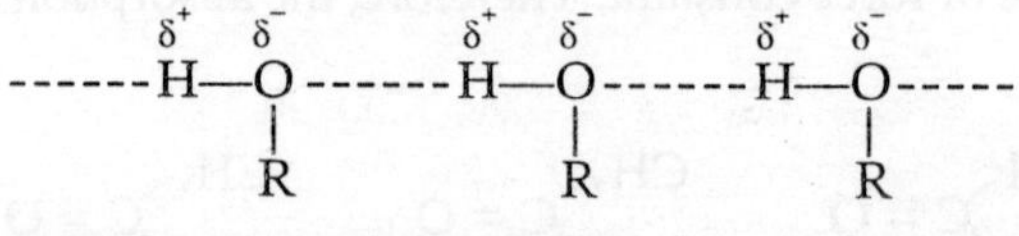

(*ii*) *Carboxylic acids.* In solution the carboxylic acids exist as dimer due to intermolecular hydrogen bonding as

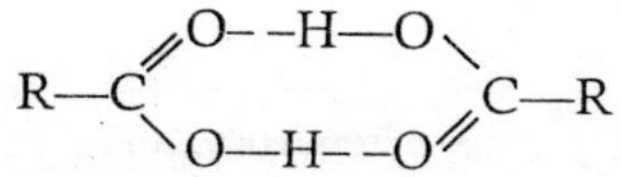

Here H-bondings are very strong, therefore, O—H stretching frequencies appear at much lower wave number (2700–2500 cm^{-1}) which are lesser than the stretching frequencies of C—H vibrations (2900–2850 cm^{-1}). Similarly, due to hydrogen bonding C = O stretching vibrations appear at lower wave number (~ 1720 cm^{-1}) which is lesser than the free C = O stretching vibrations (~ 1760 cm^{-1}).

8.9.6 Finger Print Region

The region below than 1500 cm^{-1}, is known as finger print region. Generally its range is 1500–900 cm^{-1}. It contains a number of bands ; many of which do not give exact information, but it has special characteristics. Some molecules containing the same functional group show similar absorption above 1500 cm^{-1} but their spectra differ in finger print region.

Finger print region is also very important in characterizing the compound. For example, the appearance of a doublet near 1380 cm^{-1} (*m*) and 1365 cm^{-1} (*s*) shows the presence of tertiary butyl group in the compound. Gem-dimethyl shows a medium band near 1380 cm^{-1}. Alcohols, esters, lactones, acid anhydrides show strong absorption in the finger print region due to C—O stretching. Primary alcohols form two strong bands at 1350–1260 cm^{-1} and near 1050 cm^{-1}. Phenols absorb near 1200 cm^{-1}. Esters show two strong bands between 1380–1050 cm^{-1}. Absorption in the region 1150–1070 cm^{-1} is most characteristic of ethers; *i.e.*, C—O stretching in ($>$C — O — O$<$) linkage. A band in the region 750–700 cm^{-1} shows monosubstituted benzene.

Mechanism. The IR spectrum is recorded with the help of a spectrophotometer. The sample of the substance under investigation may be either in the gaseous state or liquid state or present in the solution, is exposed to the beam of IR radiation. The relative intensity of the transmitted radiation is measured automatically. A graph is plotted between the percentages of transmission versus wave number. A common source for the infrared radiations is the Nernst glower which consists of a mixture of oxides of zirconium, yttrium and erbium and is heated electrically to about 1775 K. The sample containers are called cells and the optical parts of the spectrophotometer are made of rock salts (NaCl) or some other material. Solid samples scatter the radiations too much if placed as such directly in the IR beam. Therefore, solids are dissolved in a suitable organic solvent like $CHCl_3$, CS_2, etc. Liquid paraffin may also be used for the purpose. In some cases solids are mixed with KBr to form like a pellet by applying pressure. One thing should be noted that the sample under investigation must be taken in completely dry form. Because water gives strong absorption band near 2.7 μm (3710 cm^{-1}) and near 6.15 μm (1630 cm^{-1}) which may be interfere with the absorption due to substance under investigation.

[**Note.** During interpreting the IR spectrum, always begin at the high frequency (wave number) end of the spectrum and use the finger print region only for the confirmation. For example, the absorption of ($>$C = O) group due to stretching is 1800–1600 cm^{-1}.]

IR spectrum of some important organic compounds:

(*i*) *IR spectrum of acetic acid* (CH_3—C(=O)—OH)

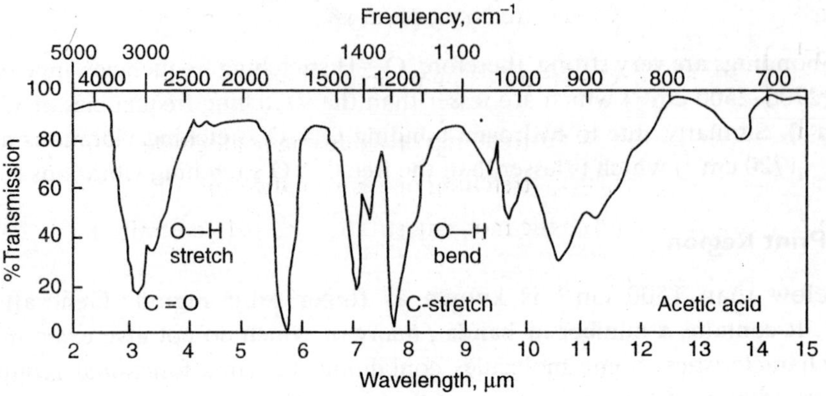

FIGURE 8.21 **IR spectrum of acetic acid, H**

The important bands are as follows :

3050 cm^{-1}	O—H stretching but hydrogen bonded
1720 cm^{-1}	(>C = O) stretching
1280 cm^{-1}	C—O stretching
950 cm^{-1}	O—H bending

(*ii*) *IR spectrum of benzaldehyde,* H—C = O

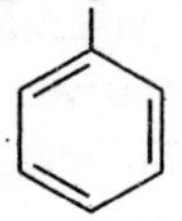

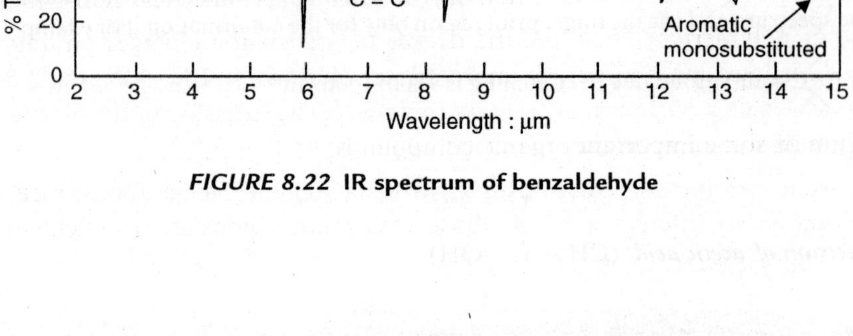

FIGURE 8.22 **IR spectrum of benzaldehyde**

Some important bands are as follows:

3080 cm^{-1}	C—H stretching of an aromatic ring
2870 cm^{-1} 2780 cm^{-1}	C—H stretching of an aldehyde
1710 cm^{-1}	C = O stretching of an aromatic aldehyde
1600 cm^{-1}	C = C stretching of an aromatic ring
680 cm^{-1}	aromatic monosubstitution (C—H deforming)

(*iii*) *IR spectra of isoborneol*

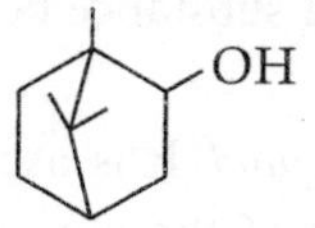

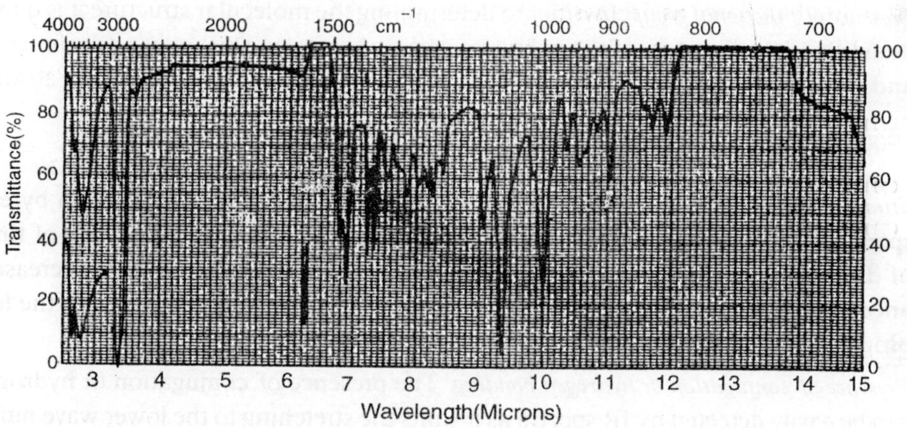

FIGURE 8.23 **Isoborneol, 8.2% w/w in carbon tetrachloride, 0.1 mm path, 2.5–3.2 μ insert, 1.5% w/w in carbon tetrachloride, 0.5 mm path**

The important bands are as follows:

3788 cm^{-1}	free O—H stretching
3571 cm^{-1}	bonded O—H stretching
1385 cm^{-1}	C—CH_3 stretching doublet of symmetrical gem-dimethyl group
1366 cm^{-1}	C—CH_3 stretching doublet of asymmetrical gem-dimethyl group

In the case of polymeric association, the degree of association changes on dilution; a larger fraction of unassociated groups is present in more dilute solutions. In Fig. 8.23, the 2.5–3.2 μ insert (isoborneol, 1.5% w/w in CCl_4, 0.5 mm cells) shows this effect. The spectrum of the more dilute solution indicates relatively freer hydroxyl groups.

Solvents also absorb the IR radiation in a particular region. Hence for correct analysis, two spectra should be run in two different solvents whose absorption regions are complementary. It is also necessary for careful analysis that the path length of the reference cell should be less than that of the solution cell for exact compensation.

Applications

It has wide applications in characterizing the functional groups. Some of the important applications are as follows:

(*i*) *Identification of substance.* It is very helpful in finding out whether a given sample of the organic substance is identical with another or not. If the IR spectra of the two compounds are same in identical conditions then the compound is identified. But this technique fails to distinguish between the two enantiomers of a substance because the IR spectra for the enantiomers are exactly the same.

(*ii*) *In determining the purity of a compound.* It is useful to check the purity of the compound by comparing its spectrum with that of the pure compound. Pure compound always shows a sharp peak whereas impure compound shows many extra poorly resolved bands.

(*iii*) *Presence of functional group.* In order to determining the molecular structure it is quite helpful in establishing the structure of the unknown organic compound by knowing the functional groups and the nature of the bond. For example, if the spectrum contains a strong absorption band 1900–1600 cm^{-1}, the presence of ($>C=O$) in a compound is suspected.

(*iv*) *Studying the progress of the reaction.* It determines the progress of a reaction by examine the spectra of small portion of compound of a reaction mixture in small interval of time. The rate of disappearance of characteristic absorption bands of a reactant and the increase of appearance of characteristic absorption bands of a product gives the exact idea of the formation of progress of a reaction.

(*v*) *Presence of conjugation or hydrogen bonding.* The presence of conjugation or hydrogen bonding can be easily detected by IR spectra as it shifts the stretching to the lower wave number. It also distinguishes between inter-molecular hydrogen bonding and intramolecular hydrogen bonding; the absorption position in the latter being independent of the change in concentration.

(*vi*) *In studying the reaction mechanism.* The reaction mechanism can be explained with the help of IR spectra. The IR spectra determine the formation of intermediates during the formation of products.

Table 8.1

Class of Compounds	*Type of Vibrations*	*Intensity*	*Approximate Range*	
			Wavelength (μ)	*Frequency* (cm^{-1})
1. Alkanes	(*i*) C—H (Stretch)	*m – s*	3.38–3.51	2960–2850
	(*ii*) $> CH_2$ (bend)	*m*	6.85–6.89	1460–1450
	(*iii*) $C—CH_3$ (bend)	*w*	~ 7.27	~1375
2. Alkenes	(*i*) = C—H (stretch)	*m*	3.29–3.32	3040–3010
	(*ii*) C = C (stretch)	*w*	6.02–6.35	1660–1600

3. Alkynes	(*i*) ≡ C—H (stretch)	S	3.05	3300
	(*ii*) C ≡ C (stretch)	*m*, *w*		
		or *v*	4.42–4.76	2260–2100
4. Aromatic hydrocarbon	(*i*) = C—H (stretch)	*w*	3.28–33.3	3050–3000
	(*ii*) C = C (stretch)	*v*	6.25	1600
		v	6.33	1580
		m	6.67	1500
	(*iii*) C—H (bending)	*m*	~ 6.09	~ 1450
	The number and position of these bonds depends upon the substitution pattern. For example	*m–s*	14.28–11.11	900–700
	Mono substituted benzenes	*s*	14.5–14.1	710–690
	(two bands)	*s*	13.7–13.0	770–730
	***o*-Disubstituted**			
	(one band)			
	***m*-Disubstituted**	*v*, *s*	13.6–13.0	770–735
	(two bands)	*m*	14.5–14.1	710–690
	***p*-Disubstituted**	*m*	13.3–12.5	800–750
	(one band)	*s*	12.5–11.9	840–800
5. Alcohols and Phenols	Free O—H (stretch)	*sh*	2.78–2.74	3650–3600
	H-bonded O—H (stretch)	*sh*	3.13–286	3500–3200
	C—O (stretch)	*s*	10–8	1250–1200
6. Ethers	C—O (stretch)	*s*	10–7.7	1300–1000
7. Amines	(*i*) N—H (stretch)			
	1° Amines			
	(two bands)	*m*	2.86–3.12	3500–3200
	2° Amines			
	(one band)	*m*	2.88–3.12	3500–3200
	(*ii*) N—H (bend)			
	1° Amines	*m–s*	6.06–6.29	1650–1550
	2° Amines	*w*	6.06–6.24	1650–1550

	(*iii*) C—N (stretch) (*a*) Aliphatic (*b*) Aromatic	 *w* *s*	 8.2–9.8 7.35–9.25	 1220–1020 1360–1080
8. Ketones	C = O (stretch) Saturated aliphatic or Saturated a cycle (*a*) Six-membered (*b*) Five-membered (*c*) Four-membered (*d*) Three-membered (As ring strain increases, absorption shifts to higher wave numbers)		 5.80–5.87 5.80–5.87 5.71–5.75 ~ 5.63 ~ 5.51	 1725–1705 1725–1705 1750–1740 ~ 1775 ~ 1815
9. Carboxylic acid	(*i*) O—H (stretch) (*ii*) C = O (stretch) (*a*) Saturated aliphatic (*b*) α, β-Unsaturated and aromatic (*iii*) C—O (stretch)	*s*, broad *s* *s* *m*	3.33–4.04 (Several bands) 5.80–5.88 5.88–5.95 7.6–8.3	300–2500 1725–1700 1700–1680 1320–1210
10. Acid chloride	C = O (stretch)	*s*	~ 5.57	~1795
11. Nitriles	C ≡ N (stretch)	*s*	4.5–4.4	2220–2260
12. Nitro-compounds	N = O (stretch) (two bands) (*i*) Antisymmetric (*ii*) Symmetric	 *s* *s*	 6.37–6.67 7.30–7.70	 1570–1500 1370–1300
s = strong *m* = medium *w* = weak *v* = variable *b* = broad *sh* = sharp				

8.10 FLAME PHOTOMETRY

Flame photometry is the type of emission spectroscopy and is termed as flame emission spectroscopy.

Principle

When certain metallic salts solution is sprayed into a flame, the solvent gets evaporated and some of the salts undergo dissociation and forms constituent atoms in vapour state. Some atoms get excited by absorbing thermal energy and the electrons get promoted from lower energy level to the higher energy level. When these electrons return to the lower (ground) energy level, they radiate energy of different wavelengths. The radiated energy, on passing through optical filter, emitted the characteristic wavelength radiation which is amplified and recorded by a suitable digital read-out system.

Instrumentation and working. The block diagram of essential components of a flame photometer is represented as in Fig. 8.24.

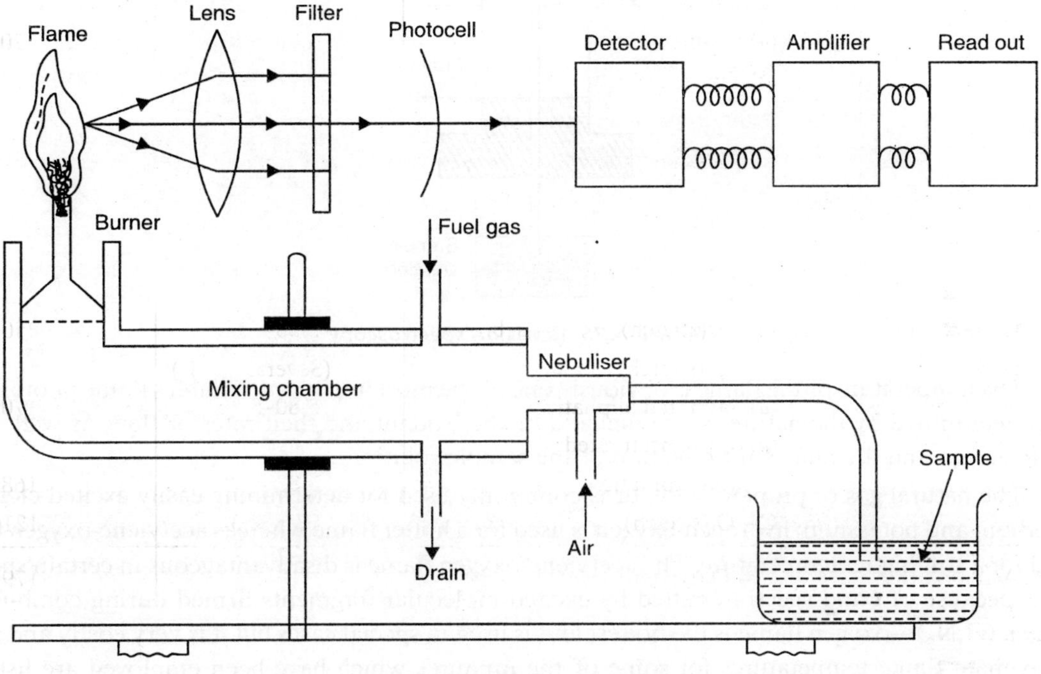

FIGURE 8.24 **Flame photometer**

At a particular pressure the air is introduced into the atomizer to produce the suction of the sample solution under examination. The solution of sample combines with air stream in the form of a fine mist is then mixed with fuel gas in the mixing chamber. The gaseous mixture (air + sample solution + fuel gas) is entered in a burner. After sample introduction, the process involves desolvation (loss and solvent by the droplet), followed by dissociation of the tiny solid particles to yield analyte atoms. A fraction of the atoms acquire sufficient energy by collision with molecules of the hot flame gases to become electronically excited. When the excited atoms revert to lower electronic energy levels, emission of radiant energy takes place. The radiations are allowed to pass through a converging lens, optical filter, photocell, and a detector. Optical filter permits the passage of characteristic radiation of the element under examination. Finally, it is amplified and read out by a digital recorder.

In early days premix nebulizer burners were used in flame emission. But now-a-days total consumption, integral aspirator-burner is used. The schematic diagram of an integral aspirator-burner is represented as in Fig. 8.25.

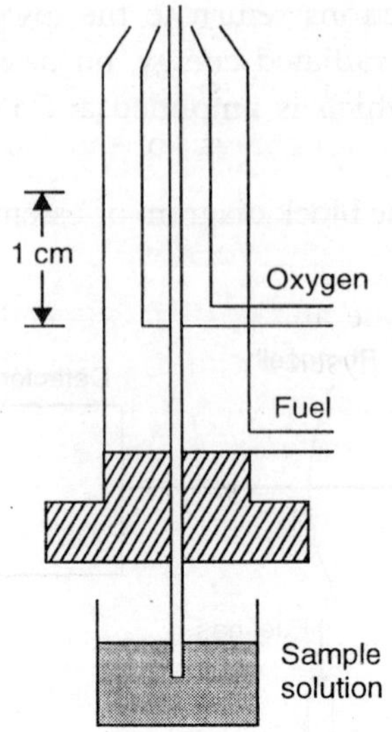

FIGURE 8.25 Emission spectroscopy

The temperature of the flame is obviously one of the most important variables flame photometry. This is determined by the nature of the fuel and of the oxidant and their rates of flow, as well as the burners design and the rate of introduction of the sample solution.

The natural gas or propane with air is commonly used for determining easily excited elements like sodium and potassium; hydrogen-oxygen is used for a hotter flame whereas acetylene-oxygen flame is used for a still higher temperature. The acetylene-oxygen flame is disadvantageous in certain spectral regions because of band spectra emitted by excited molecular fragments firmed during combustion. Cynogen (C_2N_2)—oxygen flame is the hottest and is used in special cases but it is very costly and toxic. Approximate flame temperatures for some of the mixtures which have been employed are listed in table 8.2.

Table 8.2

Mixture (Fuel-oxidant)	*Temperature (in °C)*
Natural-gas	1700
Propane-air	1800
Hydrogen-air	2000
Hydrogen-oxygen	2650
Acetylene-air	2300
Acetylene-oxygen	3200
Acetylene-nitrous oxide	2700
Cyanogen-oxygen	4800

Applications

(*i*) It is used for the analysis of Na, K, Ca, Cu, etc. in different compounds.

(*ii*) In industries, it is useful for the detection of elements in cement, glass, fuel, etc.

(*iii*) In medical science, it is used for analysing the blood and urine, etc.

(*iv*) In agriculture field it is useful for analysis of soil, water and plant materials.

Drawbacks

This technique is not applicable for the analysis of all metals. The metal which have tendency to incomplete vaporization can not be analysed because the formation of excited atoms can not be easily detected by photocell.

8.11 TITRIMETRIC ANALYSIS

It is the quantitative chemical analysis carried out by determining the volume of a solution of accurately known concentration (standard solution) which is required to react quantitatively with a measured volume of a solution of the substance to be determined. Earlier such type of analysis was known as volumetric analysis. In titrimetric analysis the reagent of known concentration is called the titrant and the substance being titrated is termed as titrand. The standard solution is usually added from a long graduated tube called a burette or micro-burette. The process of adding the standard solution until the reaction is just complete is called titrated. The point at which the reaction completes is called the equivalence end point. The completion of the titration is detected by some physical changes. For this purpose, we use an indicator. A reaction must fulfill the following conditions for using in titrimetric analysis.

(*i*) There must be a simple reaction which can be expressed by a chemical equation.

(*ii*) The substance to be determined should react completely with the reagent in stoichiometric proportions.

(*iii*) The reaction should be relatively fast. A positive catalyst may be used for this purpose.

(*iv*) There must be a alternation in some physical or chemical property of the solution at the end point.

(*v*) An indicator should sharply define the end point of the reaction by a change in physical properties like colour or precipitate formation.

(*vi*) If no such type of indicator is available then the end point may be detected by measuring :

(*a*) The potential between an indicator electrode and a reference electrode. Such type of titration is known as *potentiometric titration.*

(*b*) The change in electrical conductivity of a solution. Such type of titration is known as *conductometric titration.*

(*c*) The current which passes through the titration cell between an indicator electrode and a depolarized reference electrode at a suitable applied e.m.f. Such type of titration is known as *amperometric titration.*

(*d*) The change in absorbance of the solution. Such type of titration is called *spectrophotometric titration.*

Titrimetric methods need simple apparatus and are quickly performed. These are normally capable of high precision. That is why the methods have great advantages over gravimetric methods.

Classification of Reactions in Titrimetric Analysis

The reactions involved in the titrimetric analysis have been classified into four types:

1. *Neutralization reactions (acidimetry and alkalimetry).* These types of reactions involve the combination of hydrogen and hydroxide ions to form water. The titration of free bases or those formed from salts of weak acids by hydrolysis, with a standard acid is called acidimetry.

Similarly the titration of free acids or those formed by the hydrolysis of salts of weak bases, with a standard base is called alkalimetry.

2. *Complex formation reactions.* These types of reactions depend upon the combination of ions other than hydrogen or hydroxide ions, to form a soluble or slightly dissociated ion or compound as in the determination of hardness of water with EDTA.

3. *Precipitation reactions.* These reactions depend upon the combination of ions to form a simple precipitate as in the titration of silver ion with a chloride solution. There is no change in oxidation state occurs.

4. *Oxidation-reduction reactions.* These types of reactions depend upon the change in oxidation number of reacting substances. The standard solutions are either oxidizing or reducing agents.

In this chapter, we shall discuss only about conductometric titrations in details.

8.12 CONDUCTOMETRIC TITRATIONS

It is that type of titration in which endpoint or equivalence point is determined by measuring the conductance of the given solution. It is based on the fact that the conductance of an electrolyte solution depends upon the number of ions present in solution, their charges and mobilities. During titration the ions of one kind are replaced by ions of other kind differing in the mobilities. As a result, the conductance of solution changes with every addition of the titrant. At the equivalent point, a sharp change in conductance is observed.

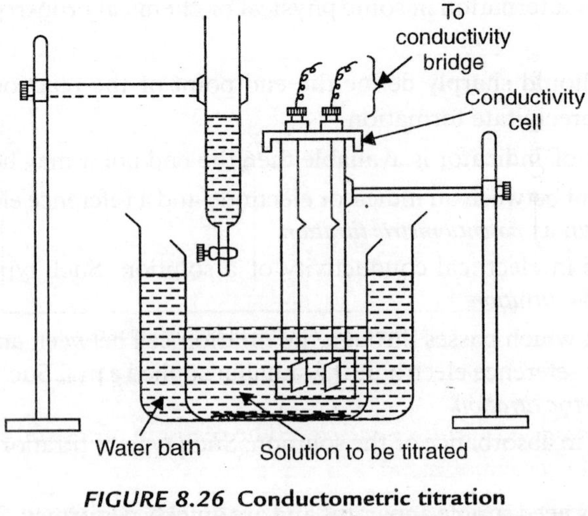

FIGURE 8.26 **Conductometric titration**

Process

Titrations are usually carried out by taking a solution to be titrated in a beaker kept in a water bath at constant temperature in which a conductivity cell is dipped and connected to a conductivity bridge circuit (Wheatstone bridge circuit). The titrant is added from a burette. The whole arrangement may be represented as in Fig. 8.26.

The solution is stirred and conductance is measured after each addition. For accuracy, there should not be an appreciable change in the volume of solution being titrated during titration. For this it is better to use a solution of titrant 20 to 100 times concentrated than the solution to be titrated using a micro burette. The conductance of solution is recorded during the course of titration and plotted against the volume of the titrant added, we get straight lines. The end point can be obtained and finding out the point of their intersection.

Types of Conductometric Titrations

(*i*) *Titration of a strong acid with a strong base.* The reaction of a strong acid (HCl) with strong base (NaOH) may be represented as:

$$\underbrace{H^+ + Cl^-}_{\text{Strong acid}} + \underbrace{Na^+ + OH^-}_{\text{Strong base}} \longrightarrow \underbrace{Na^+ + Cl^-}_{\text{Strong salt}} + H_2O$$

A standard base (whose strength is known) is taken in a burette and acid is taken in the conductivity vessel or a beaker. The electrical conductance of acid (say HCl) is measured with the help of a conductometer. The conductance is due to H^+ ions are replaced by slow moving Na^+ ions and the conductivity gets lowered. When NaOH is continuously added, the conductance goes on decreasing till the minimum and further begins to increase due to fast moving of OH^- ions. On plotting a graph between conductance and volume of NaOH used we get two straight lines AB and CD as in Fig. 8.27. The point of intersection (X) of these will be the end point (equivalence point).

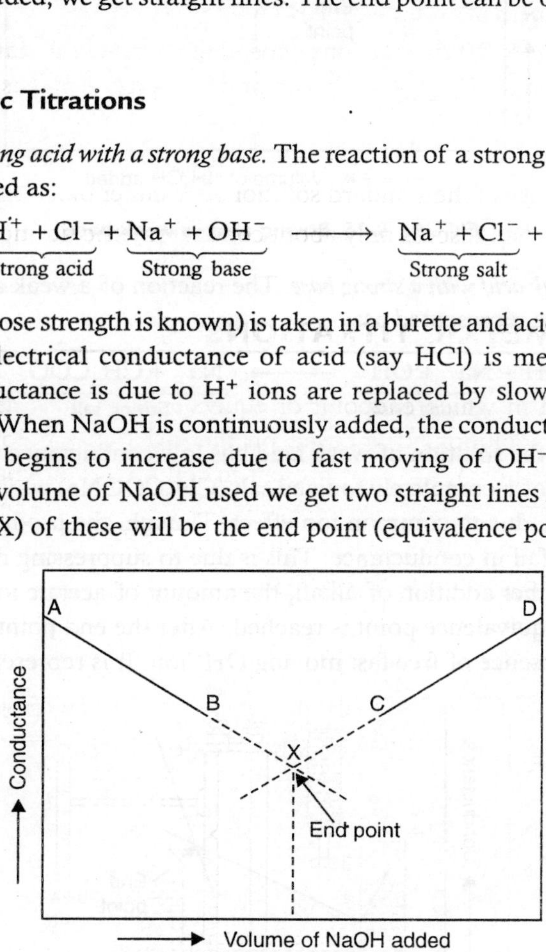

FIGURE 8.27 **Strong acid vs strong base**

(*ii*) *Titration of a strong acid with a weak base.* The reaction of a strong acid (say HCl) with a weak base (NH_4OH) may be represented as:

$$\underbrace{H^+ + Cl^-}_{\text{Strong acid}} + \underset{\text{Weak base}}{NH_4OH} \longrightarrow \underbrace{NH_4^+ + Cl^-}_{\text{Strong salt}} + \underset{\text{Unionized}}{H_2O}$$

In this conductometric titration the conductance falls at the beginning due to the replacement of fast moving H^+ ions by slow moving NH_4^+ ion. When the neutralization is complete, the further addition of NH_4OH does not vary the conductance so much as the free base behaves like a weak electrolyte and its conductance is very small as compared with that of its salt. It may be represented as in Fig. 8.28.

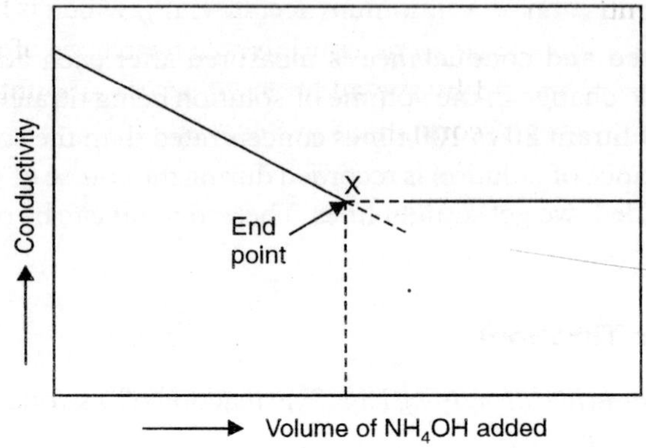

FIGURE 8.28 **Strong acid vs weak base**

(*iii*) *Titration of a weak acid with a strong base.* The reaction of a weak acid (say acetic acid) with a strong base (NaOH) may be represented as:

$$\underbrace{CH_3COOH}_{\text{Weak acid}} + \underbrace{Na^+ + OH^-}_{\text{Strong base}} \longrightarrow \underbrace{Na^+ + CH_3COO^-}_{\text{Strong salt}} + H_2O$$

In this case initially the solution of weak acid has low conductance with further falls on adding alkali due to the formation of a completely ionized salt (CH_3COONa) which suppresses the ionization of weak acid (CH_3COOH), due to common ion effect. Though, the product sodium acetate has high conductance yet there is a fall in conductance. This is due to suppressing nature of acetate ion due to common ion effect. On further addition of alkali, the amount of acetate ion increases as it is released from acetic acid until the equivalence point is reached. After the end point the conductance increases more rapidly due to the presence of free fast moving OH^- ion. It is represented as in Fig. 8.29.

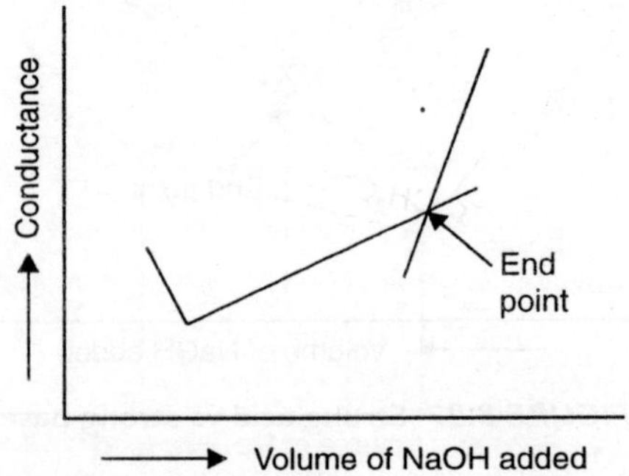

FIGURE 8.29 **Weak acid vs strong base**

(*iv*) *Titration of a weak acid with a weak base.* The reaction of a weak acid (say CH_3COOH) with a weak base (NH_4OH) is represented as:

$$\underset{\text{Weak acid}}{CH_3COOH} + \underset{\text{Weak base}}{NH_4OH} \longrightarrow \underbrace{NH_4^+ + CH_3COO^-}_{\text{Strong salt}} + \underset{\text{Unionized}}{H_2O}$$

In the beginning the conductance decreases due to the formation of ammonium acetate, which suppresses the dissociation of acetic acid. On further addition of NH_4OH, the conductance is increased due to increase in the concentration of ammonium acetate (salt), which is highly ionized due to strong salt. This continues until the equivalence point, considerable hydrolysis of the salt occurs. Beyond this point, the salt hydrolysis is suppressed by the addition of excess amount of ammonium hydroxide. Thus the conductance remains almost constant as in Fig. 8.30.

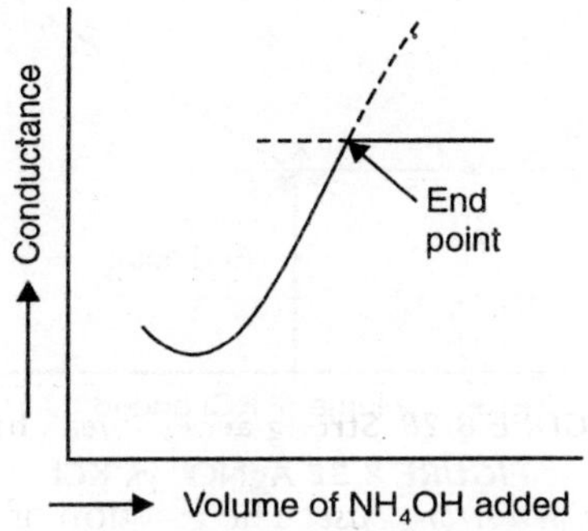

FIGURE 8.30 **Weak acid vs weak base**

(*v*) *Titration of a mixture of strong and weak acids with a strong base.* Consider a mixture of strong acid (HCl) and a weak acid (CH_3COOH) which is to be titrated with a strong base NaOH. When NaOH is added to the mixture of acid solution, the strong acid due to high ionization is neutralized first. Thus, the conductance of the solution starts decreasing. This continues till the complete neutralization of HCl. After that the conductance will start increasing due to the formation strong salt CH_3COONa, which is highly ionized. After complete neutralization of CH_3COOH, the conductance shows a sharp increase due to the addition of excess highly conducting OH^- ions. Hence two equivalence points are obtained as in Fig. 8.31.

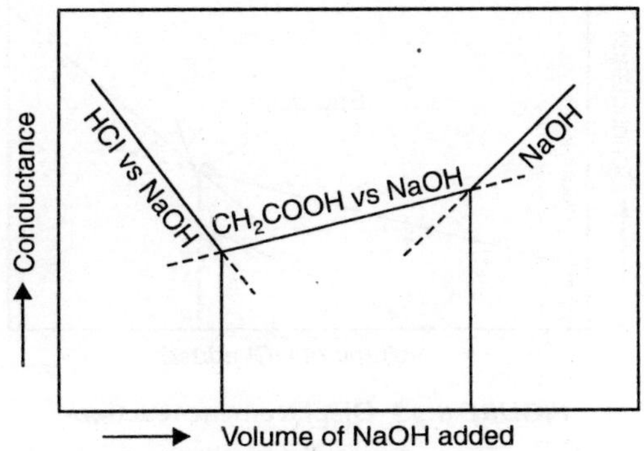

FIGURE 8.31 **Strong and weak acids vs strong base**

(*vi*) *Titration involving precipitation reactions.* Consider a precipitation reaction like $AgNO_3$ vs KCl. When KCl solution is added to a known volume of $AgNO_3$ solution, the precipitation reaction takes place as follows:

$$Ag^+ + NO_3^- + K^+ + Cl^- \longrightarrow AgCl \downarrow + K^+ + NO_3^-$$

The conductance in the initial state almost remains constant because KCl is replaced by an equivalent amount of KNO_3 and there is no appreciable difference in the mobility. After the end point, the excess amount of added salt (KCl) increases the conductance sharply due to increase in the concentration of free ions (K^+ and Cl^-) of the salt. It may be represented as in Fig. 8.32.

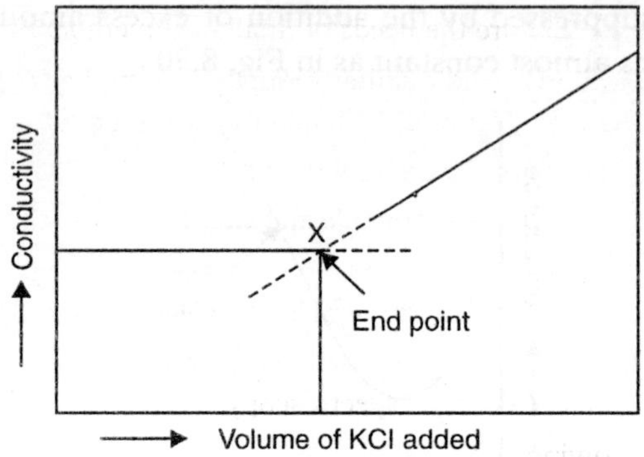

FIGURE 8.32 **$AgNO_3$ vs KCl**

(*vii*) *Titration involving displacement reaction.* Consider a displacement reaction when HCl is added to a solution of sodium acetate, the reaction takes place

$$CH_3COO^- + Na^+ + H^+ + Cl^- \longrightarrow CH_3COOH + Na^+ + Cl^-$$

In this reaction highly dissociated CH_3COONa is replaced by strongly dissociated NaCl and almost undissociated CH_3COOH. Initially the conductance increases slowly due to Cl^- has more conductance than CH_3COO^-. Beyond the end point, the conductance increases sharply due to excess of HCl. It may be shown as in Fig. 8.33.

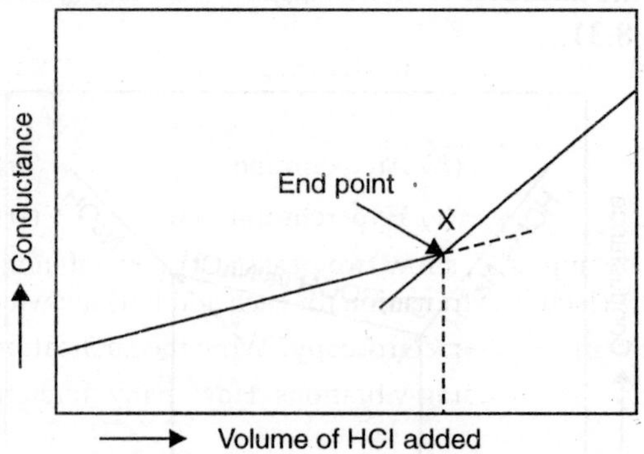

FIGURE 8.33 **Displacement reaction**

Advantages

Some important advantages of conductometric titrations are as follows:

(*i*) These titrations are very convenient for coloured solutions where the use of indicator is not possible.

(*ii*) These titrations give quite good results in case of very dilute solutions, weak acids and weak bases.

(*iii*) No observation is required near the end point because the end point is detected graphically.

EXERCISE

1. Define analytical chemistry. What are the needs of analytical methods?
2. What do you mean by quantitative and qualitative analytical methods? Explain the thermogravimetric analysis (TGA) in details. How is it applicable in quantitative analysis?
3. Discuss the principle and working of differential thermal analysis (DTA) technique. **(*M.D.U. Dec. 2002*)**
4. What is spectrophotometry? Discuss the principle and working of a spectrophotometer with the help of a schematic diagram. **(*K.U.K. Jan. 2004, June 2004*)**
5. Discuss the basic principle involved in thermogravimetric analysis. How is it applicable in the structural determination of compounds? **(*M.D.U. May 2002*)**
6. Discuss the brief introduction of electronic spectroscopy. **(*M.D.U. 1999*)**
7. Write short notes on the following:

   (*i*) Vibrational spectroscopy **(*M.D.U. 2000*)**

   (*ii*) Lambert-Beer's law

   (*iii*) Finger print region

   (*iv*) Selection rule.
8. Define spectroscopy. Which of the following molecules will show pure rotational spectra and why?

   (*i*) H_2 (*ii*) O_2 (*iii*) HBr (*iv*) N_2 (*v*) NO (*vi*) CO.
9. Define the following terms:

   (*i*) Wavelength (*ii*) Velocity (*iii*) Frequency (*iv*) Wave number.

   Which of the following will show vibrational spectra?

   (*i*) H_2 (*ii*) O_2 (*iii*) CO_2 (*iv*) NO (*v*) CO.
10. Explain 'allowed' and 'forbidden transitions'. Discuss the role of solvent in the electronic transitions.
11. (*a*) Define:

    (*i*) Chromophore (*ii*) Auxochrome (*iii*) Bathochromic shift

    (*iv*) Hypsochromic shift (*v*) Hyperchromic shift (*vi*) Hypochromic shift

    (*b*) The UV spectrum for a compound, show two peaks at λ_{max} = 280 nm, ϵ_{max} = 15 and λ_{max} = 190 nm, ϵ_{max} = 100. Identify the electronic transition for each and indicate which is more intense?
12. How UV spectroscopy differs from IR spectroscopy? Write the applications of UV and IR spectroscopy?
13. Illustrate the types stretching and bending vibrations. How many fundamental vibrational bands appear in the IR spectrum of the following?

    (*i*) CO_2 (*ii*) C_6H_6.
14. How vibrational frequencies are affected by

    (*a*) Hydrogen bonding, and (*b*) Electronic effect?
15. Why spectroscopic methods are better than classical methods, for analyzing the compounds?
16. How will you distinguish ethyl alcohol and acetic acid by IR spectra?

17. What is flame photometry? Describe its applications and drawbacks. ***(K.U.K. Jan. 2004, June 2004)***
18. What do you understand by titrimetric analysis? Why titrimetric methods have great advantages over gravimetric methods?
19. (*a*) Define the following:

    (*i*) Potentiometric titration (*ii*) Conductometric titration
    (*iii*) Amperometric titration (*iv*) Spectrophotometric titration.

    (*b*) Classify the reactions in titrimetric analysis.
20. (*a*) Describe the conductometric titration of a weak acid with a strong base. ***(K.U.K. Jan. 2004)***
    (*b*) What are the advantages of conductometric titrations over titration using indicators? ***(K.U.K. June 2004)***
21. (*a*) Name the types of conductometric titrations. ***(K.U.K. Jan. 2007)***
    (*b*) Describe the conductometric titrations of a strong acid with a strong base.
22. Describe the conductometric titrations of:
    (*i*) Strong acid with a weak base
    (*ii*) Weak acid with a weak base
    (*iii*) Mixture of strong and weak acids with a strong base.
23. (*a*) Define conductometric titration. Describe the conductometric titration involving:
    (*i*) Precipitation reactions
    (*ii*) Displacement reactions.
    (*b*) What are the advantages of conductometric titrations?
24. Explain the salient features of spectrophotometer. ***(K.U.K. Jan. 2005, June 2005, June 2006)***
25. Write an essay on
    (*i*) Flame photometry ***(K.U.K. Jan. 2005, June 2005, Jan. 2006)***
    (*ii*) Titrimetric analysis
    (*iii*) Chromophore
    (*iv*) Conductometric titration
    (*v*) Thermogravimetric analysis. ***(K.U.K. Jan. 2007)***

MODEL TEST PAPER-1

Time : 3 Hours *M. M. 100*

***Note.** Attempt Five questions selecting one question from each unit. Question 1 is compulsory. Each question carries equal marks.*

1. (*a*) Define following:

Triple point, Metastable equilibrium

(*b*) Explain homocatalysis and heterocatalysis with examples.

(*c*) Why hardness or alkalinities are expressed in terms of $CaCO_3$ equivalent?

(*d*) Write a short note on Pilling-Bedworth rule.

(*e*) How would you distinguish ethyl methyl ketone and methyl vinyl ketone by IR spectra?

5 × 4

UNIT–1

2. (*a*) Derive phase rule equation. What do you understand by reduced phase rule?

(*b*) Explain the various curves, points, areas involved in water system with a neat and sketched diagram. Why is the fusion curve in the phase diagram of water system inclined towards the pressure axis? Explain. 10 × 2

3. (*a*) Define catalysis. Write the general characteristic of catalysis. How enzymes speed up the reaction?

(*b*) Explain the effect of temperature and pH on the rate of enzyme catalysis. Write a short note on proximity and orientation. 10 × 2

UNIT–2

4. (*a*) What do you mean by hardness of water? Explain the EDTA methods in details to remove the hardness of water. What is the role of ammonium buffer solution in this process?

(*b*) 50 ml of alkaline water sample required 15 ml of N/50 H_2SO_4 using phenolphthalein indicator and another 10 ml of the same acid for complete neutralisation using methyl orange indicator. Calculate the type and amount of alkalinity producing substances in terms of $CaCO_3$ equivalent. 10 ×2

5. (*a*) Explain the ion exchange method of purifying the water. Discuss their use and regeneration with reactions involved.

(*b*) A zeolite softener was 90% exhausted by removing the hardness completely when 10^5 litres of hard water sample passed through it. The exhausted zeolite bed required 150 litres of 30% NaCl solution for its complete regeneration. Calculate the hardness of water. 10 × 2

UNIT–3

6. (*a*) Define corrosion. Explain in details the mechanism of rusting of iron.

   (*b*) Write a short note on

   (*i*) Microbiological corrosion

   (*ii*) Biodegradable lubricants. 10 × 2

7. (*a*) Compare the fluid film lubrication with that of boundary lubrication. Why molybdenum disulphide is used as solid lubricant?

   (*b*) Write a short note on viscosity index of lubricant. 10 × 2

UNIT–4

8. (*a*) What are chief physical characteristics expected of an elastomer? How they are achieved in a new product? How does the chain structure of rubbers compare with that of plastics?

   (*b*) Write a short note on 'The effect of structure on properties of polymers'. 12 + 8

9. (*a*) State and explain Lambert-Beer's law. What are limitations of Lambert-Beer's law.

   (*b*) Write a short note on

   (*i*) Finger print region

   (*ii*) Selection rule. 10 × 2

MODEL TEST PAPER-2

Time : 3 Hours *M. M. 100*

Note. *Attempt Five questions selecting one question from each unit. Question 1 is compulsory. Each question carries equal marks.*

1. (*a*) Justify the statement "The eutectic is a mixture and not a compound."

(*b*) What are the characteristics of potable water?

(*c*) Explain how can corrosion be controlled by proper designing?

(*d*) Write the significance of aniline point.

(*e*) Why spectroscopic methods are better than the classical methods? 5×4

UNIT–1

2. (*a*) What do you understand by congruent and incongruent melting point system? Give one example of each. Explain the phase diagram of Pb-Ag system in details.

(*b*) Define phase, Component and Degree of freedom. 10×2

3. (*a*) Define following:

Turnover number, Promotors, inhibitors, induced catalyst. How promotors increases the activity of enzyme catalyst?

(*b*) Derive Michaelis-Menton equation for enzyme catalysis. 10×2

UNIT–2

4. (*a*) What are scales and sludges? Write their compositions. How they are formed in the boilers? Discuss their disadvantages and methods of prevention.

(*b*) 100 ml of hard water sample required 25 ml of M/100 EDTA solution with ammonium buffer solution and EBT indicator. Calculate the hardness of water. 10×2

5. (*a*) What do you mean by demineralization and desalination? Discuss in details the electrodialysis process for desalination of sea water with the help of neat, cleaned and labelled diagram. What are its advantages and disadvantages.

(*b*) A water sample, on analysis, shows the following results

Ca^{2+} = 160 ppm CO_2 = 88 ppm

Mg^{2+} = 72 ppm HCO_3^- = 488 ppm

Calculate the amount of lime and soda required to softening 10^6 litres of water if $FeSO_4 \cdot 7H_2O$ is used as a coagulant at the rate of 139 ppm. 10×2

UNIT–3

6. (*a*) Explain the factors affecting corrosion. Write a short note on soil corrosion.

(*b*) How metal can be prevented from corrosion by galvanisation? What are the differences between galvanisation and tinning?

7. (*a*) What is grease? How is it prepared? How consistency and drop point values of lubricants are determined? Explain clearly. Give their significance. 10 × 2

(*b*) Define iodine value. Explain the method for the determination of iodine value of lubricating oil. Write its significance.

UNIT–4

8. (*a*) Explain free radical mechanism or cationic mechanism of polymerisation.

(*b*) Explain the preparation, properties and uses of Teflon.

Write a short note on biodegradable polymerisation.

9. (*a*) How UV spectroscopy differs from IR spectroscopy? Write the applications of UV and IR spectroscopy.

(*b*) Explain the principle and working of a flame photometer. Describe its applications and drawbacks.

MODEL TEST PAPER-3

Time : 3 Hours *M. M. 100*

Note. *Attempt Five questions selecting one question from each unit. Question 1 is compulsory. Each question carries equal marks.*

1. (*a*) What are the criteria for phase equilibria?

(*b*) Define catalyst. Differentiate between enzymes and inorganic catalyst.

(*c*) Write a short note on Phosphate conditioning of water.

(*d*) How much rust ($Fe_2O_3 . 3H_2O$) will be formed on completely rusting of 100 kg of iron.

(*e*) Define chromophore and auxochrome. 5 × 4

UNIT–1

2. (*a*) Write phase rule equation. Explain the phase diagram of Zn-Mg system or CO_2 system.

(*b*) Explain the terms (any two):

Phase, component, Degree of freedom

KCl-NaCl - H_2O is a three component system while NaBr-KCl-H_2O is a four component system. Explain it. 10 × 2

3. (*a*) What are active sites of enzymes? Why the reaction rate of an enzyme catalysed reaction changes from first order to zero order as the substrate concentration is increased?

(*b*) Why enzymes are known as biocatalyst? Explain the lock and key hypothesis for the activities of enzymes. Write some importance of enzymes. 10 × 2

UNIT–2

4. (*a*) Explain the method with principle of the removal of calcium and magnesium hardness of water sample. Write the reactions involved in it. What is the role of ammonium buffer solution in this process?

(*b*) 20 ml of standard hard water (SHW) containing 1.5 g $CaCO_3$ per litre, required 25 ml EDTA solution in ammonium buffer solution using EBT indicator, for end point. 100 ml of hard water sample required 18 ml EDTA solution, while same amount of water after boiling required 12 ml EDTA under same condition. Calculate carbonate and non-carbonate hardness of water. 10 × 2

5. (*a*) Explain the lime-soda process for purification of water. Why does $Mg(HCO_3)_2$ require double amount of lime for softening of hard water?

(*b*) The analytical report shows the following results of a water sample.

Ca^{2+} = 80 ppm; Mg^{2+} = 48 ppm; Na^+ = 16.10 ppm

HCO_3^- = 36.6 ppm ; SO_4^{2-} = 111.36 ppm; Cl^- = 13.49 ppm.

Express the results in terms of salts present as their $CaCO_3$ equivalents. 10 × 2

UNIT–3

6. (*a*) Explain Pitting corrosion in details. What are the effects of temperature, pH, over voltage and reactivity of metals influences the corrosion?

   (*b*) Explain the mechanism of hydrogen evolution and oxygen absorption in electrochemical corrosion. 10 × 2

7. (*a*) How lubricants are selected for used in industries? Explain the methods of determination of Iodine value of lubricants. What are its significance?

   (*b*) Define flash point and fire point of an oil. What are the factors affecting the flash point and fire point of an oil. 10 × 2

UNIT–4

8. (*a*) Explain PVC is soft and flexible whereas bakelite is hard and brittle. Write a short note on glass-resin forced plastics.

   (*b*) Explain the mechanism of coordination polymerisation. What are important features of coordination polymerisation. 10 × 2

9. (*a*) Explain the thermogravimetric analysis (TGA) in details. How is it applicable in quantitative analysis?

   (*b*) Explain 'allowed' and 'forbidden' transistions. Discuss the role of solvent in the electronic transitions. 10 × 2

INDEX